Vibration and Damping Behavior of Biocomposites

Vibration and Damping Behavior of Biocomposites

Edited by
Senthil Muthu Kumar Thiagamani,
Md Enamul Hoque, Senthilkumar Krishnasamy,
Chandrasekar Muthukumar, and Suchart Siengchin

CRC Press
Taylor & Francis Group
Boca Raton London New York

CRC Press is an imprint of the
Taylor & Francis Group, an **informa** business

First edition published 2022
by CRC Press
6000 Broken Sound Parkway NW, Suite 300, Boca Raton, FL 33487-2742

and by CRC Press
2 Park Square, Milton Park, Abingdon, Oxon, OX14 4RN

© 2022 Taylor & Francis Group, LLC

CRC Press is an imprint of Taylor & Francis Group, LLC

Library of Congress Cataloging-in-Publication Data
Names: Thiagamani, Senthil Muthu Kumar, editor.
Title: Vibration and damping behavior of biocomposites /
edited by Senthil Muthu Kumar Thiagamani, Md Enamul Hoque, Senthilkumar Krishnasamy,
Chandrasekar Muthukumar, and Suchart Siengchin.
Description: First edition. | Boca Raton, FL: CRC Press, [2022] |
Includes bibliographical references and index. |
Summary: "There has been a growing interest among research communities and industries in the use of natural fibers as reinforcements in structural and semi-structural applications, given their environmental advantages. Knowledge of the vibration and damping behavior of biocomposites is essential for engineers and scientists who work in the field of composite materials. This book brings together the latest research developments in vibration and viscoelastic behavior of composites filled with different natural fibers. This compilation will benefit academics, researchers, advanced students, and practicing engineers in materials and mechanical engineering and related fields who work with biocomposites"—Provided by publisher.
Identifiers: LCCN 2021047605 (print) | LCCN 2021047606 (ebook) |
ISBN 9781032003153 (pbk) | ISBN 9781032003122 (hbk) | ISBN 9781003173625 (ebk)
Subjects: LCSH: Fibrous composites. | Plant fibers. | Damping (Mechanics)
Classification: LCC TA418.9.C6 V49 2022 (print) | LCC TA418.9.C6 (ebook) |
DDC 620.1/18—dc23/eng/20211123
LC record available at https://lccn.loc.gov/2021047605
LC ebook record available at https://lccn.loc.gov/2021047606

ISBN: 978-1-032-00312-2 (hbk)
ISBN: 978-1-032-00315-3 (pbk)
ISBN: 978-1-003-17362-5 (ebk)

DOI: 10.1201/9781003173625

Typeset in Times
by codeMantra

I would like to dedicate this book to my late parents
Shri. P.C.G. Thiagamani
and
Smt. Prema Thiagamani
- Senthil Muthu Kumar Thiagamani

Contents

PART 1 Vibration and Damping of the Biocomposites

PART 2   *Viscoelastic Properties of the Biocomposites*

Contents

Preface

Fiber-reinforced polymer composites or biocomposites have better damping characteristics than conventional metals due to the viscoelastic nature of the polymers. There has been a mounting interest among the research communities and industries in using natural fiber as reinforcement in structural and semi-structural applications, considering the environmental implications of synthetic fibers. Other than the physical damages during the service life of parts made up of the biocomposites, their performance is also affected by the vibration that arises from the functioning of the various systems and sub-systems. Hence, the characterization of the vibration and damping behavior of the biocomposites has become essential. The conventional way of assessing vibration damping characteristics of composite material is through the free vibration and forced vibration testing methods. Furthermore, dynamic mechanical analysis (DMA) has proved to be an effective tool in characterizing the damping characteristics of the polymer-based composite, specifically due to the viscoelastic nature of the polymers.

The vibration and damping behavior of a typical composite material depend on the type of fiber reinforcement, fiber architecture, fiber loading, and the type of resin employed. In polymer-based biocomposites, natural fibers are hydrophilic in nature, while the polymers are hydrophobic, leading to inferior fiber–matrix interfacial adhesion. This, in turn, would affect the damping characteristics of the polymer-based biocomposites. Thus, in addition to exploring the vibration and damping characteristics of the biocomposites based on the above-mentioned aspects, various possible methods of improving the fiber–matrix interfacial adhesion have also been discussed.

This book is organized into two major parts, namely "Vibration and Damping of the Biocomposites" and "Viscoelastic Properties of the Biocomposites."

In Part 1, Chapter 1 presents an overview of the free vibration and damping behavior of the biocomposites. Chapter 2 discusses the various factors that affect the vibration and damping behavior, while Chapter 3 focuses on the influence of hybridization. Chapter 4 presents the free vibration and damping properties of pineapple leaf- and sisal fiber-based biocomposites. Chapters 5–9 present an overview of the influence of various aspects such as the fiber length, fiber content, organic fillers, fiber treatments, and compatibilizers on the vibration and damping characteristics of biocomposites.

In Part 2, Chapters 10 and 11 provide an overview of the characterization of viscoelastic properties of biocomposites through DMA. Chapter 12 focuses on the effects of hybridization and the addition of fillers, while Chapter 13 presents the viscoelastic properties of bionanocomposites. Chapters 14–17 provide insights on the influence of fiber treatments, uses of compatibilizers, and interphase damping modification. Chapter 17 discusses the DMA of nanomaterials-based biocomposites. Chapter 18 focuses on the recent studies in the characterization of the viscoelastic properties of the biocomposites using computational modeling.

All the chapters of this book were written by experienced researchers in the field of free vibration and viscoelastic characterization. We would like to express our sincere thanks and appreciation to all the contributing authors for their time and efforts. We would like to extend our gratitude to the publisher for their cooperation and support in successfully completing this book. We believe that this book can serve as reference material for academicians, students, researchers, and scientists looking for fundamental knowledge, recent trends in research and development in the field of vibration and damping on biocomposites.

Editors

Dr. Senthil Muthu Kumar Thiagamani is an associate professor in the Department of Mechanical Engineering at Kalasalingam Academy of Research and Education (KARE), Tamil Nadu, India. He earned his Diploma in Mechanical Engineering from the Directorate of Technical Education, Tamil Nadu in 2004; earned his Bachelor of Engineering in Mechanical Engineering from Anna University Chennai in 2007; and earned his Master of Technology in Automotive Engineering from Vellore Institute of Technology, Vellore in 2009. He completed his Doctor of Philosophy in Mechanical Engineering (specialized in Biocomposites) from KARE in 2018. He has also completed his postdoctoral research from the Materials and Production Engineering Department at The Sirindhorn International Thai-German Graduate School of Engineering (TGGS), KMUTNB, Thailand. He has 12 years of teaching and research experience. Dr. Thiagamani is also a visiting researcher at KMUTNB, Thailand. He is a member of international societies such as the Society of Automotive Engineers and the International Association of Advanced Materials. His research interests include biodegradable polymer composites and characterization. He has authored several articles in peer-reviewed international journals, book chapters, and conference proceedings. He has also published edited books in the theme of biocomposites. He is an editorial board member of the ARAI Journal of Mobility Technology. He is serving as a reviewer for various journals, including *Journal of Industrial Textiles*, *Journal of Polymers and the Environment*, *SN Applied Sciences*, *Mechanics of Composite Materials*, *International Journal of Polymer Science*, etc.

Dr. Md Enamul Hoque is a professor at the Department of Biomedical Engineering in the Military Institute of Science and Technology (MIST), Dhaka, Bangladesh. Before joining MIST, he served in several leading positions in some other global universities that include Head of the Department of Biomedical Engineering at King Faisal University (KFU), Saudi Arabia and Founding Head of Bioengineering Division, University of Nottingham Malaysia Campus (UNMC). He completed his PhD at the National University of Singapore (NUS), Singapore in 2007. He also earned his PGCHE (Post Graduate Certificate in Higher Education) from the University of Nottingham, UK in 2015. He is a Chartered Engineer (CEng) certified by the Engineering Council, UK; Fellow of the Institute of Mechanical Engineering (FIMechE), UK; Fellow of Higher Education Academy (FHEA), UK; and Member, World Academy of Science, Engineering and Technology. To date, he has published 6 books, 32 book chapters, and 170 technical papers in referred journals and international conference proceedings. His publications have elicited 1,720 citations with 19 h-index. His major areas of research interest include (but are not limited to) biomaterials, nanomaterials, nanotechnology, biomedical implants, rehabilitation engineering, rapid prototyping technology, stem cells, and tissue engineering.

Dr. Senthilkumar Krishnasamy is a research scientist at the Center of Innovation in Design and Engineering for Manufacturing (CoI-DEM), King Mongkut's University of Technology North Bangkok, Thailand. He graduated with a Bachelor's degree in Mechanical Engineering from Anna University, Chennai, India, in 2005. He then chose to continue his Master's studies and graduated with a Master's degree in CAD/CAM from Anna University, Tirunelveli, in 2009. He earned his PhD from the Department of Mechanical Engineering, Kalasalingam University (2016). Dr. Krishnasamy worked in the Department of Mechanical Engineering, Kalasalingam Academy of Research and Education (KARE), India, from 2010 (January) to 2018 (October). He completed his postdoctoral fellowship at Universiti Putra Malaysia, Serdang, Selangor, Malaysia and King Mongkut's University of Technology North Bangkok (KMUTNB) under the research topics of "Experimental investigations on mechanical, morphological, thermal and structural properties of kenaf fibre/mat epoxy composites" and "Sisal composites and fabrication of eco-friendly hybrid green composites on tribological properties in a medium-scale application," respectively. His areas of research interest include the modification and treatment of natural fibers, nanocomposites, 3D printing, and hybrid-reinforced polymer composites. He has published research papers in international journals, book chapters, and conferences in the field of natural fiber composites. He is also editing books from different publishers.

Dr. Chandrasekar Muthukumar is an assistant professor in the School of Aeronautical Sciences, Hindustan Institute of Technology & Science, Chennai, India. He graduated with a Bachelor's degree in Aeronautical Engineering from Kumaraguru College of Technology, Coimbatore, India. His Master's degree in Aerospace Engineering was from Nanyang Technological University-TUM ASIA, Singapore. He earned his PhD in Aerospace Engineering from Universiti Putra Malaysia (UPM), Malaysia, which was funded by a research grant from the Ministry of Education, Malaysia. During his association with the UPM, he obtained internal research funds from the university worth 16,000 and 20,000 MYR, respectively. He has 5 years of teaching and academic research experience. His field of expertise includes fiber metal laminate (FML), natural fibers, biocomposites, and aging and their characterization. His publications are based on the fabrication and characterization techniques of biocomposites, aging studies in biocomposites, and creep analysis of biocomposites. Dr. Muthukumar has authored or co-authored 32 research articles in SCI journals, 24 book chapters, and 5 articles in conference proceedings. He is currently co-editing six books that are to be published by

CRC Press/Taylor & Francis Group, Wiley, Springer, and Elsevier. The book *Natural Fiber-Reinforced Composites: Thermal Properties and Applications* has been submitted to Wiley and is in production. He is a peer reviewer for *Journal of Composite Materials, Polymer Composites, Materials Research Express*, and *Journal of Natural Fibres*.

Prof. Dr.-Ing. habil. Suchart Siengchin is the President of King Mongkut's University of Technology North Bangkok (KMUTNB), Thailand. He earned his Dipl.-Ing. in Mechanical Engineering from the University of Applied Sciences, Giessen/Friedberg, Hessen, Germany in 1999; MSc in Polymer Technology from the University of Applied Sciences, Aalen, Baden-Wuerttemberg, Germany in 2002; MSc in Material Science at the Erlangen-Nürnberg University, Bayern, Germany in 2004; Doctor of Philosophy in Engineering (Dr.-Ing.) from the Institute for Composite Materials, University of Kaiserslautern, Rheinland-Pfalz, Germany in 2008; and Postdoctoral Research from Kaiserslautern University and School of Materials Engineering, Purdue University, Indiana, USA. In 2016, he earned the habilitation at the Chemnitz University in Sachen, Germany. He worked as a Lecturer for Production and Material Engineering Department at The Sirindhorn International Thai-German Graduate School of Engineering (TGGS), KMUTNB. He has been a full Professor at KMUTNB and became the President of KMUTNB. He won the Outstanding Researcher Award in 2010, 2012, and 2013 at KMUTNB. His research interests are in polymer processing and composite material. He is Editor-in-Chief of *KMUTNB International Journal of Applied Science and Technology* and the author of more than 150 peer-reviewed journal articles. He has participated in presentations in more than 39 international and national conferences with respect to materials science and engineering topics.

Contributors

Md Jaynal Abedin
Department of Materials and
  Metallurgical Engineering
Bangladesh University of Engineering
  and Technology (BUET)
Dhaka, Bangladesh

Anika Anjum
Department of Biomedical Engineering
Bangladesh University of Engineering
  and Technology (BUET)
Dhaka, Bangladesh

D. Aravind
University Science Instrumentation
  Centre
Madurai Kamaraj University
Madurai, India
and
Department of Mechanical Engineering
Kalasalingam Academy of Research
  and Education
Krishnankoil, India

Yashdi Saif Autul
Department of Mechanical Engineering
Military Institute of Science and
  Technology
Dhaka, Bangladesh

S. Bolka
Faculty of Polymer Technology
Slovenj Gradec, Slovenia

Chandrasekar Muthukumar
Department of Aeronautical
  Engineering
Hindustan Institute of Technology &
  Science
Chennai, India

W.S. Chow
School of Materials and Mineral
  Resources Engineering
Universiti Sains Malaysia
Nibong Tebal, Malaysia

**Muhammad Ifaz Shahriar
Chowdhury**
Department of Mechanical Engineering
Military Institute of Science and
  Technology
Dhaka, Bangladesh

C.D. Midhun Dominic
Department of Chemistry
Sacred Heart College
Cochin, India

Dineshkumar Harursampath
Department of Aerospace Engineering
Nonlinear Multifunctional Composites:
  Analysis and Design (NMCAD)
  Laboratory
Indian Institute of Science
Bangalore, India

Md Enamul Hoque
Department of Biomedical Engineering
Military Institute of Science and
  Technology (MIST)
Dhaka, Bangladesh

Md Sarif Sakaeyt Hosen
Department of Materials and
  Metallurgical Engineering
Bangladesh University of Engineering
  and Technology (BUET)
Dhaka, Bangladesh

Naman Jain
Meerut Institute of Engineering and
    Technology
Meerut, India

Aswathy Jayakumar
Materials and Production Engineering
The Sirindhorn International Thai-
    German Graduate School of
    Engineering (TGGS)
King Mongkut's University of
    Technology North Bangkok
Bangkok, Thailand

Athul Joseph
Department of Aerospace Engineering
Nonlinear Multifunctional Composites:
    Analysis and Design (NMCAD)
    Laboratory
Indian Institute of Science
Bangalore, India
and
Department of Materials Engineering
Katholieke Universitiet Leuven
Belgium

Vinayak Kallannavar
Department of Mechanical Engineering
National Institute of Technology
    Karnataka
Surathkal, India

Jasila Karayil
Government Women's Polytechnic
    College
Calicut, India

Ashish Kasam
Department of Aeronautical
    Engineering
Hindustan Institute of Technology &
    Science
Chennai, India

Subashchandra Kattimani
Department of Mechanical Engineering
National Institute of Technology
    Karnataka
Surathkal, India

Anwar Kornipalli
Department of Aeronautical
    Engineering
Hindustan Institute of Technology &
    Science
Chennai, India

Senthilkumar Krishnasamy
Department of Mechanical Engineering
Francis Xavier Engineering College
Tirunelveli, India

Shengyu Li
Wuhan Textile University
Wuhan, China

Vinyas Mahesh
Department of Mechanical Engineering
National Institute of Technology
Silchar, India
and
Department of Mechanical Engineering
    and Aeronautics
City, University of London
London, England, UK

Vishwas Mahesh
Department of Industrial Engineering
    and Management
Siddaganga Institute of Technology
Tumkur, Karnataka
and
Department of Aerospace Engineering
Indian Institute of Science
Bangalore, India

Md Mahmudul Haque Milu
Department of Biomedical Engineering
Jashore University of Science and
    Technology (JUST)
Jashore, Bangladesh

Sriram Mukunda
Department of Mechanical Engineering
NITTE Meenakshi Institute of
  Technology
Bangalore, India

M. Muthukrishnan
Department of Mechanical Engineering
KIT-Kalaignarkarunanidhi Institute of
  Technology
Coimbatore, India

B. Nardin
Faculty of Polymer Technology
Slovenj Gradec, Slovenia

Lin Feng Ng
Centre for Advanced Composite
  Materials
School of Mechanical Engineering
Faculty of Engineering
Universiti Teknologi Malaysia
Johor Bahru, Malaysia

Z.E. Ooi
School of Materials and Mineral
  Resources Engineering
Universiti Sains Malaysia
Nibong Tebal, Malaysia

Jyotishkumar Parameswaranpillai
Department of Science, Faculty of
  Science & Technology
Alliance University
Bengaluru, India

Sabarish Radoor
Materials and Production Engineering
The Sirindhorn International Thai-
  German Graduate School of
  Engineering (TGGS)
King Mongkut's University of
  Technology North Bangkok
Bangkok, Thailand

Md Asadur Rahman
Department of Biomedical Engineering
Military Institute of Science and
  Technology (MIST)
Dhaka, Bangladesh

Md Zillur Rahman
Department of Mechanical Engineering
Ahsanullah University of Science and
  Technology
Dhaka, Bangladesh

G. Rajeshkumar
Department of Mechanical Engineering
PSG Institute of Technology and
  Applied Research
Coimbatore, India

M. Ramesh
Department of Mechanical Engineering
KIT-Kalaignarkarunanidhi Institute of
  Technology
Coimbatore, India

Dharma Raghu Raj Reddy
Department of Aeronautical
  Engineering
Hindustan Institute of Technology &
  Science
Chennai, India

Renuka Sahu
Department of Aerospace Engineering
Indian Institute of Science
Bangalore, India

M.R. Sanjay
King Mongkut's University of
  Technology North Bangkok
Bangkok, Thailand

S. Arvindh Seshadri
Department of Mechanical Engineering
PSG Institute of Technology and
  Applied Research
Coimbatore, India

Sameena Shaik
Department of Aeronautical
    Engineering
Hindustan Institute of Technology &
    Science
Chennai, India

Arjun Siddharth
Department of Aerospace Engineering
Nonlinear Multifunctional Composites:
    Analysis and Design (NMCAD)
    Laboratory
Indian Institute of Science
Bangalore, India

Suchart Siengchin
Materials and Production Engineering
The Sirindhorn International Thai-
    German Graduate School of
    Engineering (TGGS)
King Mongkut's University of
    Technology North Bangkok
Bangkok, Thailand
and
Institute of Plant and Wood Chemistry
Technische Universität Dresden
Tharandt, Germany

Xiaoning Tang
Wuhan Textile University
Wuhan, China

Senthil Muthu Kumar Thiagamani
Department of Mechanical Engineering
Kalasalingam Academy of Research
    and Education
Krishnankoil, India

Theivasanthi Thirugnanasambandan
International Research Centre
Kalasalingam Academy of Research
    and Education
Krishnankoil, India

Akarsh Verma
University of Petroleum and Energy
    Studies
Dehradun, India

Part 1

*Vibration and Damping
of the Biocomposites*

1 Free Vibration and Damping Characterization of the Biocomposites
An Overview

M. Ramesh and M. Muthukrishnan
KIT-Kalaignarkarunanidhi Institute of Technology

CONTENTS

1.1 INTRODUCTION

Composites are referred to materials having distinct components that are different both in physical and chemical forms that can be modified for tailor-made applications. Composites are classified into various categories based on the composition of primary material called matrix and the secondary material referred to as reinforcement. Biocomposites in recent years have taken center stage in research to replace non-fossil fuel composites like plastics due to environmental concerns. Biocomposites are composite materials developed from the combination of a matrix like resin and a reinforcement material like natural fibers. The matrix phase can be derived from both renewable and nonrenewable resources. The reinforcement in biocomposites comprises biofibers such as crops, reprocessed wood, waste paper, and by-products from crops processing. These developed biocomposites possess various properties such as lightweight, suitable variable designer properties, cheap,

DOI: 10.1201/9781003173625-2

3

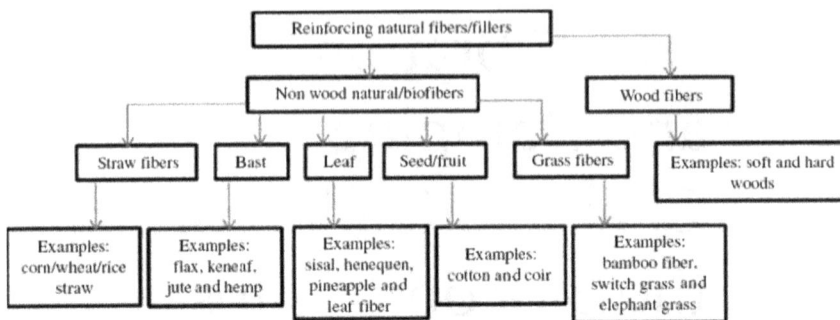

FIGURE 1.1 Types of natural fibers used as reinforcement in biocomposites [4].

biodegradable, and recyclable. Therefore, more research interests have been shown in developing biodegradable polymer composites from biological and other renewable resources. However, they exhibit poor mechanical properties, low chemical resistance, and poor durability. In order to overcome these disadvantages, some of the biocomposites are reinforced with nanofillers, thus improving its properties such as strength, fatigue life, wear resistance, stiffness, minimal material frequency, and high damping, which makes it suitable for wider applications in medicine, food packaging, cosmetics, construction, transport, agriculture, etc.

Biocomposites are generally classified into nonwood fibers (natural fibers) and wood fibers as shown in Figure 1.1. Natural fibers possess high cellulose content and have long fibers which give them increased tensile strength. Also, they possess superior vibration and damping properties when compared to synthetic fibers [1–4]. However, these fibers possess hydroxyl (OH) groups which have affinity toward water, and as a result, the increased water content results in swelling of the fiber, which affects the mechanical properties and dimensional stability [5,6] (Table 1.1).

Some of the prominent biocomposites as natural fibers are kenaf, jute, hemp, and sisal which have wide industrial applications. Other plant fibers in use are cotton, flax, recycled wood, and recycled paper [11]. The binder that holds the reinforcement in place and carries the load through it is represented by the matrix. Most widely used matrices are epoxy, polyester, polyurethane, and vinyl ester. Furthermore, the basic mechanical properties of prominent biocomposites are listed in Table 1.2. Before proceeding to the study of the biocomposites, some of the basic concepts and principles involved in the measurement of vibration and damping are necessary to be discussed.

1.2 VIBRATION AND DAMPING TECHNIQUES

Vibration is a mechanical phenomenon of a periodic motion of a particle or a body whereby oscillations occur from the equilibrium pivot point in the opposite directions. All physical bodies that have mass and elasticity are subjected to vibration. Mechanical vibrations are detrimental to the mechanical systems, and these dynamic properties contribute to higher noise, premature wear, structural instability, performance and fatigue failure, operator discomfort, and unsafe working conditions.

TABLE 1.1
Various Biocomposite Applications [3,7–10]

S. No.	Biocomposite	Applications
1	Flax and hemp fibers	Linen, nonwovens, insulation, paper industry, fiberboards, and cellular panels
2	Rice straw	Fiber boards
3	Cereal straw	Fire-resistant panels
4	Rice husks	Fiber cement blocks and applications involving improving acoustic and thermal properties
5	Bamboo fibers	Composition panels
6	Bagasse and soybean stalks	Particle boards
7	Coconut coir, banana fiber, and pineapple fibers	board manufacturing, concrete panels, floor furnishing materials, and seat supports
8	Sugar cane processing (bagasse)	Particle boards, fiber boards, and composition panel production
9	Cotton fibers	Sound absorption either on their own or by mixing with other fiber types and the different blending ratios

TABLE 1.2
Mechanical Properties of Different Biofibers [3]

S. No.	Material	Density (g/cm³)	Tensile Strength (MPa)	Young's Modulus (Gpa)	% of Elongation
1	Flax	1.45	500–900	50–70	1.5–4
2	Hemp	1.48	350–800	30–60	1.6–4
3	Kenaf	1.3	400–700	25–50	1.7–2.1
4	Jute	1.3	300–700	20–50	1.2–3
5	Bamboo	1.4	500–740	30–50	2
6	Cotton	1.51	300–600	6–12	3–10
7	Sisal	1.5	300–500	10–30	2–5
8	Coconut	1.2	150–180	4–6	20–40

Vibration is a serious concern in automobile, aviation, and electronics industry. The detrimental effects of vibration are reduced by using composite materials in noncritical structural components that have superior damping properties. All vibrations over a period of time come to rest if additional energy is not supplied where energy is dissipated over a period of time. Dissipation of the energy or damping is the important property of the tribological behavior of the composite materials, and it depends on the viscoelastic behavior of the matrix/reinforcement, stacking sequence, and the orientation of the reinforcements [12,13]. For example, transverse shear stress in the composites is determined by the quality of the fibers as reinforcement present in the biocomposites where these shear deformations become the primary cause of dissipating vibrational energy in the biocomposite materials. As a result, determining optimum structural properties without losing the biocomposites' strength and stiffness is critical.

Vibration is a complex cyclic process that involves periodic cyclic displacement of a particle around an equilibrium position. Vibration is generally classified as free vibration and forced vibration. Both these types of vibration dissipate energy due to a change of energy from the kinetic potential energy of vibration in every successive cycle. Vibration analysis is mainly done to check for the anomalies caused by vibration in the machinery using piezoelectric sensor or accelerometer. The main parameters involved in any study of vibrations are frequency, amplitude, and acceleration. Most damping vibration analyses fall under the following techniques: (a) determination of amplitude decay in free vibration, (b) determination of resonance curve during forced vibration, (c) forced vibration hysterics loop study in the stress–strain curve, (d) calculating the amount of energy absorbed during forced vibration, and (e) determination of sound wave propagation constants. Method "a" and "b" are the most preferred methods for testing among the other techniques.

Natural frequency is the frequency in which the body oscillates on its own after the initial disturbance. Generally, all the vibrational systems will have "n" degrees of freedom, which in turn have "n" distinct natural frequencies. On the other hand, damping has a reducing effect on the vibration. In physical systems, damping involves processes that result in the dissipation of stored energy due to oscillation or under cyclic stress. Generally, all composite materials possess high damping capacity due to their viscoelastic phenomena that are associated with the material properties. As a result, the composite materials' internal damping capacity is determined by (a) composition of matrix and reinforcement, (b) orientation of the reinforcements, and (c) surface treatment of the composite material.

Damping is classified as (a) viscous damping where the resisting force will be viscous in a fluid medium in the structures, (b) coulomb damping where the resisting force is due to dry friction between contact surfaces. It involves constant dissipation of force in the direction opposite to that of the velocity, and (c) solid or surface damping where the resistive force is due to the internal friction in the material. It is proportional to the displacement and is independent of frequency. In addition, damping ratio is the measure of decay of oscillations in successive cycles. The damping ratio is a system parameter, denoted by ζ (zeta) or ξ (Xi), that varies from undamped ($\xi = 0$), underdamped ($\xi < 1$) through critically damped ($\xi = 1$) to overdamped ($\xi > 1$) [14].

The other prominent testing parameters under dynamic analysis are dynamic modulus, storage modulus (E'), loss modulus (E''), damping efficiency or loss factor (tan δ), and logarithmic decrement (δ). The dynamic modulus represents the stress–strain ratio under vibratory conditions (free or forced vibrations). The storage modulus gives details about the amount of structure that has the capacity to store the input mechanical energy in a material. The storage modulus, which reflects the composite structure's elastic properties, generally show a decrease in values as the temperature rises. The loss modulus represents the viscous properties of a material. It determines the flow material under deformation and the amount of energy lost or heat dissipated in one cycle. Tan δ represents the conversion of vibration energy to heat, and thus, the energy dissipated per radian to the peak potential energy is the cycle. δ represents the natural logarithm of the rate of decrement of any two successive amplitudes under free damped vibration. The study of controlling vibration has made research-ers explore new composite viscoelastic and other polymeric materials owing to their high loss factor and high dissipation of energy and better damping capacity [15,16].

For analyzing the vibration and damping properties of composite panels, there are many vibration analysis techniques available today. General methods used for vibra-tion analysis are (a) dynamic mechanical analysis (DMA) which is used to calculate low-strain and low-frequency damping, i.e., up to 100 Hz and (b) experimental modal analysis (EMA) is used to determine vibration parameters such as natural frequency and modal damping frequency up to 1,000 Hz. Also, nondestructive technique (NDT) like impulse excitation technique (IET) is also in use. The samples for vibration test are made as per ASTM E756 and ISO 20816-1. The procedure is to excite the surface of the composite panel by a hammer (PCB Piezoelectric model 086C03) that is connected to the force sensor which will start responding to its natural frequency, and by their general geometric dimension details, the modulus of elasticity can be determined. The vibration response caused by the hammer method can be recorded using accelerometer or micro-phone (Bruel & Kjaer model 352B10). The data acquisition system is generally from Bruel and Kjaer with Photon software where data are collected from frequency response and time-dependent signals using the fast Fourier transform algorithm. Also, the nat-ural frequency of the composite beam is represented by the peaks in the frequency response function (FRF). Other methods of testing are load decrement free vibration decay method, frequency response plots, nonresonant method, tuned damped resonant method, dynamics shear nonresonant method, and coated cantilever method [17].

General design criteria of physical systems have primary focus on mechanical strength and weight reduction, whereas secondary design criteria always focus on vibration and damping properties, which involves high cost in labor and material. In recent years, many researchers worked on the vibration and damping nature of several polymer composite materials that are lightweight and environmentally friendly. An ideal structural design comprises high damping capacity along with good mechani-cal properties. Fiber architecture, fiber length, fiber loading, filler contents, fiber ori-entations, stacking sequences, and fiber chemical treatment are the parameters that affect vibration and damping in biocomposites. This chapter focuses on the vibration and damping properties of different biodegradable, high-strength, and sustainable natural fiber-reinforced biocomposites [18].

1.2.1 HEMP FIBER BIOCOMPOSITES

Hemp plant (*Cannabis sativa*) is an inexpensive tall and annual crop plant widely cultivated in hot and tropical countries. It is widely used in various industries such as paper, textiles, ropes, fiberboard, and heat-insulating materials [19,20]. Hemp fibers are known for their low density, high strength, and stiffness, which are similar to glass fibers. Hemp fibers are multicelled structures and irregular-shaped. The pri-mary bundle of bast fibers named phloem comprises mainly cellulose, hemicellulose, lignin, pectin, and wax [21]. The composition of hemp fiber is presented in Figure 1.2.

Hemp fibers are used as reinforcement material from preindustrial era and in terms of current industrial needs. Epoxy, polypropylene (PP), polyethylene (PE), polysty-rene (PS), and polylactic acid (PLA) are among the thermoplastic and thermosetting

Stem	Bundle	Elementary fiber
Diameter from 0.6 to 4.5 cm [32]	50 μm	10 μm [32]
(a)	(b)	(c)

Lumen
S3
S2 Secondary cell wall
S1
Primary cell wall
Middle lamella

Epidermis
Cortex
Phloem
Cambium
Xylem
Pith

Lumen

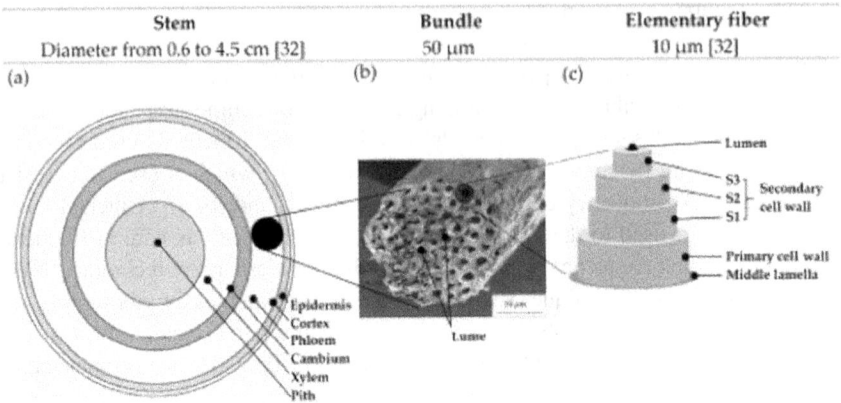

FIGURE 1.2 Hemp fiber composition: (a) transverse hemp stem segment [18], (b) bundle cross-section morphology [19], and (c) primary fiber [21].

matrices used for hemp fibers [8]. Recent research in hemp fiber composites involves its applications in sports materials and musical instruments due to its high vibration and damping capacity. They are also researched as reinforcement in brake pad applications. In addition, these natural fibers have high damping capacity and are widely used in various industrial applications for vibration and noise reduction. Ettati et al. analyzed the vibration damping properties and viscoelastic behavior of short hemp fibers by creating composite panels with a PP matrix as a matrix. As a resin matrix compatibilizer, maleic anhydride-grafted polypropylene (MA-g-PP) and maleic anhydride-grafted polyethylene octane (MA-g-PEO) are used. The biocomposite specimens thus fabricated using hemp fibers with 0–60 wt.% fiber content are subjected to vibration damping test. The samples are attached to accelerometer in the cantilever position where a rubber hammer is used for initiating the vibration [22,23]. A data acquisition program is used to record the vibration history of the test. The schematic setup is shown in Figure 1.3. Using the decay rate process, the damping ratio of the resulting cantilever beam hemp specimen is determined (Table 1.3).

The results indicated that as the percentage of fiber content in composite structures increases, the characteristics of the composite structures change from viscoelastic to elastic nature, which drastically reduces the damping ratio. But in the case of uncoupled hemp fiber, the damping ratio increases up to 30 wt.% and it decreases with increasing wt.% as shown in Figure 1.4. In the absence of compabilizer, adhesion between matrix and fiber is low which created more fiber–matrix interfacial area where more dissipation of energy takes place. Since the energy dissipation is influenced by the smaller fiber–matrix interfacial field, the effect of the compatibilizer has a decreasing effect on the damping ratio.

1.2.2 BANANA AND SISAL FIBERS-REINFORCED COMPOSITES

Sisal fibers are extracted from the leaves of the *Agave sisalina* plant which grows in Mexico's hot climate regions and in all soil types except clay. The leaves are

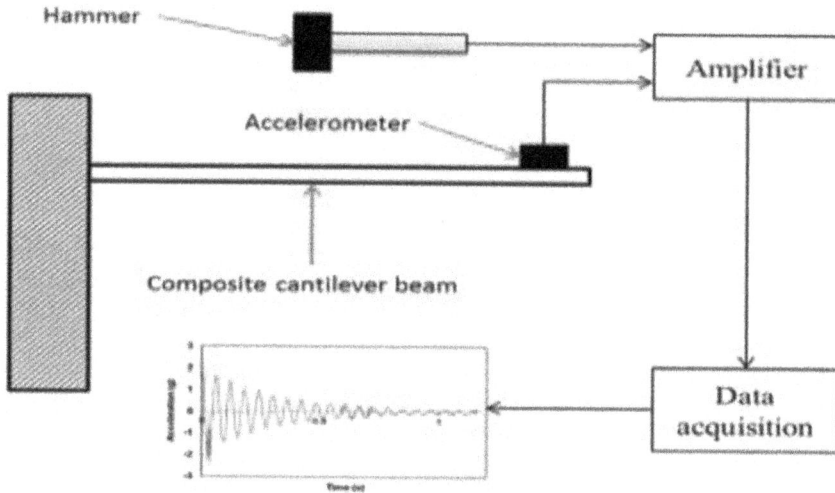

FIGURE 1.3   Schematic representation of the vibration testing system [22].

TABLE 1.3
Composition and Calculated Natural Frequency and Damping Ratio of Short Hemp Fiber Reinforcement Composite Samples under Various wt.% [22]

Specimen	Hemp (wt.%)	PP (wt.%)	MA-g-PP (wt.%)	MA-g-POE (wt.%)	Natural Frequency (Hz)	Damping Ratio
PP	0	100	0	0	10.50	0.053
10 H	10	90	0	0	12.52	0.097
20 H	20	80	0	0	12.66	0.120
30 H	30	70	0	0	11.68	0.140
40 H	40	60	0	0	11.60	0.136
50 H	50	50	0	0	11.11	0.068
60 H	60	40	0	0	11.81	0.046
30-H2.SPP-MAH	30	67.5	2.5	0	11.72	0.083
40H2.SPP-MAH	40	57.5	2.5	0	13.21	0.053
30HSPP-MAH	30	65	5	0	13.88	0.055
40HSPP-MAH	40	55	5	0	13.30	0.049
30H2.SPOE-MAH	30	67.5	0	2.5	12.47	0.064
40H2.SPOE-MAH	40	57.5	0	2.5	12.19	0.053
30HSPOE-MAH	30	65	0	5	12.50	0.052
40HSPOE-MAH	40	55	0	5	12.53	0.052

crushed to a pulp, the resultant pulp is washed, and the fibers are extracted from machine decortications. The fibers are then dried under sun and dyed further. The extracted fiber is creamy in color with around 100 cm in length and 0.3 mm approximately in diameter. The chemical composition of sisal fibers in weight % is given

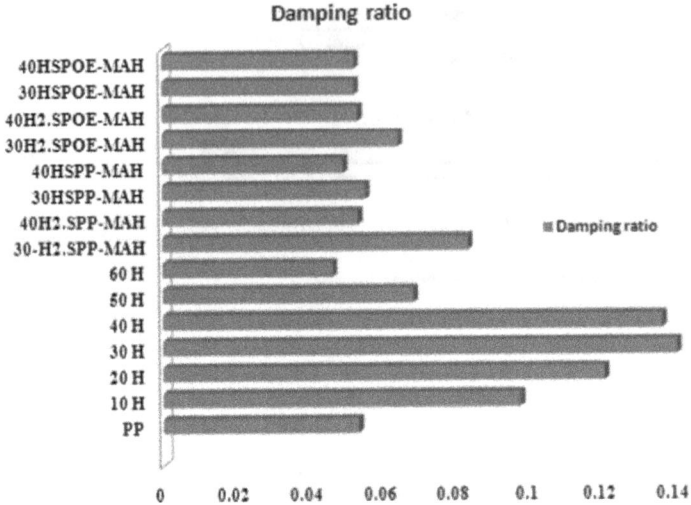

FIGURE 1.4 Damping ratio of short hemp fiber reinforcement composite samples under various wt.%.

approximately as 60% cellulose, 15% hemicellulose, 2%–4% pectin, 10%–20% lignin, 1%–4% water-soluble materials, 0.2% fat and wax, and 0.7%–1.5% ash. Various chemical treatments have been done on sisal fiber–epoxy composites, such as alkalization, silane coupling, and heating, to have better adhesion between the fiber and the matrix [24].

Banana fiber, on the other hand, is made from banana stems that have been cut longitudinally into small pieces and immersed in water for 15 days. The fibers thus extracted from the water are dried under sun for eight hours and are manually scrapped to retrieve fibers. These fibers are further treated with 50% NaOH for four hours to inhibit oxidation in banana fiber and further washed in normal running water to maintain the pH [25]. The chemical composition of banana fibers is 62%–64% cellulose, 15% hemicellulose, 3% pectin, 5% lignin, 10%–11% moisture content, 20° microfibril angle, and 11% lumen size [26–29]. Kumar et al. [30] investigated the impact of fiber length and weight percentage on the mechanical and damping properties of sisal banana polymer composites.

Rajesh et al. developed injection-molded composite panels of random sisal and banana fibers with varying weight percentages along with the resin and hardener, and tested for flexural modulus. It was found that at 50 wt.% of the banana fiber content, the flexural modulus peaked at 2,445.34 N/mm^2, and at 60 wt.%, it decreased 2,219.64 N/mm^2. Similarly, for sisal fiber, the flexural modulus peaked at 2,609.97 N/mm^2, and beyond that, the values decreased. Similarly, the 50 wt.% combination of hybrid fiber, which is a combination of banana and sisal fibers, yields high values than the other. The 50 wt.% fiber sample yields better flexural modulus results and they were also subject to similar testing with composites have NaOH treated fibers. The results have shown an improvement in the flexural modulus results as 3,071, 4,111, and 2,560 N/mm^2 for banana, sisal, and hybrids, respectively. Under free vibration test, the frequency increases up to 50 wt.% in all types of composite. On the other hand, 35 wt.% has shown a higher damping factor due to more surface contact between matrix and fiber. Similarly, chemical treatment with NaOH increases the modulus and stiffness of the composites [31].

Pothan et al. studied the dynamic analysis of hybrid composites (glass and banana) in the polyester matrix under three different temperatures and frequencies. Storage modulus, loss modulus, and damping performance are the parameters tested in dynamic analysis against the effects of temperature and fiber orientation patterns. The effect of temperature on the storage modulus of hybrid composites measured at a frequency of 10 Hz is shown in Figure 1.5a. A–F reflect various samples with a glass volume fraction ranging from 0.03 to 0.17. In these samples, glass fiber is the core, and outer skin comprises banana fiber. The other candidates considered for dynamic analysis are neat polyester and banana/polyester. Molecular motions are identified with the sudden drop in the storage modulus [32].

The plot Log E' can also be used to identify tension transfer between the matrix and the fiber. At lower temperatures, both samples have a high storage modulus, which can be due to variations in stress transfer between the two fibers. When the temperature approaches the transition range, the effect of thermal expansion on the

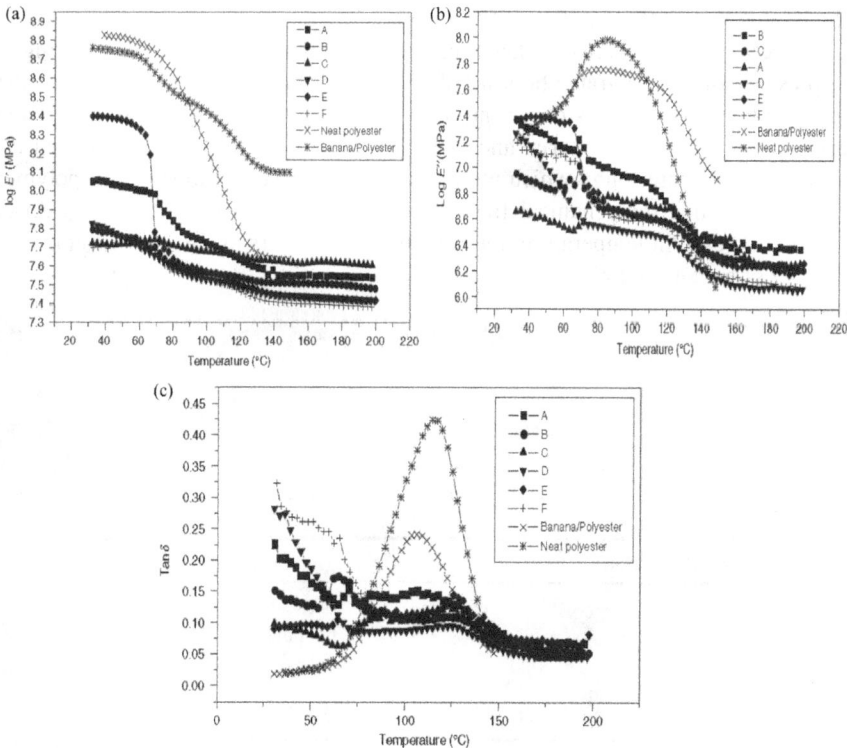

FIGURE 1.5 Effects of temperature on (a) storage modulus, (b) loss modulus, and (c) damping efficiency of hybrid composites [32].

fibers becomes more pronounced, affecting the bonding between the fiber and the matrix. As a result, all of the samples experience an initial drop in storage modulus about 55°C, followed by a second drop between 120°C and 150°C. The effect of temperature on E'' shown in Figure 1.5b is attributed to the properties of viscosity and the subsequent energy dissipation. The maximum value of E'' is around 55°C which indicates the temperature where maximum heat dissipation occurs. The loss modulus values are higher for samples having higher volume fractions [32]. Damping efficiency is dependent on the intermolecular motions within the composite, and lower tan δ values indicate the strong interactions between the fiber and the matrix, which inturn increases the damping and load-bearing capabilities of the composites (Figure 1.5c). Higher damping peaks in the sample, on the other hand, indicate a weak contact between the fiber and the matrix, allowing for more mobility and deformation before the material can return to its original shape.

The samples are prepared with varying layering patterns as shown in Figure 1.6. For varying glass volume fractions, the layering patterns are expressed by L1–L5. The findings showed that CL2 samples with an intimate mixture of glass and banana had higher tensile properties, which can be due to the high elongation of fibers and the intermingling of these fibers acting as crack arresters during tensile tests, resulting in high storage modulus values. For other layering patterns, glass fibers are taken at the periphery instead of banana skin to reduce the stress concentration on the tip which otherwise results in delamination [33].

Figure 1.7 shows the effect of layering patterns on damping coefficients. Owing to the presence of two separate fibers, all of the samples have two peaks, with the most significant change in peaks occurring in the CL3 and CL5 samples, where banana serves as the core layer material and glass serves as the outer layer. In some cases, additional peaks are found which are due to the presence of immobilized polymer which results micro-mechanical transitions at higher temperatures [34,35]. Also, with the increase in temperature, the shift in peaks occurs due to the movement of particles of banana and glass.

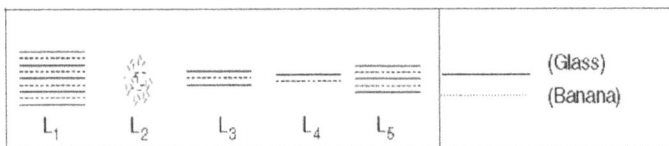

Sample marking	Layering pattern
L_1	G-B-G-B-G-B-G-B-G
L_2	Intimate mixture of G and B
L_3	G-B-G
L_4	G-B
L_5	G-B-G-B-G

G – glass; B – banana.

FIGURE 1.6   Layering pattern of glass–banana fiber composites.

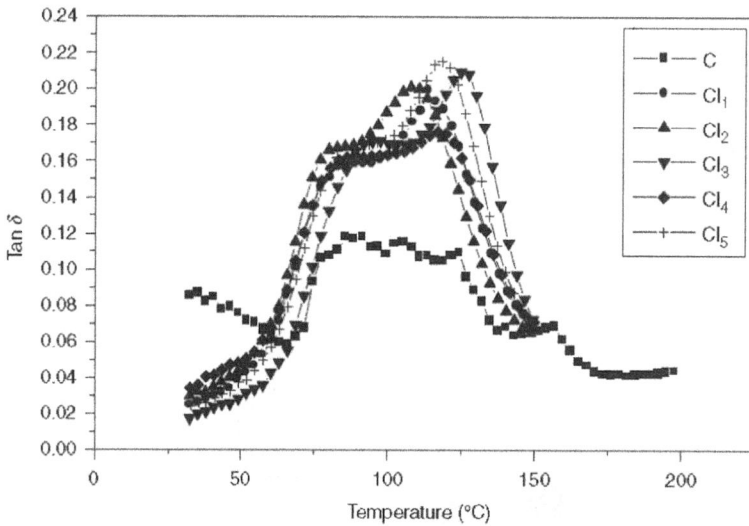

FIGURE 1.7 Effect of layering pattern on damping coefficient [32].

1.2.3 BAMBOO FIBER-REINFORCED COMPOSITES

Bamboo comprises cellulose fibers that are extracted from the natural bamboo stem. Further processing of this starchy cellulose pulp, such as alkaline hydrolysis, bleaching, and other chemical treatments, makes bamboo fibers softer than cotton. It contains on average 30% of cellulose and around 25% of lignin [36]. Bamboo fibers have high strength, elasticity, durability, tenacity, and color permeability and are easy to straighten and antibacterial in nature. These fibers are used in textile industry for intimate sweaters, socks, T-shirts, bath suits, etc. They are also used in making hygiene products like sanitary napkins, masks, food packing bags, etc. Bamboo fibers are also used in making of curtain, television covers, wall, etc. One of the prominent applications of bamboo fibers is that car-manufacturing company BMW is making use of the bamboo fiber type "Dasso Bamboo" to make car dashboard. Bamboo fibers have a high strength-to-weight ratio as compared to concrete and steel [37].

Takagi et al. [38] experimented with five different bamboo fibers: unidirectional long bamboo fiber, unidirectional short bamboo fiber, random short bamboo fiber, cotton-like bamboo fiber, and bamboo powder. The composites have bamboo fiber content from 6% to 64% and CP300 biodegradable resin. The tests were conducted on damping characteristic evaluation equipment. The specimen sizes were taken as $200 \times 10 \times 1$ mm. The test samples are shaken at the center for vibration, and half value method is used for finding the loss factor. In the graph, the effect of fiber length on the loss factor is plotted. The loss factor decreases as the fiber length increases, according to the findings shown in Figure 1.8a. The high kinetic energy coupled with low elastic modulus is due to high heat energy dissipation characteristics of the short fibers [37]. Figure 1.8b highlights the effects of bamboo fiber morphology on the loss factor. It can be seen that the short bamboo fibers or bamboo powder composites have the highest loss factor comparing to long fibers, which is again attributed to the

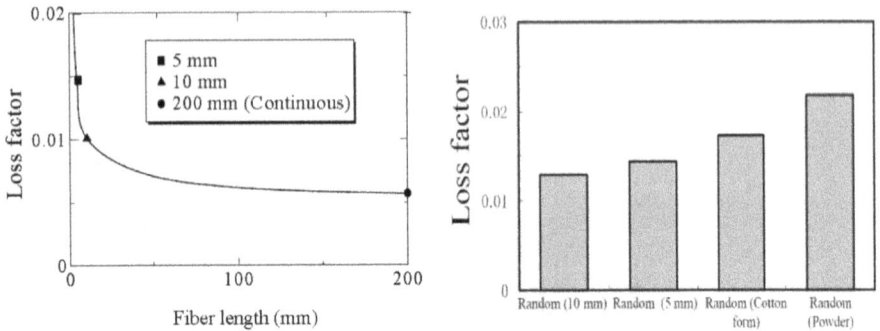

FIGURE 1.8   Effect of (a) fiber length and (b) fiber morphology on loss factor [38].

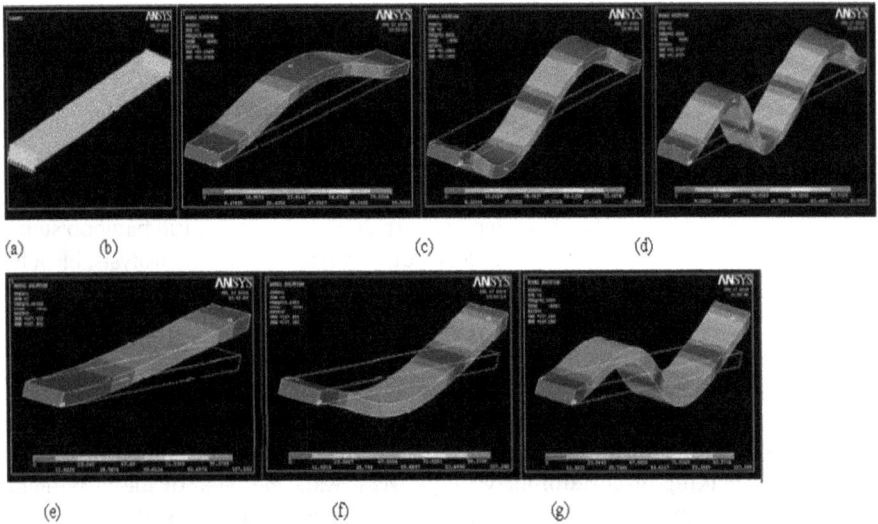

FIGURE 1.9   (a) Meshing model; (b–g) first three mode shapes in fixed-fixed and fixed-free end conditions, respectively.

high energy dissipation due to lower elastic modulus. When comparing random fiber composites to unidirectional fiber composites, it is also discovered that the loss factor decreases with increasing fiber content.

Jena et al. investigated bamboo fiber samples with different weight contents (10–15 wt.%) suspended under polyester resin matrix. Finite element analysis (FEA) using ANSYS software is used to study the frequency and modal analysis of short bamboo fiber-reinforced polymer composite beams. Figure 1.9a illustrates the meshing model of the bamboo fiber composite followed by Figure 1.9b–g, where the first three modes are under fixed-fixed and rest under fixed-free conduction. Under vibration damping test, it is found that natural frequencies tend to increase with increasing weight content in proportion to the increase in stiffness of the composite beam [41].

Krishna et al. [42] analyzed the vibration and damping behavior of the bamboo/epoxy polyurethane sandwich structures. The composite panel comprises polyurethane as a core material and bamboo/epoxy resin combinations as a face sheet. Vibration analysis was performed for in-plane and out-of-plane conditions using DEWESOFT software having impact hammer method. The results showed that as the foam density increases, the natural frequency and damping power of the composite structures increase. With increasing frequency, the damping factor is found to decrease. The authors also attribute the difference in the specimen's natural frequencies to the stiffness of the beams.

1.2.4 JUTE FIBER-REINFORCED COMPOSITES

Jute comes under bast fiber with the characteristics of long shiny soft fiber. They are lightweight, less cost, biodegradability, easy processing, and durable in nature. It is the second most abundantly produced vegetable fiber after cotton and is widely used in making sacks, curtains, wrapping bales of cotton, carpets, etc. Chemical composition of jute fiber includes cellulose, hemicelluloses, lignin, pectin, moisture, wax, etc. Rizala et al. [44] analyzed the vibration of jute fiber-based composites. The composite panels are developed using two layers of jute fibers along with three layers of epoxy resin. The goal is to use an impact hammer to analyze the composite panel's natural frequency. The vibration from the impact hammer method on the plate's surface is recorded using the PCB-1 axis accelerometer. The vibration signals are captured with an NI9250 data-capturing system and analyzed with LabView software. Similarly, the samples are placed in two supports, which are equally spaced to obtain natural frequency. Pickett's theory and ASTM norm E 1876-01 are used to investigate material properties such as modulus of elasticity ϵ, modulus of rigidity (G), and Poisson's ratio in flexural and torsional modes using natural frequency [35].

The modulus of elasticity ϵ is calculated by

$$E = 0.9465\left(mf_f^2/b\right)\left(L^3/t^3\right)T \tag{1.1}$$

where E is the modulus of elasticity (Pa), m is the mass (g), f_f is the flexural resonant frequency (Hz), b is the breadth (mm), t is the thickness (mm), L is the length of the plate specimen, and T is the correction factor for flexural mode that accounts for thickness of the plate. The frequency vibration responses of JFRPC are calculated numerically, and it is found to be compatible with the experimental results of IET. The natural frequency in flexural mode comes around 277.5 Hz, and in torsional mode, it comes around 212 Hz.

Rajesh and Pitchaimani [46] studied the effect of three different forms of warp and weft-oriented hybrid weaving patterns on inter-fly hybridization in jute and banana fiber polyester composites. Three combinations that are chosen are Type A weaving pattern where jute fiber is kept in warp direction and banana fiber is weaved in weft direction, Type B is the opposite of Type A, and Type C comprises the combination of banana and jute fiber in warp and weft directions. Epoxy resin was chosed as the matrix and was poured over the woven fiber mats to prepare the composites. The vibration damping tests were conducted by dynamic and thermal loading in nitrogen

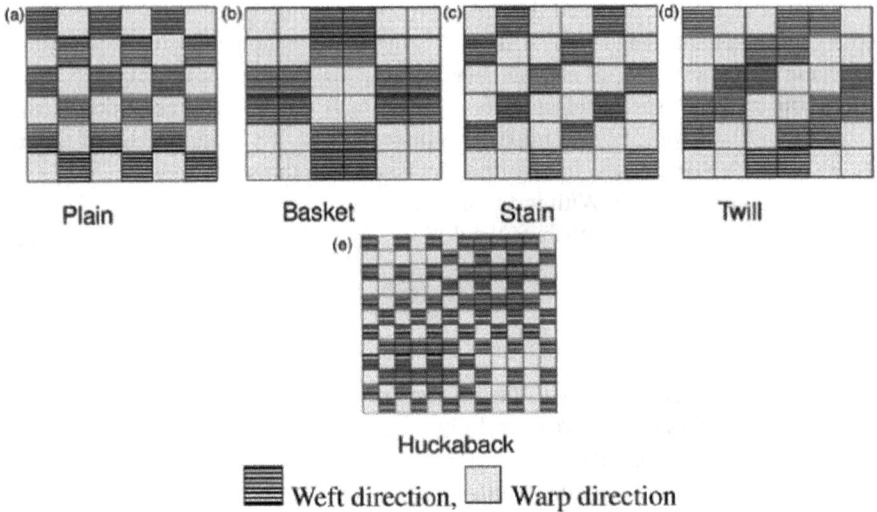

Plain   Basket   Stain   Twill

Huckaback

▤ Weft direction, ▢ Warp direction

FIGURE 1.10 (a–e) Schematic representation of weft and warp direction of woven mats [46].

atmosphere for 1 Hz under a temperature range of 0°C–200 °C. The objective was to identify the properties such as storage modulus, loss modulus, and damping factors of different types of woven mats fiber composites. Figure 1.10 represents the different types of woven mats with respect to weft and warp direction. Some of the prominent types are plain, basket, stain, twill, and huckaback [37–40].

The findings demonstrated that the basket woven mat composite sample has better dynamic mechanical properties than other woven mat forms. In the basket-type woven mats with Type A, B, and C samples, composite of jute fiber orientation in warp direction exhibits better dynamic mechanical properties. The storage modulus is improved with the woven-type fiber composite mats when compared to randomly oriented fiber samples. The effect of these woven types is also reflected with high natural frequencies for jute in warp direction-oriented basket-type samples. On the other hand, huck back-type woven sample exhibits higher damping capacity (Figure 1.11) [41–44].

1.2.5 COCONUT FIBER-REINFORCED COMPOSITES

Coconut fiber is a natural fiber derived from the coconut shell's outer husks. The coconut fibers are renewable, biodegradable, low density, nontoxic, high degree of water retention, and low cost. However, they do have disadvantages such as inflammability, dimensional instability at high temperatures, and degradability with humidity, which have made it as a difficult component as fiber in composites. Bujang et al. investigated the mechanical and vibration damping properties of a biocomposite made of randomly aligned coconut fiber and a polyester resin matrix with fiber content spanning from 5% to 15%. The results indicated that a high fiber content of 15% has poor tensile strength than 5% fiber content volume. Increased fiber content

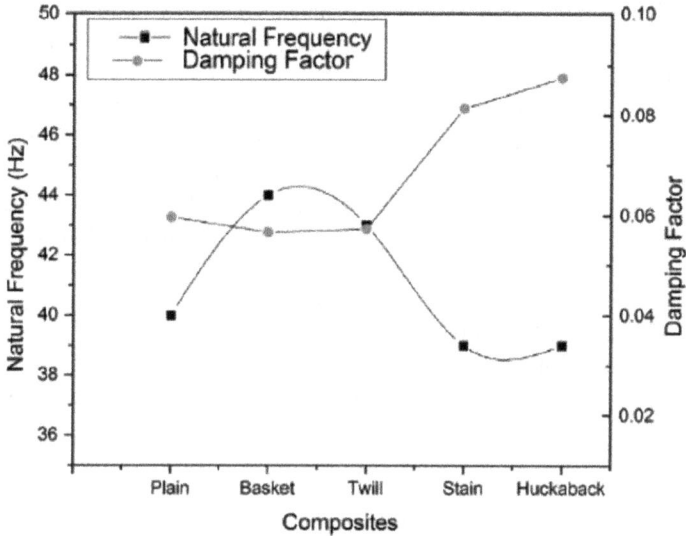

FIGURE 1.11 Natural frequency and damping power effects of weaving pattern [46].

negatively affects bonding due to a lack of interfacial adhesion between the fiber and the matrix, contributing to composite deformation [45].

Also in vibration test, composite with 5% fiber content volume shows high natural frequency than 10% and 15% fiber content when tested under different modes. According to the principles of the theoretical frequency, the natural frequency of the fiber is determined by mass and stiffness. Any increase in stiffness tends to increase the natural frequency, and any increase in mass reduces the natural frequency of the composite. Studies revealed that any increase in density of reinforcement of the fiber content in the composite will affect the mass of the composite to an increased proportion. In the current scenario, the mass of the composites remains neutral for all the samples, which proves that composites with low fiber volume content have higher stiffness than other samples with high fiber volume content. In this case, the contribution of fiber is increasing the stiffness of the composite specimen is low, which is proved by its increasing mechanical properties as well. On the other hand, samples with higher fiber content of 15% exhibit higher damping ratios that are attributed to less oscillatory motion during vibration testing thus making it suitable for applications involving vibration damping like soundproof interior structures [46,47].

It is analyzed the vibration damping properties such as loss factor, damping factor, and storage modulus over composites made of PP matrix and coconut fiber. Here, the coconut fibers are added in various proportions to characterize the viscoelastic properties of the composites. The composites have variable fiber content from 0% to 30%. The damping vibration test samples are prepared according to ASTM E756-05 standards [48]. The results indicated that natural frequency, storage modulus, loss modulus, and damping ratio decrease with an increase in fiber load. Also, the mechanical results indicated that composites with pure PP have higher values of modulus of elasticity than specimens with PP coconut fiber composites.

The results indicated the poor interfacial adhesion between the fiber and the matrix but this property is advantageous over applications where stress transfer is detrimental [49].

Few authors investigated the effects of stacking series on the tensile and vibration properties of sisal–coconut fiber hybrid composites. The sample is developed by piling pure coconut fiber (CCC) on top of pure sisal fiber (SSS). CCS, CSC, SSC, and SCS are the intermediate stacking sequences. The mechanical tests indicated that sisal–coconut polyester composites with the stacking sequences having sisal–sisal fiber together exhibit higher tensile strength compared to composites having coconut fibers together. The natural frequency of the composites was estimated using the hammer test process, which revealed that, with the exception of CCC and SSS mono fiber composites, the majority of the hybrid fiber composites have higher natural frequency values. The highest natural frequency is found in CCS composites [50,51].

1.2.6 FLAX FIBER-REINFORCED COMPOSITES

Flax (*Linum usitatissimum*) is a cellulose polymer generally called flax or linseed. It is grown for food (Linseed oil) and fiber. Flax fibers are derived from the stems of the flax bast plant. Long fine flax fibers that can be spun into yarn are used as linen in textile industries for making bed sheets, tablecloths, and underclothes. Short flax fiber is used for making kitchen towels, sails, tents, etc. Flax fiber is hydrophilic in nature and is composed of cellulose (70%), hemicellulose (16%), lignin (2%), and pectin (1.6%) as its constituents. It is studied the damping properties of the flax–epoxy composites with fiber orientation in unidirectional with variable laying angles of 0°, 45°, and 90°. Impact hammer method is used to record the natural frequency of various samples during flexural vibrations. Modal analysis helps in analyzing the dynamic characteristics of the composites in the natural frequency domain and care is taken to get the natural frequencies in pure bending mode. The damping factor decreases with increasing frequency for all fiber angles specimens, according to the findings. The maximum reduction in natural frequency occurs at 75° fiber direction, with no difference in loss factor seen at other angles. To improve the damping properties of composites, an additional viscoelastic layer is introduced. This additional layer didn't cause many changes in the natural frequencies for the first two modes, but a drastic decrease in the third and fourth modal analyses was found. The decrease in vibration amplitudes, and thus natural frequency, is due to the increase in surface mass of the material. On the other hand, the increase in surface mass affects structural rigidity but enhances flexural rigidity and increase in thickness. Overall, the addition of the viscoelastic layer shows improvement in damping coefficients [52].

The researchers analyzed the dynamic mechanical properties of flax fibers reinforced with epoxy, PP, and PLA and compared them with the carbon/epoxy and glass/epoxy composites. The composites are fabricated by resin transfer mold technology and by compression molding. The fiber volume percentage is maintained at 40% volume for all the samples. The composite plates are prepared with fiber orientation in unidirectional and twill. Temperature sweep test at specific frequencies of 0.1, 1, and 100 Hz was used to examine the composites' damping behavior. The composites are heated from 40°C to 120°C at a rate of 2 °C/min, while being deformed at each frequency [53,54]. The mechanical tensile test results indicated that carbon/epoxy and glass/epoxy composites exhibit higher tensile strength when compared to flax fiber composites. However, flax/epoxy composite shows higher elongation at the break than glass/epoxy one. The effective stiffness of the flax fiber decreases under axial loading and is lower compared to carbon and glass fibers due to the slipping of flax fiber within yarns. Overall, carbon/epoxy showed better mechanical properties compared to flax and glass fiber-reinforced composites. Compared to fiber orientations, unidirection samples showed higher tensile values of modulus of elasticity and percentage of elongation than twill samples. With regard to damping results under temperature sweep tests, the values of the loss factor and loss modulus shifted to higher temperatures. On the other hand, the storage modulus showed detrimental values at higher temperatures owing to the onset of fiber mobility in the amorphous phase. Beyond the initial peak values, the damping values increased at low frequencies. The results indicated that damping properties are more dominated by the matrix. The flax/PLA composite shows better damping and stiffness values compared to other samples.

It is explored the effect of polyols (glycerol and polyglycerol) on the vibration damping properties of flax fiber–epoxy composites [43]. In the ratio of 100:87.5:1.5, an epoxy resin system consists of diglycidyl ether of bisphenol as resin, methyl tetrahydrophthalic anhydride as curing agent, and 1-methylimidazole as a catalyst [55]. Flax is procured as 2/2 twill fabric and is used as the reinforcement for flax/epoxy composites. The flax fabric is further soaked in the aqueous solution of 2% concentrated glycerol and polyglycerol before being impregnated with epoxy resin system. Three types of flax fibers are developed: (a) plain flax without any polyol, (b) flax fiber soaked in glycerol, and (c) flax fiber soaked in polyglycerol.

The tensile results have shown a small decrement in the tensile values for polyol impregnated flax–epoxy composites than the plain composites. The values of flexural modulus remain constant for all types of composites at all types of modal analysis. With respect to the loss coefficient, the values increase with an increase in frequency due to the internal friction developed by molecules of heterogeneous compounds during vibration [56–59]. These molecules dissipate energy during vibration, which accounts for the increase in damping coefficient. The loss coefficients of the polyol-treated flax fiber composites showed improved vibration properties. Thus, polyol loadings below 5% in flax/epoxy composites have improved the loss coefficient values by 25%. Also, among the polyols, polyglycerol exhibits better vibration damping properties.

1.3 CONCLUSION

It is essential to investigate the vibrational and damping responses of fiber-based biocomposites in the context of industrial applications. This chapter attempts to review some of the published literature on biocomposites' vibration and damping properties. Fiber reinforcement's effects on the vibrational and damping responses of thermoset and thermoplastic polymer matrices are discussed. Surface treatment increases the free vibration and damping properties of biocomposites, according to the literature, by improving the interfacial bond between fiber and matrix. Because of the weak interfacial adhesion between the fibers and the matrix in the untreated composite, the damping values obtained are higher. This chapter also demonstrated that the external layers are the most significant when it comes to damping properties. The damping ratio of a biocomposite can be significantly improved by adding an additional layer, and it can be doubled by adding two layers.

REFERENCES

1. J. Flynn, A. Amiri, C. Ulven, Hybridized carbon and flax fiber composites for tailored performance. *Mater. Des.* 102 (2016) 21–29.
2. M. Ramesh, L. Rajeshkumar, Wood flour filled thermoset composites. Thermoset composites: Preparation, properties and applications. *Mater. Res. Found.* 38 (2018) 33–65. doi:10.21741/9781945291876-2.
3. M. Rueppel, J. Rion, C. Dransfeld, C. Fischer, K. Masania, Damping of carbon fibre and flax fibre angle-ply composite laminates. *Compos. Sci. Technol.* 146 (2017) 1–9.
4. K. N Bharath, S Basavarajappa, Applications of biocomposite materials based on natural fibers from renewable resources: A review. *Sci. Eng. Compos. Mater.* 23(2) (2016) 123–133.
5. W. Zhu, V. Nandikolla, B. George, Effect of bulk density on the acoustic performance of thermally bonded nonwovens, *J. Eng. Fiber. Fabr.* 10(3) (2015) 39–45.
6. M. Ramesh, L.R. Kumar, A. Khan, A.M. Asiri. (2020) Self-healing polymer composites and its chemistry. In: *Self-Healing Composite Materials*. Elsevier, pp. 415–427. doi:10.1016/b978-0-12-817354-1.00022-3.
7. M. Kucuk, Y. Korkmaz, The effect of physical parameters on sound absorption properties of natural fiber mixed nonwoven composites. *Text. Res. J.* 82(20) (2012) 2043–2053.
8. M.J. Mochane, T.C. Mokhena, T.H. Mokhothu, A. Mtibe, E.R. Sadiku, S.S. Ray, I.D. Ibrahim, O.O. Daramola, Recent progress on natural fiber hybrid composites for advanced applications: A review. *Express Polym. Lett.* 13 (2019) 159–198.
9. M. Ramesh, L. RajeshKumar, V. Bhuvaneshwari. (2021) Bamboo fiber reinforced composites. In: Jawaid M., Mavinkere Rangappa S., Siengchin S. (eds) *Bamboo Fiber Composites. Composites Science and Technology*. Springer, Singapore. doi:10.1007/978-981-15-8489-3_1.
10. L. Yu, K. Dean, L. Li, Polymer blends and composites from renewable resources. *Prog. Polym. Sci.* 31 (2006) 576–602.
11. A.M. Díez-Pascual, A.L. Díez-Vicente, ZnO-reinforced poly(3-hydroxybutyrate-co-3-hydroxyvalerate) bionanocomposites with antimicrobial function for food packaging. *ACS Appl. Mater. Interfaces.* 6 (2014) 9822–9834.
12. M. Ramesh, L.R. Kumar. (2020) Bioadhesives. In: Inamuddin R., Boddula M.I., Ahamed, Asiri A.M. (eds) *Green Adhesives*. doi:10.1002/9781119655053.
13. A. Bledzik, J. Gassan, Composites reinforced with cellulose based fibers. *Prog. Polym. Sci.* 24 (1999) 221–274.
14. J. M. Berthelot, Damping analysis of laminated beams and plates using the Ritz method. *Compos. Struct.* 74 (2006) 186–201.
15. T. Thamae, S. Aghedo, C. Baillie, D. Matovic. (2009) Tensile properties of hemp and Agave Americana fibres. In: Bunsell, A. R. (ed.) *Handbook of Properties of Textile and Technical* Fibres. Elsevier, pp. 73–99.
16. M. Ramesh, C. Deepa, L.R. Kumar, M.R. Sanjay, S. Siengchin. Life-cycle and environmental impact assessments on processing of plant fibres and its bio-composites: A critical review. *J. Ind. Text.* (2020). doi:10.1177/1528083720924730.
17. H.N. Dhakal, Z.Z. Hang, The Use of hemp fibres as reinforcements in composites. *Biofiber Reinf. Compos. Mater.* (2015) 86–103.

18. L. Nunes. (2017) *Nonwood Bio-Based Materials.* Elsevier, Amsterdam, The Netherlands.

19. G.Y. Zheng, Numerical investigation of characteristic of anisotropic thermal conductivity of natural fiber bundle with numbered lumens. *Math. Probl. Eng.* 2014 (2014) 1–8.

20. D. Balaji, M. Ramesh, T. Kannan, S. Deepan, V. Bhuvaneswari, L. Rajeshkumar, Experimental investigation on mechanical properties of banana/snake grass fiber reinforced hybrid composites. *Mater. Today: Proc.* 42 (2021) 350–355. doi:10.1016/j.matpr.2020.09.548.

21. F. Chegdani, S.T.S. Bukkapatnam, M. El Mansori, Thermo-mechanical effects in mechanical polishing of natural fiber composites. *Procedia Manuf.* 26 (2018) 294–304.

22. A Etaati, S Abdanan Mehdizadeh, H Wang, S Pather, Vibration damping characteristics of short hemp fibre thermoplastic composites. *J. Reinf. Plast. Compos.* 33 (2014) 330.

23. M. Ramesh, C. Deepa, M. Tamil Selvan, L. Rajeshkumar, D. Balaji, V. Bhuvaneswari., Mechanical and water absorption properties of Calotropis gigantea plant fibers reinforced polymer composites. *Mater. Today: Proc.* doi:10.1016/j.matpr.2020.11.480.

24. R. M. Zhi, Z. M. Qiu, L. Yuan, Y. G. Cheng, Z. H. Min, The effect of fiber treatment on the mechanical properties of unidirectional sisal reinforced epoxy composites. *Compos. Sci. Tech.* 61 (2001) 1437–1447.

25. J.O. Ajayi, K.A. Bello, S.D. Yusuf, Influence of retting media on the physical properties of bast fibers. *J. Chem. Soc. Nigeria.* 25 (2000) 112–115.

26. M. Sumaila, I. Amber, M. Bawa, Effect of fiber length on the physical and mechanical properties of random oriented, nonwoven short banana (Musa balbisiana) fiber /epoxy composite. ISSN. 2186-2476, Vol. 2, March 2013.

27. S. Pujari, A. Ramakrishna, M. Suresh Kumar. Comparison of jute and banana fiber composites: A review. *Int. J. Curr. Eng. Technol.* 2 (2014) 121–126. E-ISSN 2277-4106, P-ISSN 2347-5161. Doi: 10.14741/Ijcet/Spl.2.2014.22

28. V. Bhuvaneswari, M. Priyadharshini, C. Deepa. D. Balaji, L. Rajeshkumar, M. Ramesh. Deep learning for material synthesis and manufacturing systems: A review. *Mater. Today: Proc.* (2021). doi:10.1016/j.matpr.2020.11.351.

29. N. Venkateshwaran, A. Elayaperumal, G.K. Sathiya, Prediction of tensile properties of hybrid-natural fiber composites. *Composites: Part B.* 43 (2012) 793–796.

30. K.S. Kumar, I. Siva, P. Jeyaraj, J.W. Jappes, S. Amico, N. Rajini, Synergy of fiber length and content on free vibration and damping behavior of natural fiber reinforced polyester composite beam. *Mat. Des.* 56 (2014) 379–386.

31. M Rajesh, J. Pitchaimania, N Rajini, Free vibration characteristics of banana/sisal natural fibers reinforced hybrid polymer composite beam. *Procedia Eng.* 144 (2016) 1055–1059.

32. L.A. Pothan, C.N. George, Dynamic mechanical and dielectric behavior of banana–glass hybrid fiber reinforced polyester composites. *J. Reinf. Plast. Compos.* (2009). doi:10.1177/0731684409103075.

33. M.P. Sepe. (1998). *Properties Measured by DMA in Dynamic Mechanical Analysis for Plastics Engineering.* Plastics Design Library, New York, USA, pp. 11–14.

34. L.A. Pothan, S. Thomas, Polarity parameters and dynamic mechanical behaviour of chemically modified banana fiber reinforced polyester composites. *Comp. Sci. Technol.* 63 (2003) 1231–1240.

35. M. Ramesh, J. Maniraj, L. Rajesh Kumar. (2021) Biocomposites for energy storage. In: Khan A, Rangappa S.M., Siengchin S., Asiri A.M. (eds) *Biobased Composites: Processing, Characterization, Properties, and Applications.* Wiley Online Library, Wiley, Germany, pp. 123–142.

36. L.A. Pothen, Z. Oommen, S. Thomas, Dynamic mechanical analysis of banana fiber reinforced polyester composites. *Comp. Sci. Technol.* 63 (2003) 283–293.

37. S. Jain, R. Kumar, U.C. Jindal, Mechanical behaviour of bamboo and bamboo composite. *J. Mater. Sci.* 27 (1992) 4598–4604.
38. H. Takagi, H. Mori, M. Nakaoka, Damping performance of bamboo fibre-reinforced green composites. *WIT Trans. Eng. Sci.* 90 (2015). www.witpress.com, ISSN 1743-3533.
39. M. Ramesh, L. Rajeshkumar, D. Balaji, V. Bhuvaneswari. (2021) Green composite using agricultural waste reinforcement. In: Thomas S., Balakrishnan P. (eds) *Green Composites. Materials Horizons: From Nature to Nanomaterials.* Springer, Singapore, pp. 21–34. doi:10.1007/978-981-15-9643-8_2.
40. C. Subramanian, S.B. Deshpande, S. Senthilvelan, Effect of reinforced fibre length on the damping performance of thermoplastic composites. *Adv. Compos. Mater.* 20(4) (2011) 319–335.
41. P.C Jena, Free vibration analysis of short bamboo fiber based polymer composite beam structure. *Mater. Today.* 5(2) (2018) 5870–5875.
42. M. Krishna, B.S. Suresh, A. Joshi, A. Raj, R. Rajan, S. Bandyopadhyay, Vibration and Damping Behaviour of bamboo/epoxy polyurethane foam sandwich structures, RVJSTEAM, 1,2 (2020).
43. M. Ramesh, L. Rajeshkumar, Technological advances in analyzing of soil chemistry. *Appl. Soil Chem.* (2021) 61–78.
44. M. Rizala, A.Z. Mubarak, A. Razali, M. Asyraf, Free vibration characteristics of jute fibre reinforced composite for the determination of material properties: Numerical and experimental studies. *AIP Conference Proceedings*, 2019, vol. 2187, p. 050020.
45. S. Montecinos, S. Tognana, W. Salgueiro, Determination of the Young's modulus in CuAlBe shape memory alloys with different microstructures by impulse excitation technique. *Mater. Sci. Eng. A.* 676 (2016) 121.
46. M. Rajesh, J. Pitchaimani, Dynamic mechanical analysis and free vibration behavior of intra-ply woven natural fiber hybrid polymer composite. *J. Reinf. Plast. Compos.* 35 (2016) 228–242.
47. M. Ramesh, L. Rajeshkumar, D. Balaji. Aerogels for insulation applications. *Aerogels II: Prep., Prop. Appl.* 98 (2021) 57–76.
48. I.Z. Bujang, M.K. Awang, A.E. Ismail. Study on the dynamic characteristic of Coconut fibre reinforced composites. *Regional Conference on Engineering Mathematics, Mechanics, Manufacturing & Architecture (EMARC)* 2007, © 2007 Noise, Vibration and Comfort Research Group.
49. K. Murali, K. Mohana, Extraction and tensile properties of natural fibers: Vakka, date and bamboo. *Compos. Struct.* 77 (2007) 288–295.
50. M.V. Gelfuso, D. Thomazini, J.C. Silva de Souza, J.J. de Lima Junior, Vibrational analysis of coconut fiber-PP composites, *Mater. Res.* 17(2) (2014) 367–372.
51. M. Ramesh, L. Rajeshkumar, C. Deepa, M. Tamil Selvan, V. Kushvaha, M. Asrofi, Impact of silane treatment on characterization of ipomoea staphylina plant fiber reinforced epoxy composites. *J. Nat. Fibers.* (2021). doi:10.1080/15440478.2021.1902896.
52. D.A. Castello. (2004) *Modelagem e Identificação de Materiais Viscoelásticos do Domínio do Tempo. [Tese].* Universidade Federal do Rio de Janeiro, Rio de Janeiro.
53. S.N. Monteiro, R.J.S. Rodriguez, F.P.D. Lopes, B.G. Soares, Efeito da incorporação de fibras de coco no comportamentodinâmico-mecânico de compósitos com matriz-poliéster. *Tecnologiaem Metalurgia e Materiais.* 5(2) (2008) 111–115.
54. A. Saravana Kumar, P. Maivizhi Selvi, L. Rajeshkumar. (2017) Delamination in drilling of sisal/banana reinforced composites produced by hand lay-up process. In: *Applied Mechanics and Materials*, vol. 867. Trans Tech Publications, Switzerland, pp. 29–33.
55. H. Daoud, J.-L. Rebiere, Numerical and experimental characterization of the dynamic properties of flax fiber reinforced composites. *Int. J. Appl. Mech.* 8(5) (2016) 1650068.

56. F. Duc, P.E. Bourban, C.J.G. Plummer, J.-A.E. Manson, Damping of thermoset and thermoplastic flax fibre composites. *Composites: Part A.* (2014). doi:10.1016/j.compositesa.2014.04.016.
57. M.-J. Le Guen, R.H. Newmana, A. Fernyhough, M.P. Staiger, Tailoring the vibration damping behaviour of flax fibre-reinforced epoxy composite laminates via polyol additions. *Composites: Part A.* 67 (2014) 37–43.
58. T. Nishino, H. Naito, K. Nakamura, K. Nakamae, X-ray diffraction studies on the stress transfer of transversely loaded carbon fibre reinforced composite. *Composites Part A.* 31 (2000) 1225–1230.
59. V. Bucur. (2006) *Acoustics of wood.* 2nd ed. Springer, Switzerland.

2 Factors Affecting the Vibration and Damping Characteristics of Polymer Composites

Md Sarif Sakaeyt Hosen and Md Jaynal Abedin
Bangladesh University of Engineering
and Technology (BUET)

Md Enamul Hoque
Military Institute of Science and Technology (MIST)

CONTENTS

2.1 INTRODUCTION

Composites are multifunctional materials consisting of more than one material that are chemically different. Composites can be classified following the types of matrices and reinforcement. Fiber reinforced polymer (FRP) is one of the classifications of composites in which polymer matrix binds the fibers (Sahu & Gupta, 2017). In the composite industry, synthetic fibers are mainly used, but it is harmful to our health and environment (Sanyang et al., 2016). Therefore, we have to find some alternative superior materials, e.g., natural fiber-reinforced composites (NFRCs). Most NFRCs contain cellulose in a matrix of lignin, which shows CO_2 neutrality, ease of recycling, low-density, good thermal, mechanical, and insulation characteristics (Bajwa & Bhattacharjee, 2016;

DOI: 10.1201/9781003173625-3

Senthilkumar et al., 2018). Few fibers show some unwanted properties (low mechanical strength, less dimensional stability) because cellulose contains hydroxyl groups. These problems can be solved by making composites incorporated with a polymer (Arwinfar et al., 2016; Lakreb et al., 2015; Quiroga et al., 2016). NFRCs are used in the automobile (Senthilkumar, Siva et al., 2018), aerospace (Balakrishnan et al., 2016), building industries (Subramanian et al., 2011; Senthil Muthu Kumar et al., 2021), packaging sector (Chandrasekar et al., 2020), and so on. When fibers are reinforced with the polymeric material, the strength and stiffness of the composite will be increased. Damping properties are also affected by reinforcement. The noise control of home appliances, structural vibration, building acoustics is the anti-vibrational applications of high stiffness and good damping characteristics material (Senthil Kumar et al., 2021). Therefore, NFRCs can be a replacement for conventional materials. During service conditions, polymer composites are used to face various types of dynamic loadings. So, researches concerning the viscoelastic nature and the structure of these elements for ordaining their pertinent stiffness and damping properties for varied implementation are of great significance (López-Manchado & Arroyo, 2000). In many applications, where the material parts are subjected to dynamic stressing, it is required to control the level of stress and the amplitudes of deformation to maintain good vibrational damping properties (Sharif & Hoque, 2019).

Damping is an acoustic tool for individualizing materials. The property provides us the information about the mechanism of how the vibrational elastic energy dissipates into the material. In maintaining the impact resistance and fatigue life of a structure, the damping properties of the composites, of which the structure is made, also play a very significant role. There is no phase difference between applied stress and corresponding strain for an ideal elastic material. When materials contain a plastic or viscoelastic nature, then they would show phase lag between applied stress and strain because plastic or viscoelastic nature contributes to the stress (Lett et al., 2021). Usually, 90° phase lag is shown by the purely viscous material. Most of the polymers are viscoelastic materials and show higher damping than metallic materials. Molecular weight, degree of branching and cross-linking, number of monomeric units, and main chain stereo configuration are the factors on which the mechanical properties of polymers depend (Chauhan et al., 2009). In composite materials, reinforcement generates difficulties in the inner structure of the composite. So, the damping characteristics rely not only on the property of the constituent material, but also on the filler loading, fiber orientation, fiber loading, fiber length, fiber–matrix interface, staking sequence, hygrothermal treatment, and chemical treatment. Stress relaxation properties and static creep of materials are nearly connected with the damping phenomenon (Fong et al., 2015; Kumar et al., 2016; Senthilkumar et al., 2019). In this chapter, we report about the free vibration damping behavior and effects of various parameters on polymer composites.

2.2 DAMPING CALCULATIONS

Vibration can be classified as free vibration or forced vibration. The damping measurement technique depends on the type of vibration. Nowick and Berry explained the process of damping measurement (Nowick et al., 1975; Senthilkumar, Saba, et al.,

2021) in which the vibration produced by an impulse for free vibration method and the downfall of the amplitude are recorded. Damping is expressed as the logarithmic diminution, which is the normal logarithmic ratio of two adjacent amplitudes. For small stress wave amplitude after a cycle can be expressed as,

$$A_n = A_0 e^{-nd}$$

$$\log A_n = \log A_0 - nd$$

where A_0 is the original amplitude of the wave, A_n is the amplitude after the n-th cycle, n is the number of cycles, and d is the logarithmic diminution. The diminution can be calculated from the slope of the plot of $\log A_0$ versus n. Therefore, the damping coefficient is

$$\tan \delta = \frac{d}{\pi}$$

Spectrum analyzer could collect information about the stress wave propagation over a range of cycles.

2.3 FREE VIBRATION AND DAMPING BEHAVIOR

Many important variables affect the composite's vibration and damping properties. The following is a list of significant elements, each with a short description.

2.3.1 EFFECTS OF FILLER LOADING

Filler content in the polymer composite significantly affects its vibration and damping characteristics. Chauhan et al. (2009) observed the effect of wood flour filler loading on the damping behavior of wood-reinforced polypropylene composite with and without compatibilizer, and noticed its effect on the damping coefficient. The composite with 50% filler loading showed 33% lesser tan δ compared to the composite loaded with 10% filler content. The viscoelastic behavior of the composites turned into elastic as the filler content rise, and the percentage of energy dissipation decreased while retrievable strain energy increased (Kong et al., 2015). Because the dissipated energy of the wood is comparatively lower (tan δ of the scale of 0.01 for dry wood) than polypropylene (tan δ of the scale of 0.075), increasing the wood filler amount in the composite helps reduce energy dissipation. At low filler levels, there was a hardly noticeable difference in damping coefficients for composites produced with and without compatibilizer (up to 30%). The composites with compatibilizer showed a lesser tan δ value than the composites without compatibilizer at higher filler loadings of 40% and 50%. The dissimilation in damping coefficients with increased filler loading may be ascribed to enhance the wood particles–PP matrix interfacial adhesion, as well as exalted load transmission from matrix to fibers in compatibilizer composites. Figure 2.1 shows the effect of filler loading on damping behavior as described above.

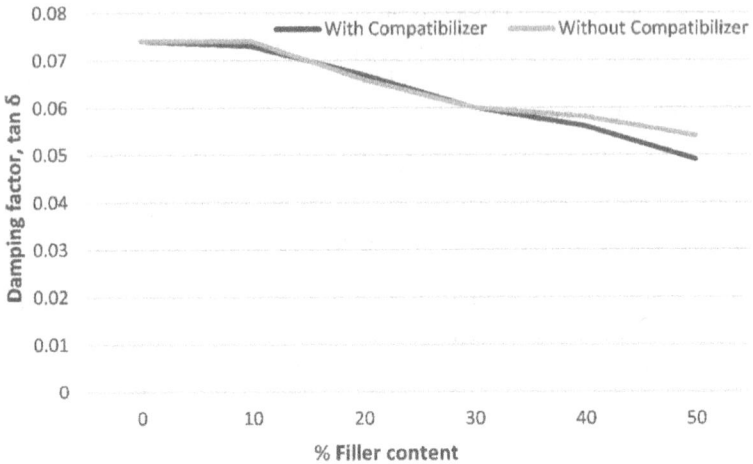

FIGURE 2.1 Effects of fiber loading on the damping behavior of wood-filled polypropylene composites with and without compatibilizer. (Reconstructed from Chauhan et al., 2009.)

Jena et al. (2014) studied the impact of chemosphere filler content on the frequency of bamboo-epoxy composite. The effect of chemosphere filler in a seven-layer composite with the highest natural frequency (NF). It has been discovered that adding chemosphere filler to a composite increases the inherent frequency of the composite. The NF of the composite with 3 wt.% chemosphere is the maximum. They discovered a reduction in NF after adding 3 wt.% chemosphere owing to agglomeration and a weak fiber–matrix contact, which reduces Young's modulus. For composites with 3 wt.% loadings, chemosphere may enhance NF by 13.92%. According to Ghofrani et al. (2016), the damping factor of plywood/waste tire rubber (WTR) rose as the WTR content increased with varied WTR levels. When 720 g of WTR was added, the damping factor reached its maximum point.

2.3.2 Effects of Fiber Orientation

The control of vibration and damping characteristics of wood-polymer composites is also influenced by fiber orientation. Bennet et al. (2014) investigated the effects of lamina fiber orientation on the free vibration properties of coconut sheath/*Sansevieria cylindrica* fiber (SCF) hybrid composites with five different intermediate lamina orientations (i.e., 0°, 30°, 45°, 60°, and 90°) using SCF, while keeping the overall wt.% of hybrid composites constant (40 wt.%), and noticed the NF and damping ratio fluctuating with increasing orientation angle. When examining the damping behavior of short carbon fiber composites, Murčinková et al. (2019) discovered that increasing the angle of the fiber orientation angle from 0° to 45° results in a reduction in the first NF, and an enhancement in the damping characteristics. The bending modulus of a beam also reduces as fiber orientation rises from 0° to 90° (giving unidirectional stress along the 0° direction), according to Berthelot and Sefrani (2004). However, in glass fiber polymer matrix composites, the loss factor reaches a maximum value of around

60°, whereas in Kevlar fiber composites, it reaches a maximum value of around 30°. Senthil Kumar et al. (2017) further investigated the impact of the orientation of the fiber on the vibration characteristics of sisal/polyester composites. Among the fiber orientations tested, 0°/45°/0° and 0°/90°/0° had the greatest NF and damping value.

2.3.3 EFFECTS OF FIBER LOADING AND FIBER LENGTH

Both fiber content and length are key factors for composite materials' vibration and damping properties. In some instances, the impact of certain factors is described below. In a study by In Eaati et al. (2014), the damping ratio of hemp fiber-reinforced polypropylene composites with different fiber concentrations (0–60 wt.%) was shown to rise with increasing hemp fiber content up to 30%, but decreases above 30%. Senthil Kumar et al. (2014) reported the impact of fiber length and stress on the vibration and damping properties of polyester composites made of short sisal fiber (SFPC) and banana fiber (BFPC). For various fiber wt.% and length, Figure 2.2a–c

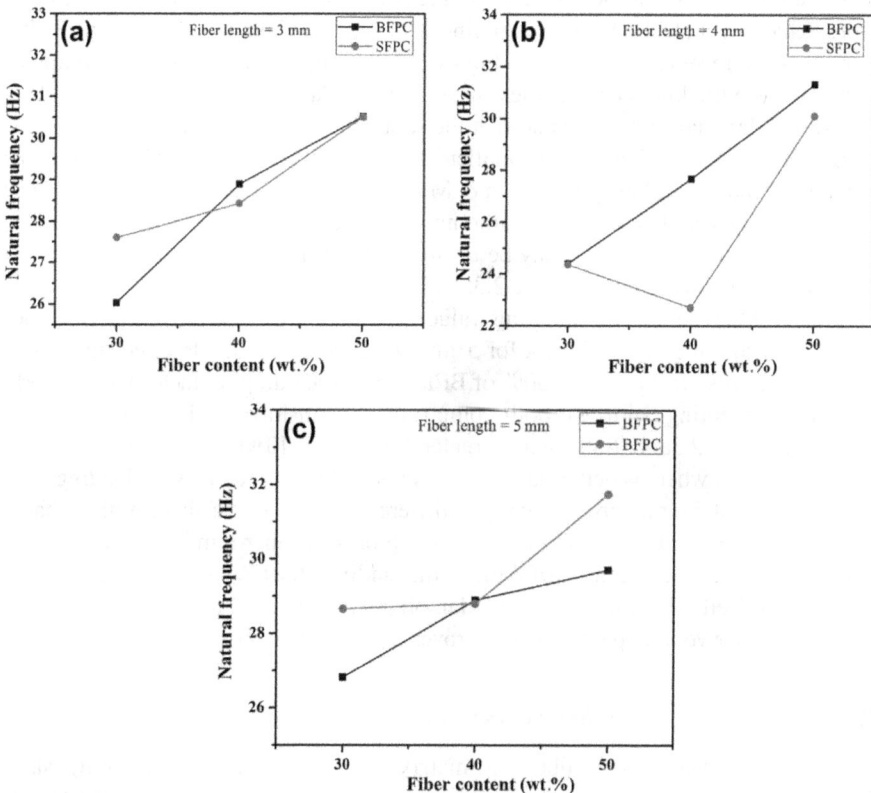

FIGURE 2.2 Effects of fiber content, length, and their types on the natural frequency of BFPC and SFPC for various fiber lengths: (a) 3 mm, (b) 4 mm, and (c) 5 mm (Senthil Kumar et al., 2014). (https://www.sciencedirect.com/science/article/abs/pii/S0261306913010881, reused with permission, License number 5110090289654.)

illustrates the first mode of NF. According to Figure 2.2a, BFPC and SFPC have distinct fiber/matrix interfaces at fiber length of 3 mm with 30 wt.% loading. The inherent frequencies of both composites blend with increasing fiber wt.%.

The shorter fiber length may enhance the contact area between the surface and matrix and also the fiber/matrix interface, improving stiffness in both composites. NF increases more linearly in BFPC than SFPC for all fiber lengths, which is likely linked to greater fiber content. Highest NF was found for banana/polyester composites with 4 mm fiber length with 50% weight. In SFPC, 30% and 50% fiber loads enhanced the NF for all fiber lengths. For all fiber lengths, the NF decreased by 40%. However, for low fiber load (30 wt.%), SFPCs had higher natural frequencies than BFPCs for all fiber lengths. This indicated that fiber stiffness, in addition to the fiber/matrix interface, affects composite NF. Figure 2.2 shows that for SFPCs with 40% fiber load, the NF decreases with fiber length. It may be owing to more widespread agglomeration, lowering NF and Young's modulus (Hoque et al., 2013).

Damping characteristics of BFPC and SFPC are shown in Figure 2.3a–c. With an increase in banana and sisal fiber content, damping firstly reduced and then increased. Due to its viscoelastic nature, greater resin concentration should dampen more (Chandra et al., 1999). This is in line with the BFPC tendency, while the SFPC trend was the reverse. According to Figure 2.3b and c, in addition to fiber composition, interface thickness and stiffness have a role in damping (Chandra et al., 1999). The BFPC dampens better than sisal at the same fiber loading (30 wt.%).

Because banana fiber has a smaller diameter (100–300 μm) (Mukherjee & Satyanarayana, 1984; Saravana Bavan & Mohan Kumar, 2010), it may have a thicker interface than sisal. However, the damping factor has reduced for banana fiber composites excepting 4 mm. This may be attributed to weaker interphase for 4 mm and 40% BFPC, as illustrated in Figure 2.3.

Banana fiber composite damping value drops abruptly compared to sisal fiber composite damping value. Except for 5 mm, the damping value dropped linearly as fiber loading rose from 40% to 50% of BFPC. And the damping factor rose linearly in SFPCs excepting 5 mm. In both composites, a transition of damping is clearly shown in Figure 2.3a–c. Banana has greater damping at 40 wt.% loading with a fiber length of 4 mm, whereas better damping was noticed in sisal at 40 wt.% loading with a fiber length of 5 mm, perhaps owing to differences in intrinsic surface fiber shape. While investigating the impact of fiber loading on sisal fiber-reinforced polypropylene composites, Munde et al. (2019) found that adding 30 wt.% sisal fibers to polypropylene improved NF while worsening damping. Hadi and Ashton (1996) found that increasing fiber volume percentage improves damping behavior.

2.3.4 EFFECTS OF FIBER–MATRIX INTERFACE

The interface area between fiber and matrix is a major issue in fiber composites. Fiber composites may have excellent or poor interfacial bonding (Enamul Hoque et al., 2014). In studies, a reduction in composite damping was linked to an increase in interface bonding (Chua, 1987; Cinquin et al., 1990; Dong & Gauvin, 1993). Even with excellent interfacial bonding, the interfacial area may have a substantial impact on the composites' mechanical characteristics and, in many cases,

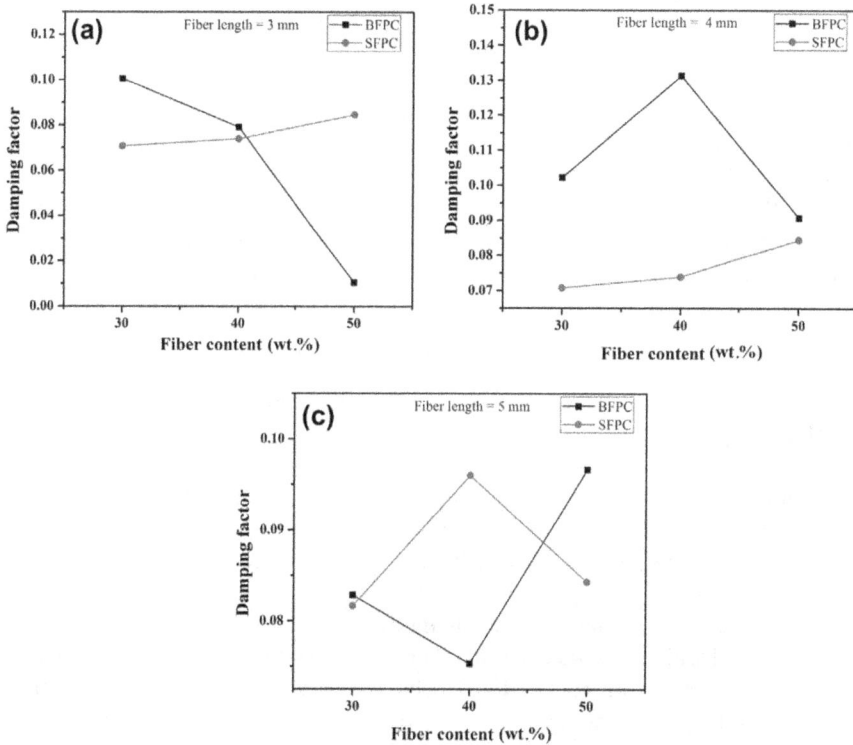

FIGURE 2.3 The effects of fiber content, length, and their types on damping factor of BFPC and SFPC for various fiber lengths: (a) 3 mm, (b) 4 mm, and (c) 5 mm (Senthil Kumar et al., 2014). (https://www.sciencedirect.com/science/article/abs/pii/S0261306913010881, reused with permission, License number 5110021157896.)

their damping level. During processing a thicker third phase termed interphase forms near the fiber/matrix interface due to the migration of some hardener molecules from the polymer matrix to the fiber sizing (He & Liu, 2005). The damping characteristics of composites are influenced by a comparative matrix to fiber shear stiffness, matrix shear modulus, and fiber diameter (He & Liu, 2005). They observed the impact of the fiber–matrix interface on damping and discovered that when adhesion between them improves (mostly to increase mechanical characteristics like strength), damping capabilities decrease. Botelho et al. (2005) mentioned also, as adhesion strength rises, damping characteristics decrease.

2.3.5 EFFECTS OF STACKING SEQUENCE

The stacking sequence of composites is an essential element to consider since the vibration and damping behavior of composites changes as the stacking order changes. Murugan et al. (2016) investigated the free vibrational characteristics of woven glass (G)/carbon (C) hybrid composite beams subjected to fixed-free boundary conditions by altering the stacking sequence. C–G–G–C had the greatest vibration response of the sequences tested. The GGGG sequence has a loss factor of 0.054, while C–C–C–C has the minimum values of 0.025, and the greatest NF of 29 Hz, 280 Hz, and 766 Hz for modes 1, 2, and 3, respectively. The stacking sequence of Sansevieria cylindrica (S) and coconut sheath (C) fiber-reinforced polyester composites impacts on its vibrational response was also studied by Bennet et al. (2015). When the results of the C–S–C stacking sequence were compared to that of the C–C–C, C–C–S, S–C–S, S–S–C, and S–S–S sequences, the NF of the C–S–C stacking sequence was determined to be better. The C–C–C sequence, on the other hand, has a greater damping ratio.

2.3.6 Effects of Hygrothermal Treatment

The most significant impact of thermal alteration of wood is that it increases dimensional stability by reducing hygroscopicity. Depending on the moisture level, reduced wood hygroscopicity stabilizes mechanical and acoustic characteristics (Arifur Rahman et al., 2015). Kubojima et al. (1998) found that in an autoclave, the damping coefficient (tan δ) value of spruce wood which is hygrothermally treated got reduced, whereas high relative humidity caused an increase. An increase in tan δ is a good thing. Such damping property reduction probably attributes to a structural alteration in the wood polymer, such as lignin cross-linking (Tjeerdsma & Militz, 2005). Crystallization of cellulose is another potential structural change response for decreased damping. Karami et al. (2020) observed that the mild hygrothermal treatment affects the physical and vibrational properties of spruce wood at temperatures ranging from 130°C to 150°C, whereas the relative humidity levels ranging from 0% to 25%, and discovered irreversible changes that resulted in an improvement in the physical properties and reduced damping (tan δ) and increased E'/d vibrational characteristics, which may be attributed to the treatment's intermediate relative humidity. Panda et al. (2013) studied the impact of temperature and moisture content on free vibration of delaminated woven fiber composite plates computationally and experimentally and discovered that NF decreases as moisture and temperature increases due to reduction of stiffness for all laminates (Panda, Sahu, and Parhi 2013). The decrease in fundamental frequencies is nonlinear in nature and depending on temperature and moisture concentration value, the issue may turn into one of the instabilities.

2.3.7 Effects of Chemical Treatment

Treating composite materials with the right chemical reagents may improve their vibration and damping characteristics, making them more useful. Ramakrishnan et al. (2021) investigated the effect of different concentrations of NaOH (0%, 2.5%, 5%, 7.5%) and wt.% of nanoclay addition (1, 3, 5, 7) on the viscoelastic and free vibration properties of Jute Fiber/Nanoclay Reinforced Epoxy Composites (JFREC), and noticed that composites incorporating 5 wt.% of NaOH and nanoclay jute fibers had the lowest tan δ as shown in Figures 2.4–2.7. Overall, it can be stated that fiber composites treated with 5% NaOH and containing 5% nanoclay perform better in terms of viscoelastic and free vibration characteristics than other natural fiber-based composites.

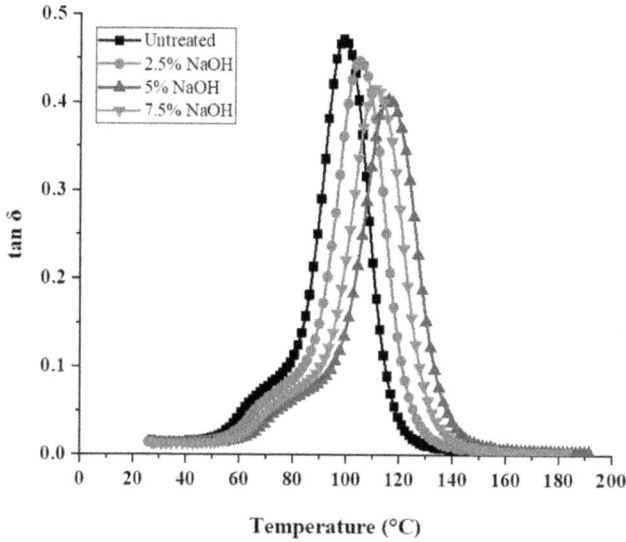

FIGURE 2.4 Effects of NaOH concentrations on tan δ (Ramakrishnan et al., 2021). (https://link.springer.com/article/10.1007/s10924-020-01945-y, reused with permission, License number 5117041305169.)

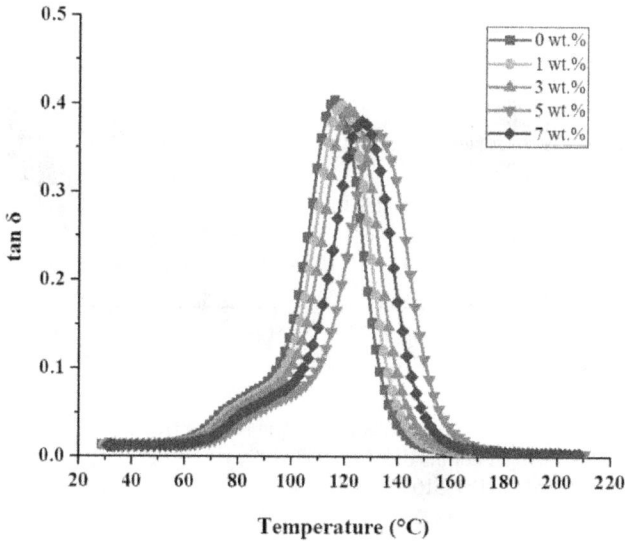

FIGURE 2.5 Effects of nanoclay concentrations on tan δ (Ramakrishnan et al., 2021). (https://link.springer.com/article/10.1007/s10924-020-01945-y, reused with permission, License number 5117041305169.)

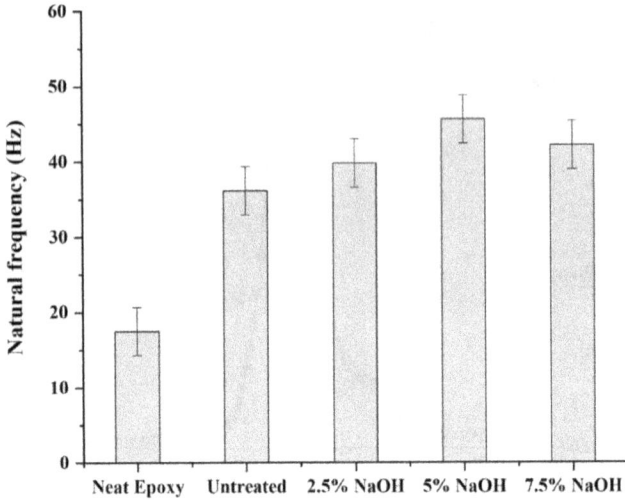

FIGURE 2.6 Effect of NaOH concentrations on natural frequency (Ramakrishnan et al., 2021). (https://link.springer.com/article/10.1007/s10924-020-01945-y, reused with permission, License number 5117041305169.)

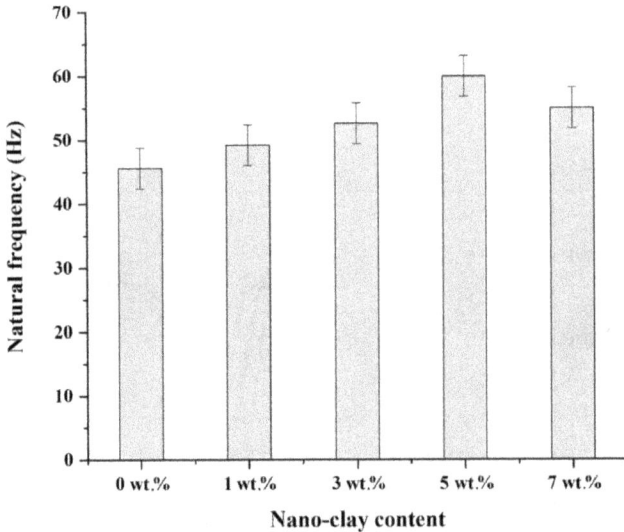

FIGURE 2.7 Effects of nanoclay concentrations on natural frequency (Ramakrishnan et al., 2021). (https://link.springer.com/article/10.1007/s10924-020-01945-y, reused with permission, License number 5117041305169.)

Arumuga Prabu et al. used an alkali solution and saline to treat banana fiber. A vibrational study was performed on prepared red mud-filled banana/polyester composites. The NF and damping are trending in opposite directions here. Untreated fiber composite has the highest NF, whereas saline-treated fiber composite has the lowest. Among the other composites, the damping of alkali-treated composites was shown to be better (Arumuga Prabhu et al., 2014). Sumesh et al. (2019) noticed the increment in NF and damping factor while increasing the alumina content in a hybrid combination of sisal/coir, sisal/banana, and banana/coir composites in all first three modes of vibration.

2.4   CONCLUSIONS

In this chapter, we reported how the free vibration and damping behavior of polymer composites changes with various parameters (filler loading, fiber orientation, fiber length, tacking sequence, hydrothermal effects, and chemical treatment). The key findings of this chapter are:

- Polymer composites will provide better vibration and damping properties if they are chemically treated with the right reagent and concentration. Reagents generate bonding with the internal polymer and create hindrance for environmental factors (air, water, etc.).
- Thermal treatment increases the dimensional stability of composites by reducing moisture content. As a result, the product shows better mechanical and acoustic properties. Hygrothermal treatment shows reduced damping and increased vibrational characteristics because of cross-linking of lignin, moisture, and crystallization of cellulose.
- Stacking sequences has an impact on damping characteristics. Among the sequences, CSC shows better NF, while CCC shows better damping properties.
- As fiber–matrix strength increases, damping decreases. Fiber length, fiber loading, and fiber orientation affect the damping and vibrational properties randomly. Different types of wood exhibit different results.

The use of composites increases day by day especially naturally found reinforcement composites because of the availability and outstanding properties of natural resources. Therefore, natural fiber can be an alternative. Very few researches have been conducted about the free vibration and damping properties of various polymer composites, and how various parameters regulate them. This chapter has given a brief idea about the impact of various parameters on the vibration and damping behavior of polymer composites, and also the manifestation of the lacking of information and researches in this field. There is a huge research opportunity in this field to come up with some extraordinary properties by analyzing with further deep investigations and advancements controlling those parameters.

REFERENCES

Arifur Rahman, M., Parvin, F., Hasan, M., & Hoque, M. E. (2015). Introduction to Manufacturing of Natural Fibre-Reinforced Polymer Composites. In M. S. Salit, M. Jawaid, N. B. Yusoff, & M. E. Hoque (Eds.), *Manufacturing of Natural Fibre Reinforced Polymer Composites* (pp. 17–43). Springer International Publishing. https://doi.org/10.1007/978-3-319-07944-8_2.

Arumuga prabu, V., Uthayakumar, M., Manikandan, V., Rajini, N., & Jeyaraj, P. (2014). Influence of redmud on the mechanical, damping and chemical resistance properties of banana/polyester hybrid composites. *Materials & Design*, *64*, 270–279. https://doi. org/10.1016/j.matdes.2014.07.020.

Arwinfar, F., Hosseini Hashemi, S. K., Latibari, A., Lashgari, A., & Ayrilmis, N. (2016). Mechanical Properties and Morphology of Wood Plastic Composites Produced with Thermally Treated Beech Wood. *BioResources*, *11*, 1494–1504. https://doi.org/10.15376/ biores.11.1.1494-1504.

Bajwa, D. S., & Bhattacharjee, S. (2016). Current progress, trends and challenges in the application of biofiber composites by automotive industry. *Journal of Natural Fibers*, *13*(6), 660–669.

Balakrishnan, P., John, M. J., Pothen, L., Sreekala, M. S., & Thomas, S. (2016). 12— Natural Fibre and Polymer Matrix Composites and Their Applications in Aerospace Engineering. In S. Rana & R. Fangueiro (Eds.), *Advanced Composite Materials for Aerospace Engineering* (pp. 365–383). Woodhead Publishing. https://doi.org/10.1016/ B978-0-08-100037-3.00012-2.

Bennet, C., Rajini, N., Jappes, J. W., Siva, I., Sreenivasan, V., & Amico, S. (2015). Effect of the stacking sequence on vibrational behavior of Sansevieria cylindrica/coconut sheath polyester hybrid composites. *Journal of Reinforced Plastics and Composites*, *34*(4), 293–306. https://doi.org/10.1177/0731684415570683.

Bennet, C., Rajini, N., Winowlin Jappes, J. T., Venkatesh, A., Harinarayanan, S., & Vinothkumar, G. (2014). Effect of lamina fiber orientation on tensile and free vibration (by Impulse Hammer Technique) properties of coconut sheath/Sansevieria cylindrica hybrid composites. *Advanced Materials Research*, *984–985*, 172–177. https://doi. org/10.4028/www.scientific.net/AMR.984-985.172.

Berthelot, J.-M., & Sefrani, Y. (2004). Damping analysis of unidirectional glass and Kevlar fibre composites. *Composites Science and Technology*, *64*(9), 1261–1278. https://doi. org/10.1016/j.compscitech.2003.10.003.

Botelho, E. C., Costa, M. L., Pardini, L. C., & Rezende, M. C. (2005). Processing and hygrothermal effects on viscoelastic behavior of glass fiber/epoxy composites. *Journal of Materials Science*, *40*(14), 3615–3623. https://doi.org/10.1007/s10853-005-0760-2.

Chandra, R., Singh, S. P., & Gupta, K. (1999). Damping studies in fiber-reinforced composites – a review. *Composite Structures*, *46*(1), 41–51. https://doi.org/10.1016/ S0263-8223(99)00041-0.

Chandrasekar, M., Senthilkumar, K., Thiagamani, S. M. K., Radoor, S., Parameswaranpillai, J., & Siengchin, S. (2020). Chitosan-Based Hybrid Nanocomposites for Food Packaging Applications. In S. M. Rangappa, P. Jyotishkumar, S. M. K. Thiagamani, S. Krishnasamy, & S. Siengchin (Eds.), *Food Packaging* (pp. 327–346). CRC Press, Boca Raton, FL.

Chauhan, S., Karmarkar, A., & Aggarwal, P. (2009). Damping behavior of wood filled polypropylene composites. *Journal of Applied Polymer Science*, *114*(4), 2421–2426. https:// doi.org/10.1002/app.30718.

Chua, P. S. (1987). Dynamic mechanical analysis studies of the interphase. *Polymer Composites*, *8*(5), 308–313. https://doi.org/10.1002/pc.750080505.

Cinquin, J., Chabert, B., Chauchard, J., Morel, E., & Trotignon, J. P. (1990). Characterization of a thermoplastic (polyamide 66) reinforced with unidirectional glass fibres. Matrix additives and fibres surface treatment influence on the mechanical and viscoelastic properties. *Composites*, *21*(2), 141–147. https://doi.org/10.1016/0010-4361(90)90006-I.

Dong, S., & Gauvin, R. (1993). Application of dynamic mechanical analysis for the study of the interfacial region in carbon fiber/epoxy composite materials. *Polymer Composites*, *14*(5), 414–420. https://doi.org/10.1002/pc.750140508.

Enamul Hoque, M., Aminudin, M. A. M., Jawaid, M., Islam, M. S., Saba, N., & Paridah, M. T. (2014). Physical, mechanical, and biodegradable properties of meranti wood polymer composites. *Materials & Design*, *64*, 743–749. https://doi.org/10.1016/j.matdes.2014.08.024.

Etaati, A., Mehdizadeh, S. A., Wang, H., & Pather, S. (2014). Vibration damping characteristics of short hemp fibre thermoplastic composites. *Journal of Reinforced Plastics and Composites*, *33*(4), 330–341.

Fong, T. C., Saba, N., Liew, C. K., De Silva, R., Hoque, M. E., & Goh, K. L. (2015). Yarn Flax Fibres for Polymer-Coated Sutures and Hand Layup Polymer Composite Laminates. In M. S. Salit, M. Jawaid, N. B. Yusoff, & M. E. Hoque (Eds.), *Manufacturing of Natural Fibre Reinforced Polymer Composites* (pp. 155–175). Springer International Publishing. https://doi.org/10.1007/978-3-319-07944-8_8.

Ghofrani, M., Ashori, A., Rezvani, M. H., & Arbabi Ghamsari, F. (2016). Acoustical properties of plywood/waste tire rubber composite panels. *Measurement*, 94, 382–387. https://doi.org/10.1016/j.measurement.2016.08.020.

Hadi, A. S., & Ashton, J. N. (1996). Measurement and theoretical modelling of the damping properties of a uni-directional glass/epoxy composite. *Composite Structures*, *34*(4), 381–385. https://doi.org/10.1016/0263-8223(96)00005-0.

He, L. H., & Liu, Y. L. (2005). Damping behavior of fibrous composites with viscous interface under longitudinal shear loads. *Composites Science and Technology*, *65*(6), 855–860. https://doi.org/10.1016/j.compscitech.2004.09.003.

Hoque, M. E., Ye, T. J., Yong, L. C., & Dahlan, K. M. (2013). Sago starch-mixed low-density polyethylene biodegradable polymer: Synthesis and characterization. *Journal of Materials*, *2013*, 1–7.

Jena, H., Pradhan, A. K., & Pandit, M. K. (2014). Effect of cenosphere filler on damping properties of bamboo-epoxy laminated composites. *Advanced Composites Letters*, *23*(1), 096369351402300103. https://doi.org/10.1177/096369351402300103.

Karami, E., Bardet, S., Matsuo, M., Bremaud, I., Gaff, M., & Gril, J. (2020). Effects of mild hygrothermal treatment on the physical and vibrational properties of spruce wood. *Composite Structures*, *253*, 112736. https://doi.org/10.1016/j.compstruct.2020.112736.

Kong, I., Tshai, K. Y., & Hoque, M. E. (2015). Manufacturing of Natural Fibre-Reinforced Polymer Composites by Solvent Casting Method. In M. S. Salit, M. Jawaid, N. B. Yusoff, & M. E. Hoque (Eds.), *Manufacturing of Natural Fibre Reinforced Polymer Composites* (pp. 331–349). Springer International Publishing. https://doi.org/10.1007/978-3-319-07944-8_16.

Kubojima, Y., Okano, T., & Ohta, M. (1998). Vibrational properties of Sitka spruce heat-treated in nitrogen gas. *Journal of Wood Science*, *44*(1), 73–77. https://doi.org/10.1007/BF00521878.

Kumar, K. S., Siva, I., Rajini, N., Jappes, J. T. W., & Amico, S. C. (2016). Layering pattern effects on vibrational behavior of coconut sheath/banana fiber hybrid composites. *Materials & Design*, *90*, 795–803.

Lakreb, N., Bezzazi, B., & Pereira, H. (2015). Mechanical behavior of multilayered sandwich panels of wood veneer and a core of cork agglomerates. *Materials & Design*, *65*(C), 627–636. https://doi.org/10.1016/j.matdes.2014.09.059.

Lett, J. A., Sagadevan, S., Fatimah, I., Hoque, M. E., Lokanathan, Y., Léonard, E., Alshahateet, S. F., Schirhagl, R., & Oh, W. C. (2021). Recent advances in natural polymer-based hydroxyapatite scaffolds: Properties and applications. *European Polymer Journal*, *148*, 110360.

López-Manchado, M. A., & Arroyo, M. (2000). Thermal and dynamic mechanical properties of polypropylene and short organic fiber composites. *Polymer*, *41*(21), 7761–7767. https://doi.org/10.1016/S0032-3861(00)00152-X.

Mukherjee, P. S., & Satyanarayana, K. G. (1984). Structure and properties of some vegetable fibres. *Journal of Materials Science*, *19*(12), 3925–3934. https://doi.org/10.1007/BF00980755.

Munde, Y. S., Ingle, R. B., & Siva, I. (2019). Vibration damping and acoustic characteristics of sisal fibre-reinforced polypropylene composite. *Noise & Vibration Worldwide*, *50*(1), 13–21. https://doi.org/10.1177/0957456518812784.

Murčinková, Z., Vojtko, I., Halapi, M., & Šebestová, M. (2019). Damping properties of fibre composite and conventional materials measured by free damped vibration response. *Advances in Mechanical Engineering*, *11*(5), 1687814019847009. https://doi.org/10.1177/1687814019847009.

Murugan, R., Ramesh, R., & Padmanabhan, K. (2016). Investigation of the mechanical behavior and vibration characteristics of thin walled glass/carbon hybrid composite beams under a fixed-free boundary condition. *Mechanics of Advanced Materials and Structures*, *23*(8), 909–916. https://doi.org/10.1080/15376494.2015.1056394.

Nowick, A. S., Berry, B. S., & Katz, J. L. (1975). Anelastic Relaxation in Crystalline Solids. *Journal of Applied Mechanics*, *42*(3), 750–751. https://doi.org/10.1115/1.3423694.

Panda, H. S., Sahu, S. K., & Parhi, P. K. (2013). Hygrothermal effects on free vibration of delaminated woven fiber composite plates – Numerical and experimental results. *Composite Structures*, *96*, 502–513. https://doi.org/10.1016/j.compstruct.2012.08.057.

Quiroga, A., Marzocchi, V., & Rintoul, I. (2016). Influence of wood treatments on mechanical properties of wood–cement composites and of Populus Euroamericana wood fibers. *Composites Part B*, *84*(Complete), 25–32. https://doi.org/10.1016/j.compositesb.2015.08.069.

Ramakrishnan, S., Krishnamurthy, K., Rajeshkumar, G., & Asim, M. (2021). Dynamic mechanical properties and free vibration characteristics of surface modified jute fiber/nano-clay reinforced epoxy composites. *Journal of Polymers and the Environment*, *29*(4), 1076–1088. https://doi.org/10.1007/s10924-020-01945-y.

Sahu, P., & Gupta, M. (2017). Sisal (Agave sisalana) fibre and its polymer-based composites: A review on current developments. *Journal of Reinforced Plastics and Composites*, *36*(24), 1759–1780. https://doi.org/10.1177/0731684417725584.

Sanyang, M. L., Sapuan, S. M., Jawaid, M., Ishak, M. R., & Sahari, J. (2016). Recent developments in sugar palm (Arenga pinnata) based biocomposites and their potential industrial applications: A review. *Renewable and Sustainable Energy Reviews*, *54*, 533–549.

Saravana Bavan, D., & Mohan Kumar, G. (2010). Potential use of natural fiber composite materials in India. *Journal of Reinforced Plastics and Composites*, *29*(24), 3600–3613. https://doi.org/10.1177/0731684410381151.

Senthil Kumar, K., Siva, I., Jeyaraj, P., Winowlin Jappes, J. T., Amico, S. C., & Rajini, N. (2014). Synergy of fiber length and content on free vibration and damping behavior of natural fiber reinforced polyester composite beams. *Materials & Design (1980–2015)*, *56*, 379–386. https://doi.org/10.1016/j.matdes.2013.11.039.

Senthilkumar, K., Siva, I., Sultan, M. T. H., Rajini, N., Siengchin, S., Jawaid, M., & Hamdan, A. (2017). Static and dynamic properties of sisal fiber polyester composites–effect of interlaminar fiber orientation. *BioResources*, *12*(4), 7819–7833.

Senthilkumar, K., Saba, N., Rajini, N., Chandrasekar, M., Jawaid, M., Siengchin, S., & Alotman, O. Y. (2018). Mechanical properties evaluation of sisal fibre reinforced polymer composites: A review. *Construction and Building Materials*. https://doi.org/10.1016/j.conbuildmat.2018.04.143.

Senthilkumar, K., Siva, I., Rajini, N., Jappes, J. T. W., & Siengchin, S. (2018). Mechanical Characteristics of Tri-Layer Eco-Friendly Polymer Composites for Interior Parts of Aerospace Application. In M. Jawaid & M. Thariq (Eds.), *Sustainable Composites for Aerospace Applications* (pp. 35–53). Woodhead Publishing. https://doi.org/10.1016/B978-0-08-102131-6.00003-7.

Senthilkumar, K., Saba, N., Chandrasekar, M., Jawaid, M., Rajini, N., Alothman, O. Y., & Siengchin, S. (2019). Evaluation of mechanical and free vibration properties of the pineapple leaf fibre reinforced polyester composites. *Construction and Building Materials*. https://doi.org/10.1016/j.conbuildmat.2018.11.081.

Senthilkumar, K., Pulikkalparambil, H., Kumar, T. S. M., Britto, J. J. J., Parameswaranpillai, J., Siengchin, S., et al. (2021). Free Vibration Analysis of Bamboo Fiber-Based Polymer Composite. In *Bamboo Fiber Composites* (pp. 97–110). Springer.

Senthilkumar, K., Saba, N., Chandrasekar, M., Jawaid, M., Rajini, N., Siengchin, S., et al. (2021). Compressive, dynamic and thermo-mechanical properties of cellulosic pineapple leaf fibre/polyester composites: Influence of alkali treatment on adhesion. *International Journal of Adhesion and Adhesives*, *106*, 102823.

Senthil Muthu Kumar, T., Senthilkumar, K., Chandrasekar, M., Karthikeyan, S., Ayrilmis, N., Rajini, N., & Siengchin, S. (2021). Mechanical, thermal, tribological, and dielectric properties of biobased composites. In *Biobased Composites* (pp. 53–73). John Wiley & Sons, Ltd. https://doi.org/10.1002/9781119641803.ch5.

Sharif, A., & Hoque, M. E. (2019). Renewable Resource-Based Polymers. In M. L. Sanyang & M. Jawaid (Eds.), *Bio-based Polymers and Nanocomposites*: Preparation, Processing, Properties & Performance (pp. 1–28). Springer International Publishing. https://doi.org/10.1007/978-3-030-05825-8_1.

Subramanian, C., Deshpande, S. B., & Senthilvelan, S. (2011). Effect of reinforced fiber length on the damping performance of thermoplastic composites. *Advanced Composite Materials*, *20*(4), 319–335.

Sumesh, K. R., Kanthavel, K., & Vivek, S. (2019). Mechanical/thermal/vibrational properties of sisal, banana and coir hybrid natural composites by the addition of bio synthesized aluminium oxide nano powder. *Materials Research Express*, *6*(4), 045318. https://doi.org/10.1088/2053-1591/aaff1a.

Tjeerdsma, B. F., & Militz, H. (2005). Chemical changes in hydrothermal treated wood: FTIR analysis of combined hydrothermal and dry heat-treated wood. *Holz Als Roh- Und Werkstoff*, *63*(2), 102–111. https://doi.org/10.1007/s00107-004-0532-8.

3 Influence of Hybridization on the Free Vibration and Damping Characteristics of Bast Fiber-Based Polymer Composites

Lin Feng Ng
Universiti Teknologi Malaysia

CONTENTS

3.1 INTRODUCTION

Composite materials have been extensively applied in load-bearing and outdoor applications over the past decades (Al-Oqla and Sapuan 2014; Ng et al. 2017). These materials offer a wide range of advantages compared to conventional metal alloys. Lightweight and high specific strength of composite materials are particularly fascinating to the aerospace and automotive industries to enhance energy efficiency. The composite materials are composed of two or more distinct phases having their own unique characteristics. The combination of these distinct phases leads to new material properties. In the composite materials, the two phases are merely combined physically rather than dissolving and blending together. Although composites have been well-known materials among research communities, continuous effort is still given to these materials to optimize their material properties. Chemical treatments and the addition of fillers are among the techniques used to improve the miscellaneous

DOI: 10.1201/9781003173625-4

properties of composite materials. In the 1970s, the idea of combining the two competing materials, polymer composites and metal alloys, had been realized in aerospace industries. The intention of combining these two competing materials is to tackle the poor fatigue performance of metal alloys. Later, it was found that the addition of metal layers can resolve the weak impact resistance of composites, developing sandwich materials with excellent fatigue resistance and superior toughness.

In recent years, composite materials are still under continuous evolution in order to satisfy the demand and requirements as requested by the industries. Following the contemporary trend in the field of composites, lightweight and environment-friendly materials are particularly desirable to the research communities. In the market, composite materials are mainly dominated by artificial fibers such as carbon, glass, and aramid fibers. Among the three types of artificial fibers, glass fiber has prominently attracted the attention of researchers and scientists, mainly because of its high-strength and low-cost characteristics. The high mechanical strength and stiffness of the carbon fiber are the primary advantages that make such fiber commonly used in load-critical applications. Nevertheless, the high cost of carbon fiber limits its large-scale applications. In contrast, aramid fiber, with trade name of Kevlar, is being commonly used in military applications. However, the photodegradation may occur when exposing Kevlar fiber to ultraviolet irradiation.

The demand for composite materials has been massively increasing over the past few decades, particularly for glass fiber-based composites. Meanwhile, the ever-growing issues related to environmental degradation due to several key factors such as waste accumulation, the depletion of natural resources, and a high level of pollution are alarming. Instead of solving the problems, prevention is essential and more effective. Prevention should be the main focus in the environmental management hierarchy (Elleuch et al. 2018). In composite materials, the extensive use of artificial fibers contributes a significant amount of waste accumulation since they are not biodegradable. It is estimated that glass fiber--reinforced polymer (GFRP) composites account for 95% of the end-of-life composite material flow (Diani and Colledani 2020). Recycling composite materials is an alternative method to handle a large amount of GFRP composites to prevent waste accumulation. In addition to the waste accumulation, producing artificial fibers requires high energy consumption, which indirectly results in greenhouse gas emissions. In this context, the substitution of artificial fibers with environment-friendly materials is considered as the long-term solution to avoid a large amount of waste accumulation and negative impacts to the environment.

In order to prevent environmental degradation, more renewable, recyclable, and biodegradable natural resources should be employed to develop green and sustainable materials. Natural fibers have gained popularity due to their several promising advantages such as low energy consumption, biodegradability, high availability, low cost, and carbon dioxide neutral. These advantages make natural fibers become the prior choice as the potential reinforcements in composite materials. However, the replacement of artificial with natural fibers is not ideal due to the relatively low strength, low fiber–matrix compatibility, and hydrophilic nature of natural fibers. Therefore, a new concept associated with incorporating both natural and artificial fibers within a single matrix is highlighted. The hybridization concept aims at reaching the balance point between mechanical strength and environmental friendliness. It is anticipated that the combination of natural and artificial fibers could result in the synergetic effect that could improve the overall mechanical performance of the materials.

3.2 HYBRIDIZATION

Hybrid composites are composed of multiple types of reinforcement embedded within a single matrix. It is commonly known that plant and artificial fibers have their respective demerits. To resolve the demerits of the fibers, hybridizing both types of fibers in the composites has been proven to have a positive effect. The demerits of one fiber could be offset by the benefits of the other fiber, resulting in a positive hybrid effect. Feng et al. (2020a) reported that the hybridization between kenaf and pineapple leaf fiber can balance the mechanical properties and moisture sensitivity of the hybrid composites. It is undoubtedly that hybridization can improve mechanical properties (Sanjay et al. 2018). In the hybrid composites, the stress can still be bridged by the surrounding high elongation fiber after the fracture of the low elongation fiber, and thus improving the mechanical properties. However, combining two strain-compatible fibers in the composites gives the optimum hybrid effect. It is expected that lightweight, high-strength and eco-friendly materials can be successfully developed through the hybridization concept. In fact, hybrid composites can be grouped into three major categories, including natural/natural, natural/artificial, and artificial/artificial hybrid composites. Although the natural/natural fiber-reinforced hybrid composites could have their distinctive properties, their mechanical strength is apparently lower than artificial fiber-based composites. Over the years, natural/artificial hybrid composites have drawn significant attention due to their balance in mechanical performance and environmental friendliness. In principle, the mechanical properties of hybrid composites can be predicted using the rules of a hybrid mixture (RoHM) as shown in Equation 3.1. Nonetheless, the properties of hybrid composites may not precisely follow the RoHM. This is because RoHM is only applicable when the hybrid composites are free from any defects. The fiber–matrix adhesion and the interface quality have a very significant effect on the mechanical properties of hybrid composites (Banerjee and Sankar 2014). The experimental investigation had revealed that any microscopic imperfection in the composites could lead to the failure of the prediction using RoHM (Mirbagheri et al. 2007):

$$P = P_1V_1 + P_2V_2 + P_mV_m \tag{3.1}$$

where P is the property of hybrid composites to be determined; P_1, P_2, and P_m are the properties of first fiber, second fiber, and matrix, respectively; V_1, V_2, and V_m are the volume fractions of first fiber, second fiber, and matrix, respectively.

In addition to the environmental friendliness, the increasing interest in incorporating plant fibers in composite materials is also due to its lightweight characteristic. By partially substituting the artificial with plant fibers, the overall weight of the structures can be reduced. The density of plant fibers ($\sim$1.4 g/cm^3) is approximately half of the glass fibers ($\sim$2.5 g/cm^3). The low-density characteristic enhances the energy efficiency in transportation sectors when plant fiber-reinforced composites are used as structural components. Plant fibers also possess adequate energy absorbing capacity during an impact event (Ahmed et al. 2021). These properties are particularly critical for transportation sectors. Besides that, certain types of bast fiber even have higher specific properties than glass fiber when weight is taken into consideration. Subramaniam et al. (2019) studied the influence of hybridization on the tensile and indentation properties of composites with weight normalization. They reported that kenaf/glass fiber-reinforced hybrid composites had the highest tensile properties and indentation resistance in comparison with non-hybrid GFRP composites. Meanwhile, plant fibers are highly abundant, and they are cheaper than artificial fibers. Hence, it is expected that the hybrid composites with their unique features could attain high mechanical performance economically. Today, hybrid composites consisting of plant and glass fibers have been used as roofing panels, decks, bridges, and off-shore deck platforms (Shahzad and Nasir 2017).

3.2.1 Reinforcement

Basically, reinforcement can be grouped into four major categories, including particulate, fiber, fillers, and flakes. The role of reinforcement is to provide rigidity, mechanical strength, and stiffness to the composites. The selection of reinforcement mainly depends on the applications in particular sectors. The shape, size, and mechanical properties of the reinforcement are taken into consideration during the material selection process. Among the few categories of reinforcement, fiber is the most extensively used material in composites because of its superior mechanical properties. Its high aspect ratio (length to diameter) provides a better stress transfer efficiency, from the matrix to the fiber through the effective interfacial region, improving the mechanical strength and stiffness of the composites. Fiber-reinforced polymer (FRP) composites indeed offer several virtues to the structural components (Ng et al. 2020). Moreover, fiber addition may reduce the use of plastic materials that are generally derived from petroleum which causes a negative impact on the environment.

When further looking into the fiber types, artificial and plant fibers are the two major types of reinforcement used extensively in composite materials. Artificial fibers have attested their high mechanical strength and stiffness over those of plant fibers. However, there is an increasing trend of using plant fibers as potential reinforcement in composite materials. To date, plant fiber-reinforced composites have been used for non-load-critical applications in automotive industries. Nonetheless, plant fibers are limited to non-structural applications because of their low mechanical strength compared to artificial fibers. In addition, the hydrophilic nature of plant fibers entails the mechanical degradation of the composites, deteriorating their long-term performance. Therefore, it is necessary to hybridize artificial and plant fibers to obtain the desired properties.

In general, natural fibers can be obtained from three different sources, including animals, minerals, and plants. The classification of natural fibers and their respective origins are shown in Figure 3.1. In recent years, plant fibers have magnetized considerable attention because they have higher mechanical strength than most animal

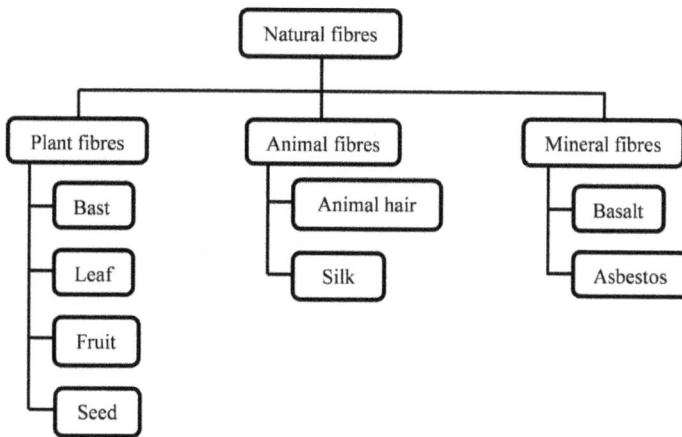

FIGURE 3.1   Classification of natural fibers and their origins.

fibers. Mineral fibers such as basalt and asbestos fibers are associated with health issues, which are considered as their main limitation (Sadrmanesh and Chen 2019). Besides, plant fibers are available in abundance, increasing their economic value. Truthfully, plant fibers can be derived from various sources, including bast, leaves, fruits, and seeds. However, most of the bast fibers such as kenaf, flax, jute, and hemp. exhibit higher mechanical strength than leaf, fruit, and seed fibers. It is undeniable that leaf fiber such as pineapple leaf fiber possesses remarkable mechanical strength because of its high cellulose content.

Cellulose is the main chemical constituent that provides mechanical strength to the plant fibers. Thus, those fibers with high cellulose content and low lignin tend to have high mechanical strength. Apart from the cellulose content, the microfibrillar angle also has a significant effect on the mechanical strength of the fibers. Low microfibrillar angle endows the plant fibers with high mechanical strength. It is worth mentioning that high cellulose content and low microfibrillar angle are generally observed in those of bast fibers (Pickering, Efendy, and Le 2016). Some bast fibers have shown comparable mechanical properties to glass fiber. This indicates that the fiber strength is highly dependent on the chemical composition. However, the plant fiber strength is also greatly influenced by the extraction methods, environment, and harvesting time (Feng, Malingam, and Irulappasamy 2019). Generally, fiber can be in different forms such as long fiber, short fiber, and woven fabric. Figure 3.2 shows the pineapple leaf fiber in three different forms.

All plant fibers have the demerits of weak compatibility with most polymer matrices and high moisture sensitivity. Hence, surface pre-treatments are required to alter the fiber surface structure, lowering their hydrophilicity and improving their fiber–matrix compatibility. Alkali and silane treatments are the most commonly applied techniques that tackle the disadvantages of plant fibers. Alkali treatment alters the fiber surface structure by eliminating a certain amount of lignin, hemicellulose, wax, and impurities, exposing more reactive groups of cellulose to interact with the polymer matrix (Feng et al. 2020b). The fiber surface becomes rougher upon the alkali

FIGURE 3.2   Plant fiber (a), short fiber (b), and long fiber (c) woven fabric.

treatment, allowing the mechanical anchoring between the fiber and polymer matrix, thereby enhancing the fiber–matrix adhesion. In contrast, silane is a bi-functional molecule that can simultaneously react with polymer matrix and plant fibers, forming the siloxane bridge across the interface, and therefore enhancing the fiber–matrix adhesion. Instead of alkali and silane treatments, Werchefani et al. (2020) suggested the use of biological technologies to resolve the limitations of plant fibers. The enzymatic treatment improves the fiber–matrix interface through the degradation of the pectin layers. It was found that the enzymatic treatment imparts the same effect as chemical treatments to develop composites with strong interfacial shear strength.

3.2.2 POLYMER MATRIX

The polymer matrices for FRP composites can be either thermoplastic or thermoset, which have different processing conditions and preparation methods. The selection of polymer matrix affects the composite performance as well, and hence the polymer matrix should be appropriately selected in accordance with the area of applications.

The main difference between thermoset and thermoplastic polymers is that the thermoset polymer is not recyclable. Thermoset polymers cannot be remolded or reshaped upon curing. The formation of three-dimensional cross-link networks held by strong covalent bonds after the curing of thermoset limits its recyclability (Post et al. 2020). In contrast, thermoplastic polymers can be remolded and reshaped easily by applying heat up to their melting temperature. Moreover, thermoplastic polymers have better ductility, inferring that they can be formed into any complex three-dimensional shape. Even though thermoset polymers have difficulty in recycling, the mechanical properties, thermal stability, chemical resistance, and durability of thermoset outperform thermoplastic polymers (Yan, Chouw, and Jayaraman 2014).

Epoxy is one of the most commonly used thermoset polymers in FRP composites. Typically, the majority of the artificial fiber-reinforced composites are based on epoxy polymers. Good corrosion resistance, high mechanical strength and stiffness, high compatibility with fiber, and good thermal stability are the benefits of epoxy polymers (Ma et al. 2011; Khan et al. 2016). The high-density cross-link networks of epoxy polymers allow them to have high mechanical strength and rigidity. Nevertheless, the high-density cross-link networks also result in the brittleness of epoxy polymers. In the aerospace industries, epoxy polymers are employed to bind the artificial fibers together, forming lightweight and high-strength FRP composites. Most of the commercial epoxy resins are derived from petroleum. Unfortunately, the functional groups of petroleum-based epoxy will be hydrolyzed and bio-accumulated in the water, and they are considered harmful to aquatic organisms (Sogancioglu, Yel, and Ahmetli 2020). Due to environmental issues, bio-epoxy resins have been developed to replace commercial epoxy. These bio-epoxy resins can be synthesized from plant oils, lignin, lipids, sugars, etc. Plant oils are among the most attractive natural resources to yield bio-epoxy, but the mechanical properties of plant oil-based epoxy are lower than commercial epoxy. The blending of plant oil-based with commercial epoxy has been identified as an effective method to maintain high mechanical properties. Niedermann, Szebényi, and Toldy (2017) reported that the jute fiber-reinforced composites based on 25 wt.% plant oil-based/commercial epoxy provided comparable mechanical properties to the composites based on commercial epoxy.

3.3 PARAMETERS AFFECTING THE MISCELLANEOUS PROPERTIES OF FRPs

The current focus has been shifted to plant fiber-based hybrid composites in recent years. Undoubtedly, hybridizing two types of fibers with different diameters has shown additional advantages as the fiber aspect ratio might be improved, which, in turn, enhances the mechanical properties of the hybrid composites. Similar to those conventional composites, the mechanical, free vibration, and damping properties of hybrid composites are influenced by several factors such as fiber length, weight composition, fiber orientation, and adhesion level between fiber and matrix. The evaluation of these parameters is particularly critical when involving plant fibers in the hybrid composites. These parameters are paramount to be evaluated comprehensively in order to develop cost-effective hybrid composites.

Innumerable intensive research studies have concluded that the increase of fiber length improves the mechanical properties of the composites. It is expected since the load is conveyed from the matrix to the fiber through the shear mechanism at the interface. There is no stress at the fiber tips, but the load tends to increase along the fiber length. For this reason, the fiber length needs to be greater than the critical fiber length in order to have the optimum reinforcing effect. Nevertheless, it should be noted that the long fiber has the tendency to agglomerate owing to its hydrogen bonds. The critical fiber length can be determined through shear-lag analysis, as shown in Equation 3.2.

$$\frac{L_c}{D} = \frac{\sigma_{fibre}}{2\tau_i} \tag{3.2}$$

where L_c is the critical fiber length, D is the diameter of the fiber, σ_{fibre} is the tensile strength of the fiber, and τ_i is the interfacial shear strength.

As shown in Equation 3.2, the critical fiber length can be decreased by improving the interfacial shear strength. Once the hybrid composites involve the incorporation of plant fibers, it becomes a challenging issue as the plant fibers have very weak compatibility with most of the polymer matrices. Therefore, it is necessary to implement chemical treatments to the plant fibers to alter the surface structure so as to improve the interfacial shear strength and reduce the tendency of agglomeration.

The stress transfer efficiency between the two phases in the composites has a strong relationship with the fiber length, weight composition, fiber–matrix compatibility, and fiber orientation. It has been well-known that the optimum fiber weight composition is in the range of 30–40 wt.%. Since reinforcement is the main load-carrying constituent, the increase of fiber content certainly increases the mechanical properties. Once the fiber content surpasses the critical level, the mechanical properties drop due to insufficient matrix material in FRP composites, which retards the stress transfer efficiency. Besides, excellent stress transfer efficiency should be supported by a robust fiber–matrix adhesion since the load is transmitted through the interfacial shear mechanism. When scrutinizing the effect of fiber orientation on the mechanical strength of composite materials, it had been identified that the FRP composites with fiber orientation, which is parallel to the loading direction, have high mechanical strength.

When looking into the free vibration and damping properties of FRP composites, it had been found that fiber length, fiber content, and fiber orientation could also have remarkable effects on the dynamic stability of FRP composites. Senthil Kumar et al. (2014) studied the free vibration and damping properties of sisal and banana fiber-reinforced polyester composites. They concluded that the increase in fiber content improved the free vibration and damping properties. However, the increase in fiber length improved the damping but did not significantly affect the natural frequency of the composites. Daoud et al. (2016) reported that the natural frequency of flax/epoxy composites was decreased when the fiber orientation was changed from 0° to 90°. Thus, the fiber parameters should be appropriately selected to optimize the free vibration and damping properties of the FRP composites.

3.4 FREE VIBRATION AND DAMPING CHARACTERISTICS OF HYBRID FRP COMPOSITES

Vibration can be classified as repetitive or periodic motion over a specific time interval. It can be induced in many engineering applications such as the engine of a vehicle. The materials with excellent damping characteristics are particularly desirable in automotive industries as these materials can control and suppress vibration and noise, improving the comfort level of the driver and passenger. In many industrial applications, the resonant amplitude of vibration should be minimized or eliminated to maintain structural integrity, position control, and durability (Karthik et al. 2016). High strength, stiffness, and damping behaviors are those characteristics that are desirable for structural applications. In general, the damping properties of FRP composites are greater than conventional metal alloys because of the viscoelastic behavior, interaction between reinforcement and matrix, and the damage mechanism of FRP composites (Rajesh, Pitchaimani, and Rajini 2016). Due to the viscoelastic behavior of the matrix, the increase of matrix content leads to the higher damping of FRP composites. FRP composites have been widely used in structural applications in aerospace and automotive industries where it involves vibration. Thus, FRP composites usually are vulnerable to dynamic loads during their service life (Cheng et al. 2015). In this case, FRP composites are required to have excellent damping characteristics to attain vibration reduction. Practically, increasing the storage modulus or damping capacity of FRP composites is able to suppress the vibration. Mechanical properties and dynamic mechanical properties are of the utmost importance when designing composite materials for certain applications. Dynamic stability of FRP composites is an indispensable criterion during the material selection for particular applications as it prominently influences the vibration of the structures. The free vibration and damping properties of the materials are widely studied because the majority of industrial applications induce vibration. Since FRP composites are involved in the primary structure of aerospace industries, it is especially pivotal to evaluate both the natural frequency and damping characteristics of the materials. Intensive research studies have been performed to determine the vibration properties of FRP composites.

Several techniques have been developed to enhance the mechanical, thermal, free vibration, and damping properties to widen the use of plant fiber as potential reinforcement in FRP composites. In plant fiber-reinforced composites, it was found that the addition of nanoparticles has been shown to have a beneficial effect on the vibration, mechanical, and thermal properties (Biswal, Mohanty, and Nayak 2011). Apart from the addition of nanoparticles, hybridization is another technique that can considerably enhance the free vibration and damping properties of FRP composites. In this regard, hybridization of glass with natural bast fibers could be advantageous to the damping improvement of FRP composites. Among natural fibers, bast fibers have been found to have high damping properties. It is expected that the incorporation of high damping bast fibers could improve the free vibration properties. A recent study has been performed by Ramraji, Rajkumar, and Sabarinathan (2020) on the free vibration properties of basalt/flax/vinyl ester hybrid composites. The results displayed that the increase of flax content increased the natural frequency of the hybrid composite laminates due to the better damping of flax fiber. However, the hybrid composites with basalt as the skin layers had the highest natural frequency because the basalt skin fiber conveyed the vibration energy to the flax fiber, where the energy was dissipated internally. Furthermore, it was noted that the increase of flax content enhanced the damping of the hybrid composites, indicating the potential of hybrid composites for structural damping applications. For woven fabrics, two types of fiber can be hybridized in two different forms which are intra-ply and inter-ply. Rajesh and Pitchaimani (2016) performed an interesting research study on the free vibration and damping properties of jute/banana fiber-reinforced hybrid composites. They showed that the intra-ply hybridization offered higher damping and free vibration properties over short and random fiber-based hybrid composites. To date, the exploration of the free vibration properties of hybrid composites is still very limited. The ongoing investigation of hybrid composites is paramount to expanding the usage of bast fibers in structural damping applications.

The dynamic mechanical properties of FRP composites are commonly investigated, including the storage modulus and damping ratio. A number of research studies have proven that the addition of bast fibers can increase the tan delta of the hybrid composites, indicating better damping of the materials. Tan delta indicates the damping behavior of the materials, and therefore the higher the tan delta, the more efficient the materials to absorb the energy and dissipate it safely. Assarar et al. (2018) evaluated the damping properties of flax/carbon hybrid composites. The positive hybrid effect was found in which the addition of flax fiber augmented the damping properties of the hybrid composites. Selver, Ucar, and Gulmez (2018) experimentally investigated the damping properties of flax/glass and jute/glass hybrid composites. As expected, non-hybrid GFRP composites evidenced the highest storage modulus. Nonetheless, the highest damping was noticed in those of non-hybrid flax and jute fiber-reinforced composites. Overall, improvement in the damping was observed in the hybrid composites compared to glass fiber-based composites. Mazlan et al. (2020) obtained similar findings in which the addition of kenaf fiber in the hybrid composites enhanced the tan delta. However, the addition of glass fiber increased the storage modulus of the materials due to the high stiffness of the fiber. Safri et al. (2019) evidenced promising results on the damping properties of sugar palm/glass fiber-reinforced hybrid composites. They reported that the addition of glass fiber increased the storage modulus of the composites. Figure 3.3 shows the damping factor of non-hybrid and hybrid sugar palm/glass fiber composites. As observed in Figure 3.3, the hybrid composites with 30% sugar palm and 70% glass fibers had outstanding tan delta values. These findings imply that the hybridization improves the storage modulus while enhancing the damping factor of the composites.

3.5 CONCLUSION

Lightweight bast fiber-reinforced hybrid composites are gaining their acceptance from the industries due to their distinctive properties, making these materials have a great potential to reduce the reliance on artificial fibers. The difficulty in recycling

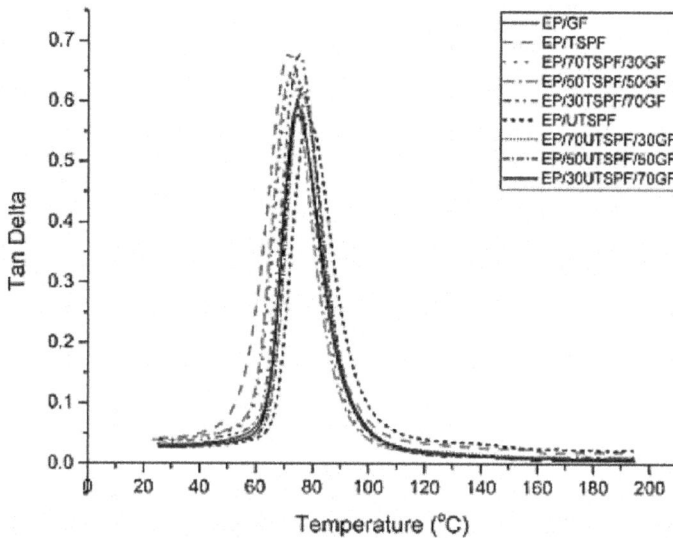

Legend:
- EP/GF
- EP/TSPF
- EP/70TSPF/30GF
- EP/60TSPF/50GF
- EP/30TSPF/70GF
- EP/UTSPF
- EP/70UTSPF/30GF
- EP/50UTSPF/50GF
- EP/30UTSPF/70GF

FIGURE 3.3 Tan delta of sugar palm/glass fiber-reinforced hybrid composites (Safri et al. 2019). (License number: 4993540672588.)

and the non-biodegradable characteristic of artificial fiber-based composites spur the researcher to employ bast fiber in the FRP composites. In practice, it is not realistic to use bast fiber alone in FRP composites owing to the low mechanical strength and hydrophilic nature of plant fibers. Therefore, hybridization of artificial and bast fibers is regarded as an alternative way to combine the benefits from both types of fiber to reach the balance properties. Similar to the conventional FRP composites, the mechanical, free vibration, and damping properties of hybrid composites are governed by several factors, including fiber length, fiber content, fiber orientation, and relative fiber ratio. In the viewpoint of free vibration and damping properties, incorporating bast fibers in the hybrid composites generally increases the natural frequency of the composites, mainly due to their excellent damping behaviors. However, the addition of artificial fibers augments the storage modulus of the composites. Overall, hybrid composites consisting of bast fibers exhibit better damping properties than the non-hybrid artificial fiber-reinforced composites. This implies that the hybrid composites have the ability to absorb energy and dissipate it safely. Meanwhile, the resonant amplitude of vibration can be drastically reduced by using the high damping bast fiber-based hybrid composites, which, in turn, enhances the structural stability and durability.

REFERENCES

Ahmed, M. M., H. N. Dhakal, Z. Y. Zhang, A. Barouni, and R. Zahari. 2021. "Enhancement of Impact Toughness and Damage Behaviour of Natural Fibre Reinforced Composites and Their Hybrids through Novel Improvement Techniques: A Critical Review." *Composite Structures* 259: 113496. doi:10.1016/j.compstruct.2020.113496.

Al-Oqla, F. M., and S. M. Sapuan. 2014. "Natural Fiber Reinforced Polymer Composites in Industrial Applications: Feasibility of Date Palm Fibers for Sustainable Automotive Industry." *Journal of Cleaner Production* 66: 347–54. doi:10.1016/j.jclepro.2013.10.050.

Assarar, M., W. Zouari, R. Ayad, H. Kebir, and J. M. Berthelot. 2018. "Improving the Damping Properties of Carbon Fibre Reinforced Composites by Interleaving Flax and Viscoelastic Layers." *Composites Part B: Engineering* 152: 248–55. doi:10.1016/j.compositesb.2018.07.010.

Banerjee, S., and B. V. Sankar. 2014. "Mechanical Properties of Hybrid Composites Using Finite Element Method Based Micromechanics." *Composites Part B: Engineering* 58: 318–27. doi:10.1016/j.compositesb.2013.10.065.

Biswal, M., S. Mohanty, and S. K. Nayak. 2011. "Mechanical, Thermal and Dynamic-Mechanical Behavior of Banana Fiber Reinforced Polypropylene Nanocomposites." *Polymer Composites* 32: 1190–1201. doi:10.1002/pc.21138.

Cheng, T. H., M. Ren, Z. Z. Li, and Y. D. Shen. 2015. "Vibration and Damping Analysis of Composite Fiber Reinforced Wind Blade with Viscoelastic Damping Control." *Advances in Materials Science and Engineering* 2015. doi:10.1155/2015/146949.

Daoud, H., J. L. Rebière, A. Makni, M. Taktak, A. El Mahi, and M. Haddar. 2016. "Numerical and Experimental Characterisation of the Dynamic Properties of Flax Fiber Reinforced Composites." *International Journal of Applied Mechanics* 8. doi:10.1142/S175882511650068X.

Diani, Marco, and Marcello Colledani. 2020. "Energy Consumption Assessment and Modeling of a Comminution Process: The Glass Fibers Reinforced Composites Case-Study." *Procedia CIRP* 90: 483–87. doi:10.1016/j.procir.2020.01.117.

Elleuch, B., F. Bouhamed, M. Elloussaief, and M. Jaghbir. 2018. "Environmental Sustainability and Pollution Prevention." *Environmental Science and Pollution Research* 25: 18223–25. doi:10.1007/s11356-017-0619-5.

Feng, N. L., S. D. Malingam, and S. Irulappasamy. 2019. "Bolted Joint Behavior of Hybrid Composites." In *Failure Analysis in Biocomposites, Fibre-Reinforced Composites and Hybrid Composites*, 79–95. doi:10.1016/B978-0-08-102293-1.00004-8.

Feng, N. L., S. D. Malingam, C. W. Ping, and N. Razali. 2020a. "Mechanical Properties and Water Absorption of Kenaf/pineapple Leaf Fiber-Reinforced Polypropylene Hybrid Composites." *Polymer Composites* 41 (4): 1255–64. doi:10.1002/pc.25451.

Feng, N. L., S. D. Malingam, N. Razali, and S. Subramonian. 2020b. "Alkali and Silane Treatments towards Exemplary Mechanical Properties of Kenaf and Pineapple Leaf Fibre-Reinforced Composites." *Journal of Bionic Engineering* 17: 380–92. doi:10.1007/s42235-020-0031-6.

Karthik, K., R. Rohith Renish, I. Irfan Ahmed, and T. Niruban Projoth. 2016. "Free Vibration Test for Damping Characteristics of Hybrid Polyester Matrix Composite with Carbon Particles." *Nano Hybrids and Composites* 11: 1–6. doi:10.4028/www.scientific.net/nhc.11.1.

Khan, R., M. R. Azhar, A. Anis, M. A. Alam, M. Boumaza, and S. M. Al-Zahrani. 2016. "Facile Synthesis of Epoxy Nanocomposite Coatings Using Inorganic Nanoparticles for Enhanced Thermo-Mechanical Properties: A Comparative Study." *Journal of Coatings Technology and Research* 13: 159–69. doi:10.1007/s11998-015-9736-6.

Ma, J., L. T. B. La, I. Zaman, Q. Meng, L. Luong, D. Ogilvie, and H. C. Kuan. 2011. "Fabrication, Structure and Properties of Epoxy/Metal Nanocomposites." *Macromolecular Materials and Engineering* 296: 465–74. doi:10.1002/mame.201000409.

Mazlan, N., T. C. Hua, M. T. H. Sultan, and K. Abdan. 2020. "Thermogravimetric and Dynamic Mechanical Analysis of Woven Glass/Kenaf/Epoxy Hybrid Nanocomposite Filled with Clay." *Advances in Materials and Processing Technologies*. doi:10.1080/2374068X.2020.1755114.

Mirbagheri, J., M. Tajvidi, J. C. Hermanson, and I. Ghasemi. 2007. "Tensile Properties of Wood Flour/kenaf Fiber Polypropylene Hybrid Composites." *Journal of Applied Polymer Science* 105: 3054–59. doi:10.1002/app.26363.

Ng, L. F., D. Sivakumar, K. A. Zakaria, and M. Z. Selamat. 2017. "Fatigue Performance of Hybrid Fibre Metal Laminate Structure." *International Review of Mechanical Engineering* 11 (1): 61–68. doi:10.15866/ireme.v11i1.10532.

Ng, L. F., S. Dhar Malingam, M. Z. Selamat, Z. Mustafa, and O. Bapokutty. 2020. "A Comparison Study on the Mechanical Properties of Composites Based on Kenaf and Pineapple Leaf Fibres." *Polymer Bulletin* 77: 1449–63. doi:10.1007/s00289-019-02812-0.

Niedermann, P., G. Szebényi, and A. Toldy. 2017. "Effect of Epoxidised Soybean Oil on Mechanical Properties of Woven Jute Fabric Reinforced Aromatic and Aliphatic Epoxy Resin Composites." *Polymer Composites* 38: 884–92. doi:10.1002/pc.23650.

Pickering, K. L., M. G. A. Efendy, and T. M. Le. 2016. "A Review of Recent Developments in Natural Fibre Composites and Their Mechanical Performance." *Composites Part A: Applied Science and Manufacturing* 83: 98–112. doi:10.1016/j.compositesa.2015.08.038.

Post, W., A. Susa, R. Blaauw, K. Molenveld, and R. J. I. Knoop. 2020. "A Review on the Potential and Limitations of Recyclable Thermosets for Structural Applications." *Polymer Reviews*. doi:10.1080/15583724.2019.1673406.

Ramraji, K., K. Rajkumar, and P. Sabarinathan. 2020. "Mechanical and Free Vibration Properties of Skin and Core Designed Basalt Woven Intertwined with Flax Layered Polymeric Laminates." *Proceedings of the Institution of Mechanical Engineers, Part C: Journal of Mechanical Engineering Science* 234: 4505–19. doi:10.1177/0954406220922257.

Rajesh, M., and J. Pitchaimani. 2016. "Dynamic Mechanical Analysis and Free Vibration Behavior of Intra-Ply Woven Natural Fiber Hybrid Polymer Composite." *Journal of Reinforced Plastics and Composites* 35: 228–42. doi:10.1177/0731684415611973.

Rajesh, M., J. Pitchaimani, and N. Rajini. 2016. "Free Vibration Characteristics of Banana/Sisal Natural Fibers Reinforced Hybrid Polymer Composite Beam." *Procedia Engineering* 144:1055–59. doi:10.1016/j.proeng.2016.05.056.

Sadrmanesh, V., and Y. Chen. 2019. "Bast Fibres: Structure, Processing, Properties, and Applications." *International Materials Reviews* 64: 381–406. doi:10.1080/09506608.2018.1501171.

Safri, S. N. A., M. T. H. Sultan, M. Jawaid, and M. S. Abdul Majid. 2019. "Analysis of Dynamic Mechanical, Low-Velocity Impact and Compression after Impact Behaviour of Benzoyl Treated Sugar Palm/Glass/Epoxy Composites." *Composite Structures* 226: 111308. doi:10.1016/j.compstruct.2019.111308.

Sanjay, M. R., P. Madhu, M. Jawaid, P. Senthamaraikannan, S. Senthil, and S. Pradeep. 2018. "Characterisation and Properties of Natural Fiber Polymer Composites: A Comprehensive Review." *Journal of Cleaner Production*. doi:10.1016/j.jclepro.2017.10.101.

Selver, E., N. Ucar, and T. Gulmez. 2018. "Effect of Stacking Sequence on Tensile, Flexural and Thermomechanical Properties of Hybrid Flax/Glass and Jute/Glass Thermoset Composites." *Journal of Industrial Textiles* 48: 494–520. doi:10.1177/1528083717736102.

Senthil Kumar, K., I. Siva, P. Jeyaraj, J. T. Winowlin Jappes, S. C. Amico, and N. Rajini. 2014. "Synergy of Fiber Length and Content on Free Vibration and Damping Behavior of Natural Fiber Reinforced Polyester Composite Beams." *Materials and Design* 56: 379–86. doi:10.1016/j.matdes.2013.11.039.

Shahzad, A., and S. U. Nasir. 2017. "Mechanical Properties of Natural Fiber/Synthetic Fiber Reinforced Polymer Hybrid Composites." *Green Energy and Technology* 355–96. doi:10.1007/978-3-319-46610-1_15.

Sogancioglu, M., E. Yel, and G. Ahmetli. 2020. "Behaviour of Waste Polypropylene Pyrolysis Char-Based Epoxy Composite Materials." *Environmental Science and Pollution Research* 27: 3871–84. doi:10.1007/s11356-019-07028-3.

Subramaniam, K., S. Dhar Malingam, N. L. Feng, and O. Bapokutty. 2019. "The Effects of Stacking Configuration on the Response of Tensile and Quasi-Static Penetration to Woven Kenaf/Glass Hybrid Composite Metal Laminate." *Polymer Composites* 40: 568–77. doi:10.1002/pc.24691.

Werchefani, M., C. Lacoste, H. Belguith, A. Gargouri, and C. Bradai. 2020. "Effect of Chemical and Enzymatic Treatments of Alfa Fibers on Polylactic Acid Bio-Composites Properties." *Journal of Composite Materials*. doi:10.1177/0021998320941579.

Yan, L., N. Chouw, and K. Jayaraman. 2014. "Flax Fibre and Its Composites – A Review." *Composites Part B: Engineering* 56: 296–317. doi:10.1016/j.compositesb.2013.08.014.

4 Free Vibration and Damping Properties of the Pineapple Leaf Fiber- and Sisal Fiber-Based Polymer Composites

Muhammad Ifaz Shahriar Chowdhury,
Yashdi Saif Autul, and Md Enamul Hoque
Military Institute of Science and Technology

CONTENTS

DOI: 10.1201/9781003173625-5

4.1   INTRODUCTION

The availability of petroleum-based materials is thought to be limited and unpredictable. As a result, a low-cost, long-lasting, and the readily available raw material is needed. Natural fiber reinforcement to manufacture bio-based composites for different applications has piqued interest in recent decades as a way to combat the rising cost of energy and the environmental effects of polymer composites reinforced by synthetic fibers [1,2]. Composite materials have mainly two distinct constructing components – matrix and fiber. A polymer matrix composite is plastic with incorporated fibers. Here, plastic is the matrix, and fiber is the reinforcement [3]. So, various components of fruits and plants such as banana, jute, flax, hemp, and kenaf can be used as raw-fiber materials in the manufacturing of these composites. Aside from the natural fibers mentioned above, pineapple leaf fibers (PALFs) and sisal fibers are gaining popularity as green reinforcement material (green reinforcement material refers to a composite constructed of natural fibers). The PALF and sisal fibers also have excellent mechanical characteristics, and their biodegradability is environmentally sustainable. Thus, sisal fibers and PALF can be used in coming structural, automotive, airplane, and non-structural industry applications in place of synthetic fiber, according to recent studies [4]. Vehicle parts, machine part architecture, machine supports, and construction sectors all rely heavily on dynamic properties [5]. It is essential to reduce the resonant amplitude in the process of designing various machine parts. The natural fibers-reinforced composites typically have better damping properties than metal matrix composites [6]. It is the interfacial debond and viscoelastic nature of natural fibers-reinforced composites that make molecular mobility occur freely. PALF and sisal fibers are used individually or in hybrid composition as the free vibration properties are not readily found in polymer composites. This chapter primarily discusses the vibration and damping properties of PALF and sisal fibers-based polymer composites based on the works of many researchers. This chapter also briefly describes the origins, compositions, and mechanical properties of PALF and sisal fiber. How various chemical treatments influence the damping characteristics and natural frequency of these polymer composites are also discussed.

4.2   NATURAL FIBERS

Fibers that are extracted from natural sources such as animals, plants, and geological substances are called natural fibers. These fibers are biodegradable, organic, low-cost, recyclable (to some extent), and environmentally friendly. They are promising and appealing substitutes to carbon, glass, and other synthetic fibers due to their availability, low density, and low cost along with reasonable mechanical characteristics [7,8]. They can be categorized according to their origins – (a) plant fiber, (b) animal fiber, and (c) mineral fiber [9].

Plant-based fibers					
Bast/Stem fibers	Leaf fibers	Seed/Fruit fibers	Stalk fibers	Grass/Reed fibers	Wood fibers
Abaca	Agave	Borassus	Barley	Bamboo	Hard wood
Banana	Caroa	Coir	Corn	Bagasse	Soft wood
Flax	Harakeke	Cotton	Maize	Esparto	
Hemp	Henequen	Kapok	Oat	Sabei	
Jute	Palm	Milk weed floss	Rey	Phragmites	
Kenaf	Pineapple	Rice husk	Rice	Communis	
Ramie	Sisal		Sugarcane	Canary grass	
Soyabean			Sunflower	Elephant grass	
			Wheat	Snake grass	

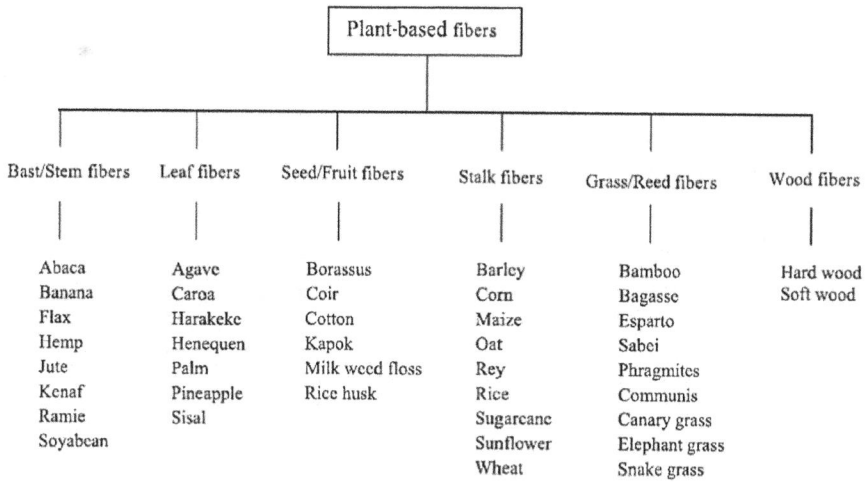

FIGURE 4.1 Classification of plant fibers. (Reprinted from [10] with permission from Elsevier.)

4.3 PLANT FIBERS

Plant fibers are often made up of cellulose, which is also combined with other materials, including lignin. Examples include pineapple leaf, sisal, flax, and jute hemp. Plant fibers can be categorized based on their source in plants (shown in Figure 4.1).

4.3.1 STEM/BAST FIBERS

Bast fibers are known for their long fibers and high mechanical strength. Examples include kenaf, soybean, banana, hemp flax, and jute. [10].

4.3.2 LEAF FIBERS

Leaf fibers are rough fibers removed from the plant's leaf by mechanical extraction or hand scraping after the retting process. They possess comparatively high strength. Examples: sisal, PALF, agave, palm, henequen, and carol. [10].

4.3.3 SEED/FRUIT FIBERS

The distinguishing features of the fiber are its strength and lightweight. The fibers are collected from the external shell of related fruit (example: coir fiber [CF]). Another type of seed fiber is formed from the pod or boll of respective plant seeds (example: milky weed floss, rice husk, and cotton.) [10].

4.3.4 STALK FIBERS

The fibers are collected from corn, wood, sunflower, sugarcane, maize, etc. Extraction of fiber is done from the plant stalks [10].

4.3.5 REED/GRASS FIBERS

The tall grasses, e.g., bamboo plants, switchgrass, elephant grass, and ryegrass, are the sources of these fibers. Snake grass, canary grass, phragmites, esparto, bagasse, and other crop fiber remnants are used as reinforcements in composites based on cement [10].

4.3.6 WOOD FIBERS

Wood fibers come in two varieties – hardwood and softwood. They are collected from different trees. The hardwood is usually shorter than softwood. The regulated processing of the fibers greatly minimizes the wide variation in mechanical characteristics and dimensional stability related to unprocessed fibers based on plants [10].

4.4 COMPOSITION AND MECHANICAL PROPERTIES OF PALF AND SISAL FIBERS

Natural plant fibers have low density, high stiffness, and high strength. An extended chain of about 30–100 cellulose molecules makes up cellulose fibrils. The diameter of the in-plant fibers is approximately 10–33 mm. It is the cellulose fibrils that deliver mechanical strength to fibers [11]. The mechanical and physical characteristics of natural fibers also rely on fiber–matrix adhesion, orientation, stress transfer efficiency at the interface, volume fraction, and aspect ratio of fiber [12]. The chemical composition and mechanical properties of PALF and sisal fibers are given in Table 4.1.

Because of its low cost and renewable nature, PALF has proven as a good substitute for synthetic fibers. Moreover, PALF-based polymer composites exhibit outstanding strength and stiffness relative to cellulose-based composite materials [17]. The low microfibrillar angle (14°) and high alpha-cellulose content of PALFs contribute to their superior mechanical properties. Owing to these exceptional properties, PALF can be used as a reinforcing composite matrix [18]. However, when PALF bundles are made wet, they show 50% lower strength, although when they are converted to yarn, they exhibit 13% higher strength [19].

4.5 PINEAPPLE PLANT

Pineapple (shown in Figure 4.2) (*Ananas comosus* L. Merr.) is a herbaceous perennial plant with edible fruit and included in the Bromeliaceae family monocotyledons subclass, *Pseudoananas*, and *Ananas genera*, and *A. comosus* species [20]. It is mainly grown in tropical and coastal areas for its fruits [21]. It is originated in Brazil but now widespread in tropical countries such as China, India, Indonesia, the Philippines, Thailand, Brazil, Nigeria, Costa Rica, and South Africa [22]. Pineapple varieties are primarily categorized into three groups – Cayenne, Spanish, and Queen. Since 1990, world pineapple production has risen to 23.33 million MT, and the region of cultivation has augmented as well. With 10.88 million MT, Asia takes the lead (46.62% share) in pineapple production.

TABLE 4.1

Chemical Composition and Mechanical Properties of PALF and Sisal Fiber

Composition and Properties		Pineapple Leaf Fiber (PALF)	Sisal Fiber	References
Chemical composition	Cellulose (%)	80.5	60	[13]
	Hemicellulose (%)	17.5	11.5	
	Lignin (%)	8.3	8	
	Pectin (%)	4	1.2	
	Wax (%)	–	–	
Mechanical properties	Tensile Strength (MPa)	413–1,627	350–7,000	[14–16]
	Density (g/cm³)	1.32	1.45–1.5	
	Moisture Constant (%)	11.8	11	
	Young's Modulus (GPa)	60–82	09–22	
	Failure Strain (%)	0.0–1.6	02–14	
	Elongation at break (%)	2.4	2–7	
	Microfibrillar spiral angle	14	20	

FIGURE 4.2 Pineapple plant. (Reprinted from [10] with permission from Elsevier.)

4.5.1 PINEAPPLE LEAF

Growing with a shorter stem, the pineapple plant generates a rosette of leaves that elongates as it matures. A full-fledged plant has about 80 leaves of different lengths (about 3 ft in length and 2–3 in. wide) and shapes (sword-like shapes). Leaves are generally softer-surfaced, stiff, sharp-pointed, waxy and thin, and good absorber [23,24].

4.5.2 EXTRACTION PROCESS OF PALF (SCRAPPING METHOD)

The scrapping machine consistunf of three rollers: (a) feed roller, (b) leaf scratching roller, and (c) serrated roller is used to scrape PLF [25]. The leaves are fed into feed roller; then, it passes through the scratching roller where the upper layer of leaves gets scratched and becomes wax-free. The leaves finally get crushed by the serrated roller, which introduces many breaks for the retting microbes to enter.

4.5.3 RETTING PROCESS OF PINEAPPLE LEAVES

In the retting process, a water tank is used for the immersion of scratched pineapple leaves. To accelerate retting reactions, liquor is held in the tank in a 1:20 ratio, diammonium phosphate, or urea of 0.5%, as substrate. After being processed, the fibers are washed in the pond, then separated, and at last dried in air [21].

4.5.4 CELL STRUCTURE OF PALF

PALF primarily consists of lignin and polysaccharides along with other chemical components such as inorganic substances, coloring matter, pentosan, uronic anhydride, pectin, wax, and fat [26]. It has an elliptical shape, and its area is determined by the concept of the minor and major axes of an ellipse [27].

The fiber comprises a middle lamella, primary wall, and secondary wall with S1, S2, and S3 layers, as shown in Figure 4.3. The stem xylem cells are divided and grown into thin layers, which make the primary wall. The S2 layer of the secondary wall contains the maximum amount of cellulose, making it the thickest (40 times more thick than other layers) of the layers. It is at about 5°–30° angled to the microfibrils-reinforced axis. Most of the layers are S3 layers. The middle lamella, a region of a high amount of lignin, is dark in the shade as it remains with cells glued together. A unique feature of every PALF is the dark spots on the fiber [28].

There is a similarity between PALF and cotton fibers as PALF is made of approximately 70%–82% cellulose. The fiber possesses a mesh-less structure that makes it a fiber of great quality and strength. The filaments are more extensible and well segregated. There are parallel, 3D arrangements of molecular chains of cellulose in the crystalline area of fibers [23].

4.6 VIBRATION ANALYSIS OF COMPOSITES

The natural frequency of composites relies on several factors, e.g., fiber–matrix bonding, fiber orientation, fiber loading, fiber length, chemical treatment, density, tensile modulus, and moment of inertia.

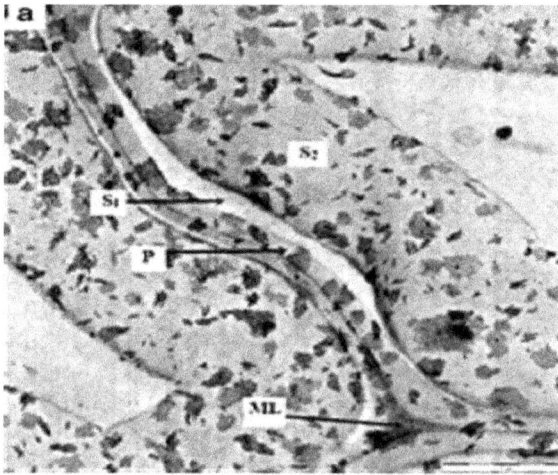

FIGURE 4.3 Highly magnified cell structure of a transverse section of PALF. (Reprinted by permission from Springer Nature Customer Service Centre GmbH: Springer [28], Copyright 2012.)

A number of researchers have studied vibration analysis of composites. Luo and Hanagud [29] experimented on the dynamics of delaminated beams using a nonlinear modal analysis that can anticipate nonlinear dynamic responses.

Voyiadjis et al. [30] studied vibrational analysis of the beams made of composites with random longitudinal, lateral, and both multiple delaminations predicting an analytical formulation that exhibited how the location and size of multiple delaminations affected the natural frequency of the beams. This study showed as the number of delamination increased, the frequency decreased linearly (in the case of multiple lateral delaminations where the length of delamination was kept constant). Della and Shu [31] analyzed the free vibration of beams made of composites with overlapping delamination.

Pothan et al. [32] performed a dynamic mechanical analysis (DMA) of composites reinforced by banana fiber taking the influence of temperature, frequency, and fiber loading under consideration. Incorporating banana fiber caused the damping peaks and loss modulus to be lowered. One contributing factor in improving the loss modulus is allowing the relaxation process to occur, which can be caused by interference of neighboring chains [33].

4.6.1 Dynamic Mechanical Analysis

It is a method for determining a material's mechanical responses by measuring dynamic property variations at a set frequency or over several frequencies at a set temperature [34]. It is a tool to gauge the viscoelastic properties and the morphology of composites and crystalline polymers. These properties are associated with essential parameters and primary relaxations, e.g., the non-Arrhenius variation of relation times with temperature, stress–relaxation modulus/creep compliance, storage/loss compliance, dynamic/complex viscosity, dynamic fragility, cross-linking density, etc. [35]. This method can be used to determine the stiffness properties and damping factors of fiber-filled reinforced and unfilled composite materials for different purposes [5]. The tests related to DMA are used to determine on how a material responds to sinusoidal or various periodic stress. Thus, DMA is quite helpful in determining the response of polymer under cyclic deformation [36].

4.6.2 DAMPING FACTOR

Damping factor (tan δ) (shown in Figure 4.4) is defined as the ratio of loss modulus (E'') and storage modulus (E') (Equation 4.1). It is a dimensionless number. The relationship between tan δ and E', E'' in a DMA graph is also shown in Figure 4.5.

$$\tan \delta = E''/E' \tag{4.1}$$

The damping factor (tan δ) is linked to viscoelasticity and molecular motions, as well as defects such as step boundaries, various interfaces, grain boundaries, and dislocations that contribute to damping [37]. The stability between the elastic phase and viscous phase in a polymer can be known with the help of the damping factor [38,39].

A low tan δ value indicates high elasticity, whereas a high tan δ specifies the high, non-elastic strain component of the material. A reduction in damping factor is a result of an increase of interface bonding of matrix and fiber, which is due to the decrease in the movement of molecular chains at the matrix and fiber. Therefore, if the loss of energy related to the storage capacity is low, the system will have a high value of damping factor [5]. Besides, the maximum value of tan δ occurs at a temperature called glass transition temperature (T_g) [40].

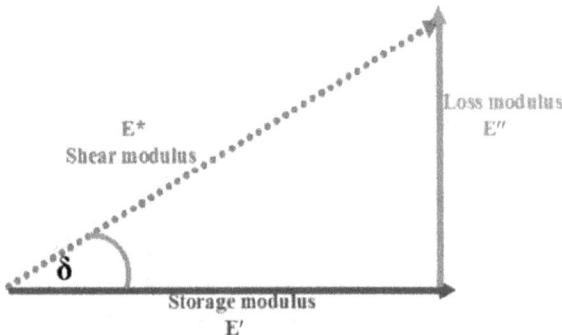

FIGURE 4.4 Relationship between tan δ and E', E''. (Reprinted from [5] with permission from Elsevier.)

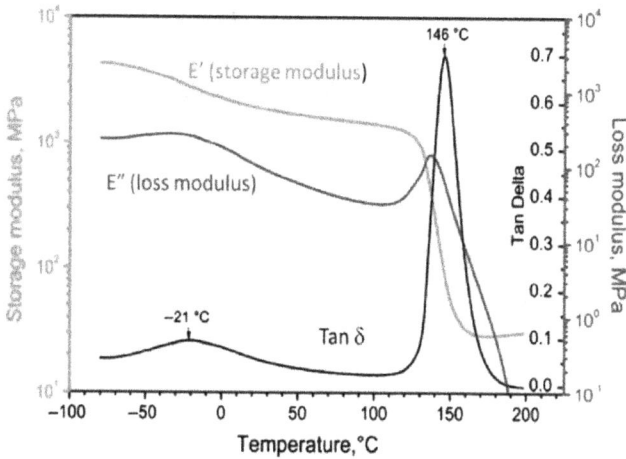

FIGURE 4.5 Relationship between tan δ and E', E'' in a DMA graph. (Reprinted from [5] with permission from Elsevier.)

4.7 VIBRATION AND DAMPING PROPERTIES OF PALF-BASED COMPOSITES

It is crucial while designing various machine components to reduce the vibrational amplitude. This vibrational amplitude is mostly influenced by structures' modal damping per mode [6]. Due to the high bonding of fiber/matrix and viscoelastic characteristics, fiber-based composite materials have a high damping rate [41,42].

Very few studies on vibration damping applications for composites reinforced by natural fibers have been conducted [43]. Several studies have shown that if the PALF is added with polyester (PE)-based composites, the natural fiber composites gain a good damping ratio and natural frequencies [44].

George et al. [45] employed thermal and dynamic mechanical analyses to assess the thermal behavior of PALF/polyethylene composites. They found that the damping factor and storage modulus increased as the fiber loading increased and decreased as the temperature increased.

Devi et al. [46] found that the PE reinforced with PALF that had a length of 30 mm and an aspect ratio of 600 had an increased storage modulus which has the highest value at 40 wt.% (Figure 4.6). The loss tangent, storage modulus, and other dynamic mechanical properties of PALF-reinforced PE were determined at temperatures varying from 30°C to 200°C. The frequencies were fixed at 0.1, 1, and 10 Hz. There was a positive shift of T_g and additional relaxation as well as PALF at 40 wt.% is incorporated in PE resin, which indicates constrained mobility of polymer molecules at the interface. The damping factor, the peak height of the composite decreased, but storage modulus increased with the rise of frequency. There was a maximum peak width in this instance. The results were compared to morphological data. Therefore, this study concluded that PALF could be used in an optimal concentration to reinforce the PE matrix efficiently.

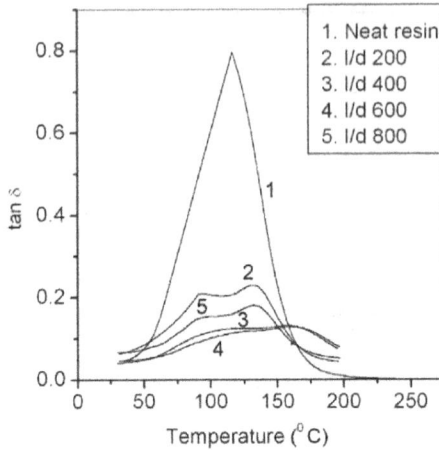

FIGURE 4.6   tan δ vs. Temperature graph showing the effect of fiber aspect ratio on PALF/ PE composites (fiber loading 40 wt.%, frequency of 10 Hz). (Reprinted with permission from [46], Copyright (2011) John Wiley and Sons.)

Devi et al. [47] inspected the dynamic mechanical characteristics of hybrid PALF/ glass fiber (GF) in PE matrix, which was unsaturated. The PALF and GF were arbitrarily oriented and were mixed intimately. They reported that for the hybrid composites if the V_f (volume fraction) of GF is enhanced to 0.2, the damping values decreases, and the material stiffness increases (Figure 4.7).

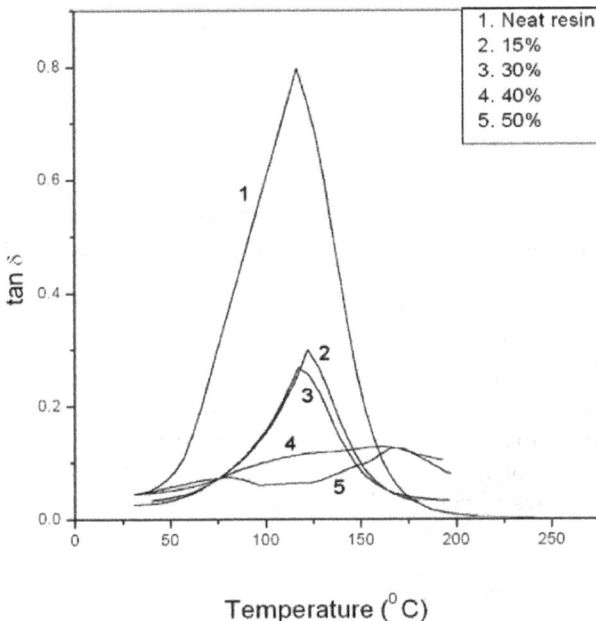

FIGURE 4.7   tan δ vs. Temperature graph showing the effect of fiber loading on PALF/PE composites (fiber aspect ratio 600, frequency of 10 Hz). (Reprinted with permission from [46], Copyright (2011) John Wiley and Sons.)

Another study on hybrid kenaf/PALF-reinforced high density polyethylene (HDPE) composites concluded that the increase of fiber length caused an escalation of composites' storage modulus. Also, the damping peak gets elevated as the fiber of the hybrids is increased, Aji et al. [48].

Senthilkumar et al. [44] inspected the influence of PALF loading in PE composite materials on damping, free vibrational, morphological, and mechanical properties. Hand lay-up was applied to form the composites, which were made with arbitrary directed PALF in a PE matrix squeezed at 17 MPa with the application of compression molding. As the PALF loading increased in the composites, the natural frequency improved (45 wt.% PALF/PE showed the maximum natural frequency). Still, there was a decrease in the damping ratio (at 25 wt.% PALF loading, the highest damping value of 0.1939 is found). The results showed that biocomposites with 35 wt.% PALF fiber loading had improved dynamic and mechanical properties, making them a good substitute for cementitious composites in low-strength structural applications. The three mode shapes (peaks) – Mode-1 (bending), Mode-2 (twisting), and Mode-3 (secondary bending) were also observed in this study (as shown in Figure 4.8 and Table 4.2) [44].

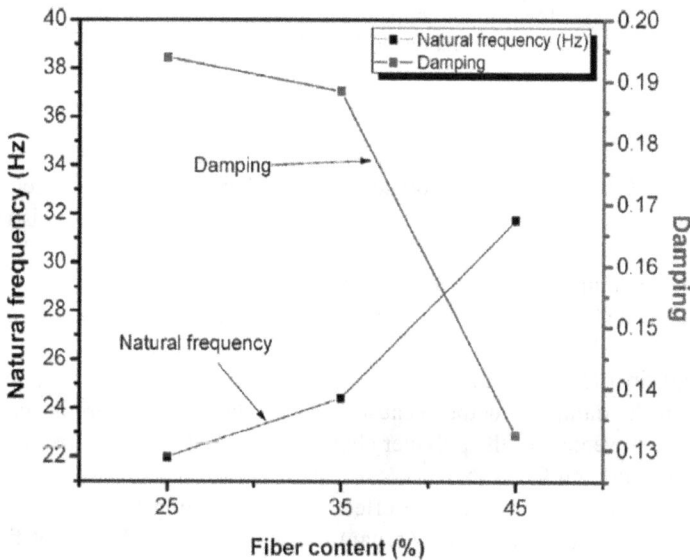

FIGURE 4.8 Damping and natural frequency of PALF/PE composite materials at Mode-1. (Reprinted from [44] with permission from Elsevier.)

TABLE 4.2

Damping and Natural Frequency of PALF/PE Composite Materials (Mode-2 and Mode-3) (Reprinted from [44] with permission from Elsevier)

| PALF Loading (wt%) | Natural Frequency (Hz) | | Damping | |
	Mode 2	Mode 3	Mode 2	Mode 3
25	175.78	351.56	0.0242	0.0121
35	119.63	195.31	0.0431	0.0255
45	195.31	383.90	0.0215	0.0109

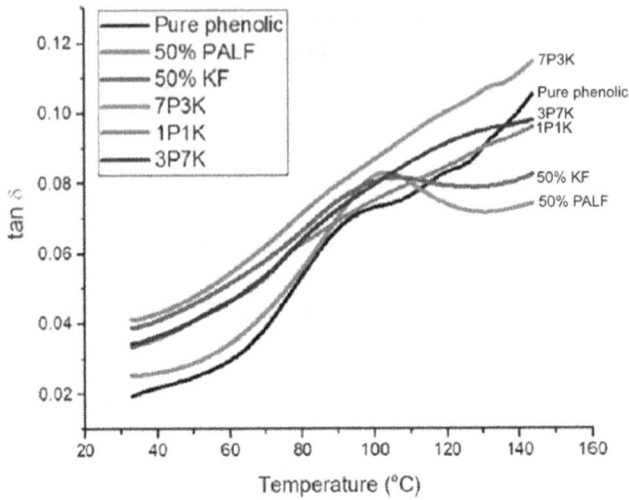

FIGURE 4.9 tan δ vs. Temperature graph of PALF, KF, and hybrid composites. (Reprinted with permission from [49], Copyright (2019) John Wiley and Sons.)

Mohammad Asim et al. [49] worked on hybrid kenaf/PALF-reinforced phenolic composites to observe the thermo-mechanical and dynamic characteristics of the composites. The curve (shown in Figure 4.9) generated from the research indicated that in all the samples, the tan δ values were low below the glass transition temperature (T_g) but started to rise in response to the temperatures, hitting a peak in the transition zone [50].

As kenaf fiber (KF) was introduced into PALF composite materials, there was a reduction of the damping factor because this introduction of fibers in the polymer hindered the movement of the polymer chain and lowered the flexibility, resulting in lowering the damping factor [51].

Doddi et al. [52] determined the effect of incorporating PALF into basalt and gauged the mechanical properties (dynamic and static) of PALF/basalt fiber composites. Unidirectional basalt layers having high modulus were introduced to PALF composite to retain damping characteristics and attain better specific characteristics. The result showed PALF/basalt hybrid had improved dynamic mechanical, flexural, and tensile properties.

The experiment exhibits that the PBP hybridized composite (hybridized composite polymer with basalt as core and PALF as skin) has lower storage modulus and tan δ values than BPB hybridized composite (hybridized composite polymer with PALF as core and basalt as skin), which indicates that the dynamic mechanical characteristics in hybrid composite polymers are influenced by the layering sequence. The BPB has improved dynamic, flexural, and tensile properties than PBP. The graphs of the dynamic factor of the composites at three frequencies (0.1, 1, and 10 Hz) are shown in Figure 4.10 [52].

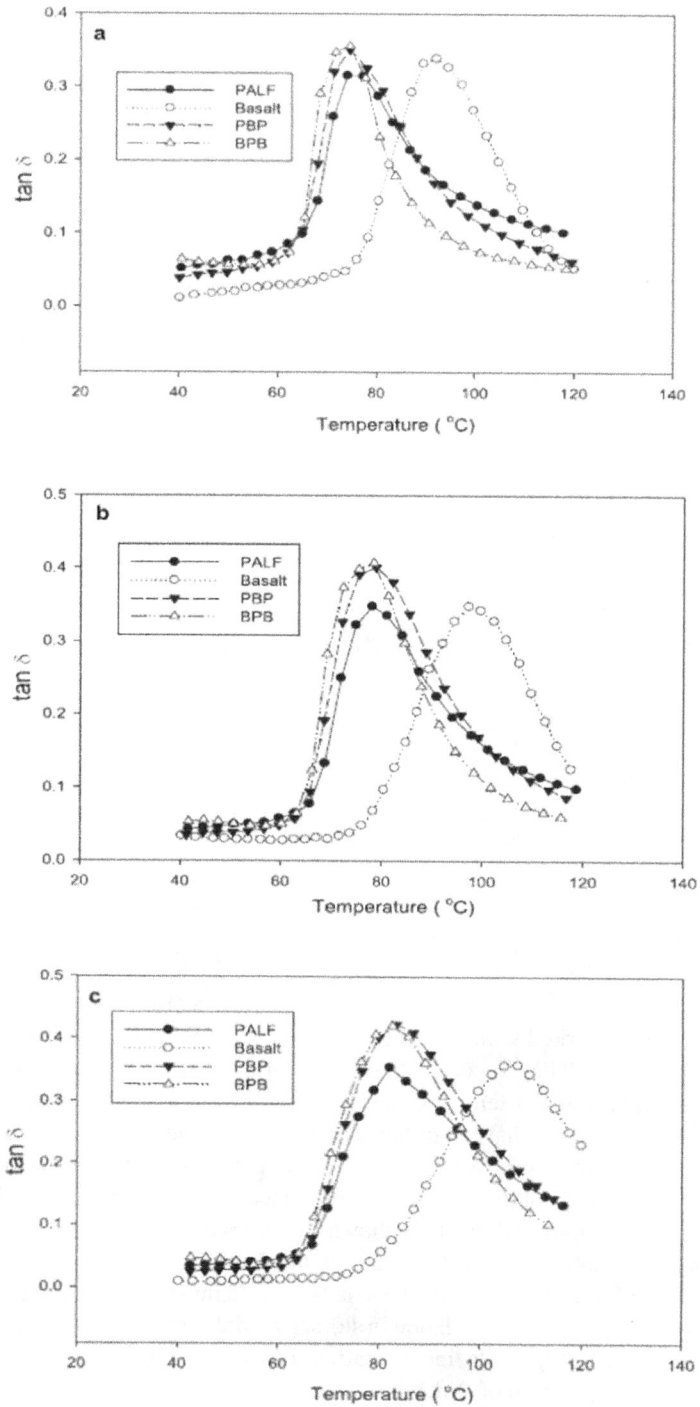

FIGURE 4.10 Changes of tan δ of PALF, basalt, PBP, and BPB at three frequencies (0.1, 1, and 10 Hz). (Reprinted with permission from [52] copyright 2019, IOP Science.)

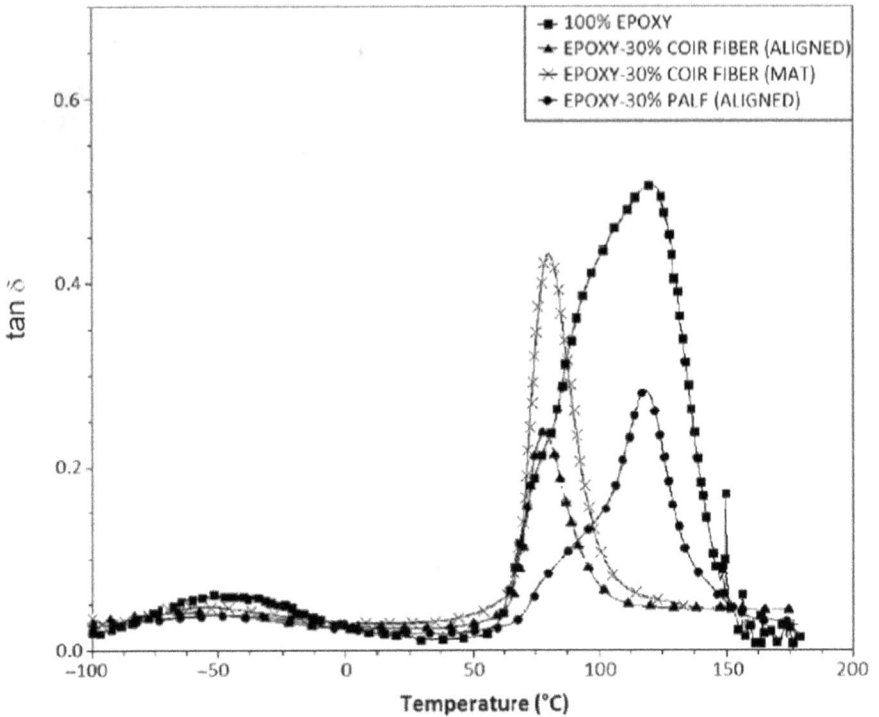

FIGURE 4.11   tan δ vs. Temperature graph with different fibers (PALF, coir fiber). (Reprinted with permission from [53], An open access article distributed under the terms of the Creative Commons Attribution License.)

    Luz et al. [53] worked with two fibers – PALF and CF for reinforcing the epoxy matrix composites. They reported that incorporation of PALF in epoxy matrix increased glass transition temperature (T_g) and also increased the moduli values (E', E''). The composites have superior dynamic mechanical properties relative to the neat epoxy resin. Moreover, it was found that PALF/epoxy interacted interracially more strongly than did CF/epoxy. The effect of temperature on the damping factor for different reinforcements is shown in Figure 4.11.

    Siakeng et al. [54] inspected the dynamic mechanical properties of biocomposites that are reinforced by PALF and CFs incorporated into the polylactic acid (PLA) matrix. This study found an escalation in storage modulus and decline in tan δ values (shown in Figure 4.12) while incorporating PALF and CF in the PLA matrix, although the incorporation of PALF is more effective.

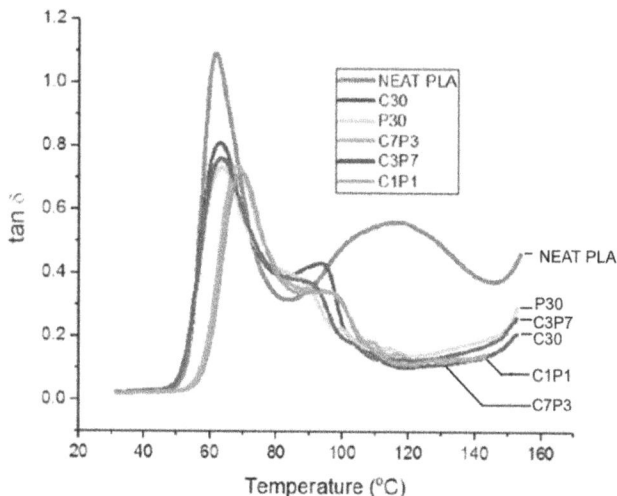

FIGURE 4.12 tan δ vs. Temperature graph of PALF- and CF-reinforced and hybrid composites. (Reprinted with permission from [54], Copyright (2019) John Wiley and Sons.)

4.8 CONSEQUENCE OF TREATMENT ON PROPERTIES OF PALF-REINFORCED COMPOSITES

Lopattananon et al. [18] used a NaOH solution (concentrations of 1%, 3%, 5%, and 7% w/v) to treat PALF and incorporate it in the rubber matrix. The treated fiber composites have greater elongation due to the increase in the molecular direction in the microfibrils.

Asim et al. [55] analyzed the thermal, dynamic mechanical, and flammability characteristics and features of untreated and treated PALF/KF phenolic composites. They noticed that the peak of loss modulus was higher and moved toward a high temperature after PALF/KF was added with phenolic composites. In addition, the peak height of tan δ had a maximum value for 60% PALF phenolic composites and the lowest value for pure phenolic composites (Figure 4.13).

Efforts had been put to upgrade the mechanical and vibrational properties of PALF, sisal, and banana fibers by using TiO_2 nanopowder, which is synthesized chemically. It was observed that natural frequency was maximum in PALF–sisal hybrid with 3% TiO_2 filler. Since the good contact of surface between filter/resin and resin/hybrid natural fibers makes the composites highly stiff, the natural frequency is improved. Factors such as energy dissipation at maximum strain area and frictional characteristics of nanofiller mixture improved damping properties of the composites. The PALF-based hybrid composites at 35 wt.% provide a better damping rate than sisal and banana fibers. However, the natural frequency is reduced for the accumulation at 4% nanofiller substitution in hybrid natural fibers

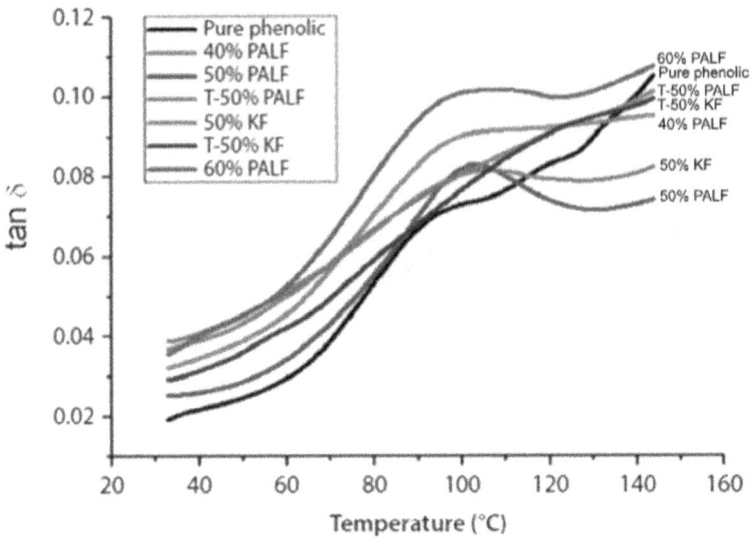

FIGURE 4.13   Damping factor curves of pure phenolic and PALF/KF fiber loading composites. (Reprinted with permission from [55], An Open Access article licensed under a Creative Commons Attribution 4.0 International License.)

of BP (Banana/PALF), PS (PALF/Sisal), and SB (Sisal/Banana) because of the weaker interface [43]. The damping ratio and natural frequency for different combinations are shown in Figures 4.14–4.16.

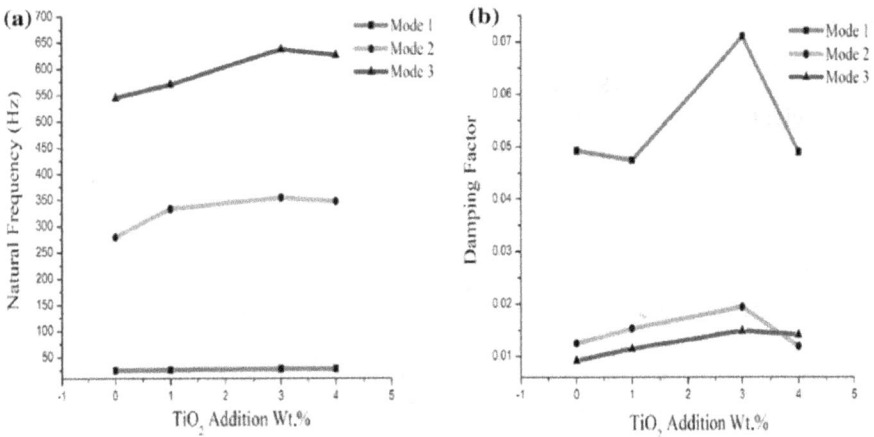

FIGURE 4.14   (a) Natural frequency, (b) the damping ratio of sisal/banana/TiO$_2$ combinations. (Reprinted by permission from Springer Nature Customer Service Centre GmbH: Springer [43], Copyright 2020.)

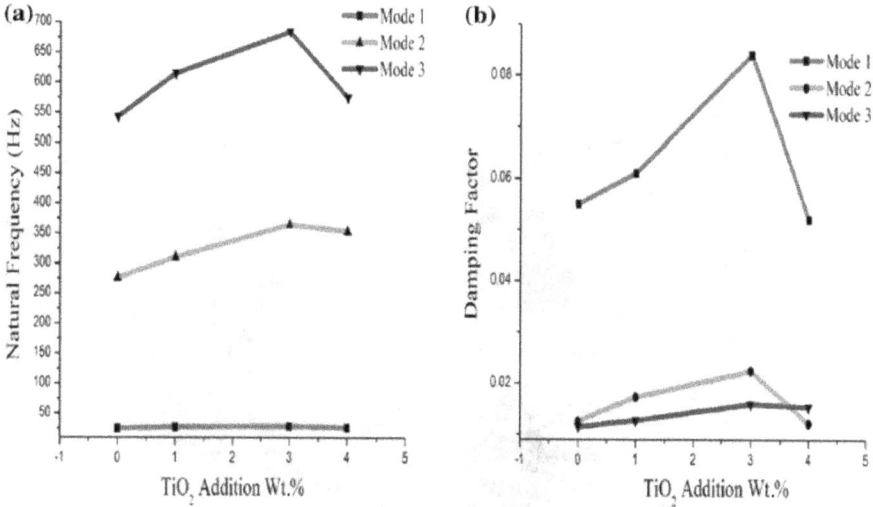

FIGURE 4.15 (a) Natural frequency, (b) the damping ratio of banana/pineapple/TiO$_2$ combinations. (Reprinted by permission form Springer Nature Customer Service Centre GmbH: Springer [43], Copyright 2020.)

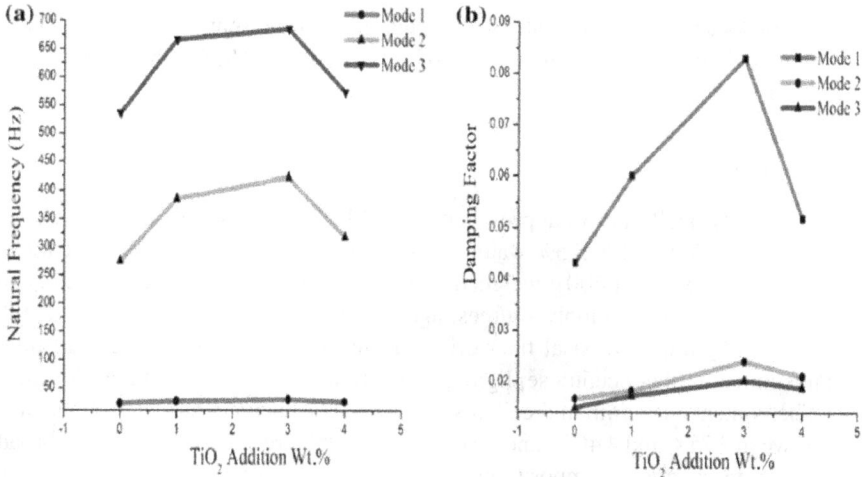

FIGURE 4.16 (a) Natural frequency, (b) the damping ratio of pineapple/sisal/TiO$_2$ combinations. (Reprinted by permission form Springer Nature Customer Service Centre GmbH: Springer [43], Copyright 2020.)

4.9 SISAL PLANT

Plant fibers are divided into various groups and can be classified according to their plant derivations and origins, among which sisal is under the leaf fibers category [56,57]. Sisal can be quickly grown in a limited amount of time. Field hedges and

FIGURE 4.17   Sisal plant. (Reprinted from [10] with permission from Elsevier.)

railway tracks are natural habitats for the plant [58]. At present, sisal plants (*Agave sisalana*) are being cultivated in tropical countries in Africa, the Caribbean, and Asia [23] (Figure 4.17).

4.9.1 SISAL LEAF

A sisal plant typically grows approximately 200 leaves. The leaves contain around 1,000 fibers made up of 87.25% water, 8% dry matter, 4% fiber, and 0.75% cuticle. Therefore, on average, a 600 g weighed leaf gives around 3% of its total weight in fiber [58]. Based on the supply sources, age factor, measuring techniques, etc., the chemical configuration of sisal fiber differs from one place to the next. Sisal fiber contains moisture, hemicellulose, lignin, and cellulose, much as most natural fibers [59]. The contents of lignin and cellulose in sisal fiber fluctuate based on the plant's age, between 3.75% and 4.40%, and 49.62% and 60.95%, respectively [60]. The broad differences in chemical compositions of sisal fiber are due to its various sources, ages, measuring processes, and other factors [58].

4.9.2 EXTRACTION OF SISAL FIBER

Mechanical, xylem, and ribbon are the three varieties of sisal fiber derived from the leaves. Mechanical fibers are harvested from the leaf's periphery. They can be separated by the extraction process and resemble a horseshoe. Xylem fibers possess an odd structure that quickly breaks up through production. The longest kind of fiber is ribbon, which can be cut in a longitudinal direction during processing [59].

TABLE 4.3
Mechanical Properties of Sisal-Reinforced Composites

Properties	Sisal/Epoxy	Sisal/PE	Sisal/Benzene/Epoxy
TS (MPa)	83.96	65.5	64
TM (GPa)	1.5	1.90	1.4
FS (MPa)	252.39	99.5	75
FM (GPa)	11	2.49	3
IS	2 (kJ/m²)	–	22.5 (J/m)

4.9.3 SISAL FIBER-REINFORCED COMPOSITES

Thermoplastics (PVC, polystyrene, polypropylene, polyethylene, etc.), rubber (styrene-butadiene rubber, natural rubber, etc.), thermosets (PE, epoxy, etc.), cement, and gypsum are among the matrix materials usually incorporated in sisal fiber-reinforced composites [58]. Senthilkumar and Saba found the mechanical properties for the following sisal-reinforced composites as shown in Table 4.3 [59].

4.10 VIBRATION AND DAMPING PROPERTIES OF SISAL FIBER-BASED COMPOSITES

Zaman Abud Ali investigated the effect of Sisal natural fiber reinforcement on vibration using experimental and numerical methods, preparing samples of composite material in the form of plates, with volume fractions of sisal–resin ranging from 35% to 65% (sample A), 30% to 70% (sample B), 25% to 75% (sample C), 20% to 80% (sample D), 15% to 85% (sample E), 10% to 90% (sample F), and 5% to 90% (sample G). Vibration tests were performed on specimens as forms of specimen fixations. The result revealed that at 202.73 Hz and for four edges (CCCC) arrangement supported by clamp, the natural frequency was high compared to each of another specimen as the sisal plant fiber had a high volume fraction and the PE resin {(35–65)% sisal-PE} had a low volume fraction in the specimen (CCCC). Again, the sample (CCCC) had high material strength as plant sisal fibers were applied before the full volume fraction (35%) of sisal was applied in specimen construction. The better specimen composite material had high natural frequency and a high volume fraction sisal, and according to the numerical findings shown in Figure 4.18. Figure 4.19 shows that when the ratio of fibers is high, the results provide the best vibration test. Figure 4.20 shows a comparison of natural frequency effect with volume fraction sisal at each level of assisted used in vibration testing in both numerical and experimental work [61].

Senthilkumar and Shiva examined the influence of length sisal fiber on free vibration characteristics in banana PE composites made of sisal fibers. The modal analysis was used to determine the natural frequencies and corresponding modal damping values of the composite laminates. The first mode of damping of the various sisal and banana PE composites is seen in Figure 4.21. With an increase in banana and sisal fiber content, two types of damping patterns were observed for a constant fiber length: damping decreased for the former and increased for the latter. The best

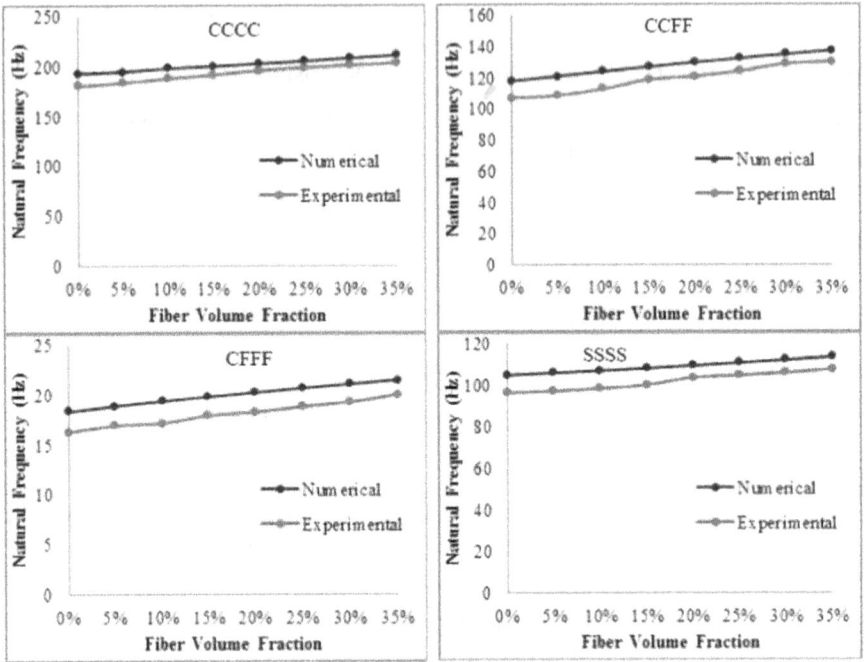

FIGURE 4.18   Relationship between volume fraction and natural frequency. (Reprinted with permission from [61], Open Access.)

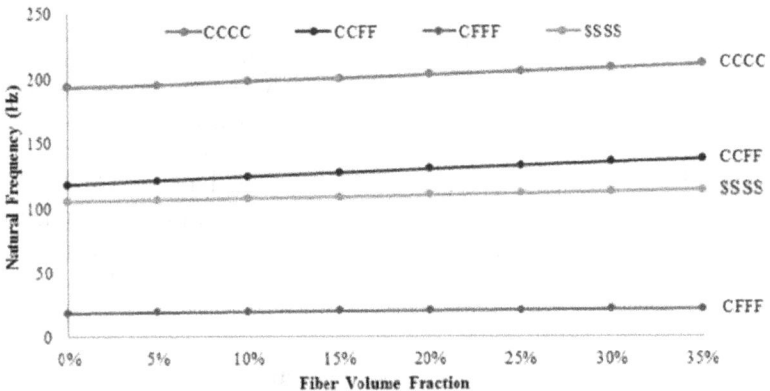

FIGURE 4.19   Influence of the type of support for specimens on volume fraction. (Reprinted with permission from [61], Open Access.)

combination for sisal fiber polyester composite (SFPC) was 50 wt.% fiber material and 3 mm fiber length and while the best blend for banana fiber polyester composite (BFPC) was 50 wt.%, fiber content, and 4 mm fiber length. The interfacial process was investigated using scanning electron microscopy [6].

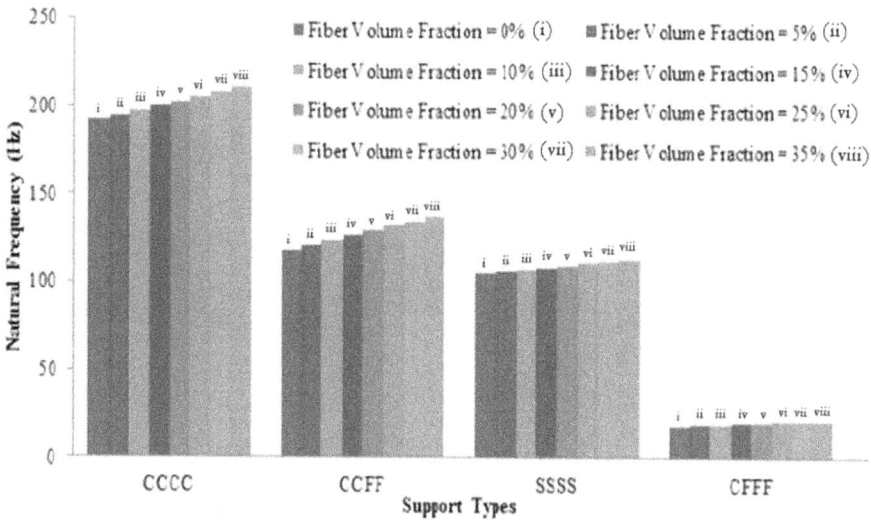

FIGURE 4.20 Natural frequency vs. support types graph. (Reprinted with permission from [61], Open Access.)

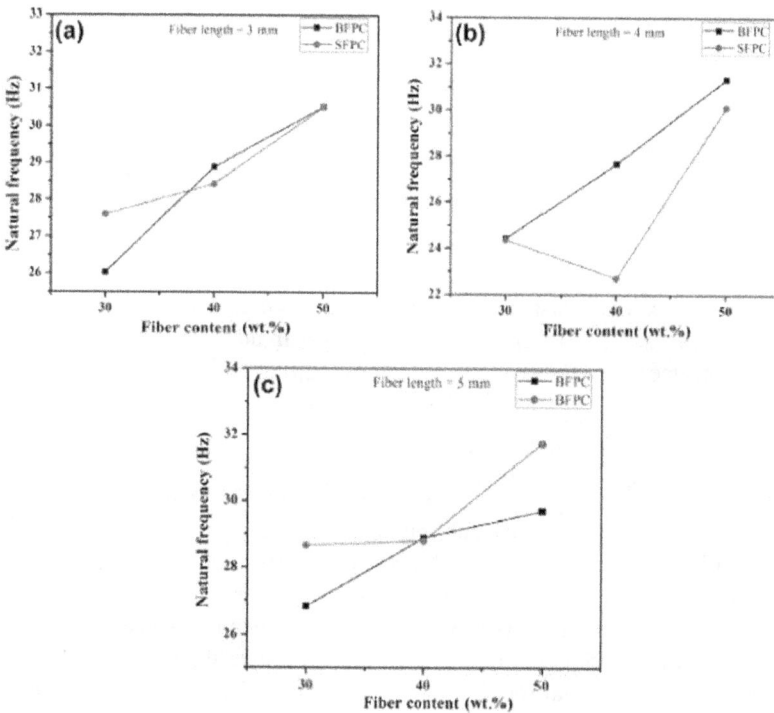

FIGURE 4.21 Influence of fiber form, material, and length on the natural frequency of composites (Mode-1) for 3 mm (a), 4 mm (b), and 5 mm fiber lengths (c). (Reprinted from [6] with permission from Elsevier.)

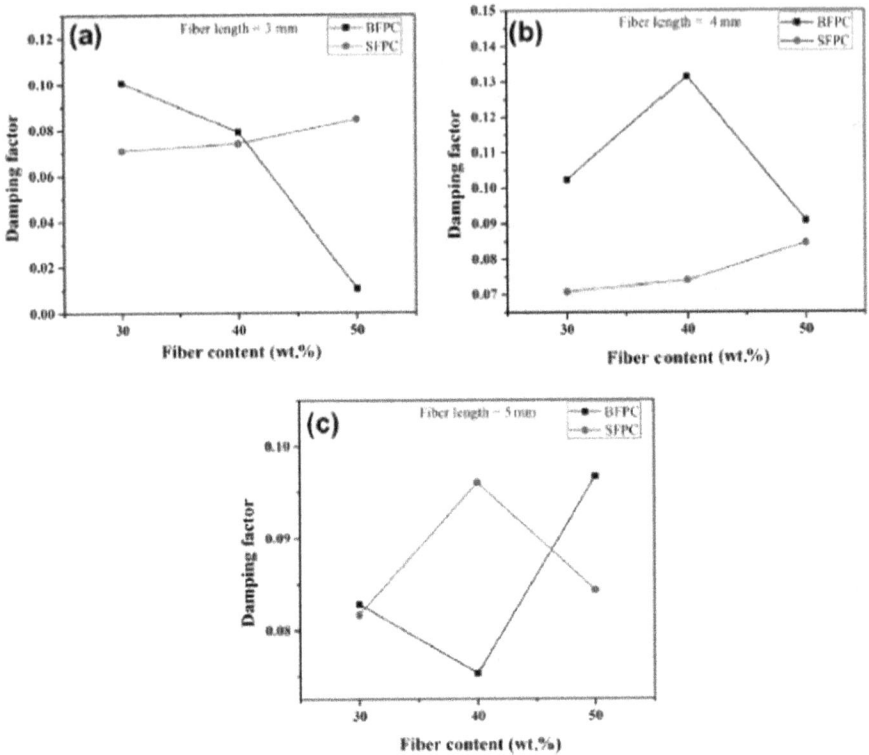

FIGURE 4.22   Influence of fiber type, content, and length on damping of composite (Mode-1) for 3 mm (a), 4 mm (b), and 5 mm (c) fiber length. (Reprinted from [6] with permission from Elsevier.)

Because of its viscoelastic nature, higher resin content should result in higher damping. It can be deduced from Figure 4.22 that, in addition to the fiber material, stiffness, interface thickness contribute to the damping mechanism [62].

4.11 CONCLUSIONS

Industries today are trying to introduce biofibers in several technological sectors for a variety of applications. Better damping quality, reduced vibrational property, low cost, high-specific modulus, lightweight, easy availability, and biodegradability of the natural fibers are the main factors that are making industries enthusiastic about using natural fibers in making important composites [63]. The dynamic mechanical properties such as natural frequency, damping factor, and mode shapes of PALF and sisal fibers can play an essential role in aerospace, automobile, and packaging industries where many vibration inputs can cause resonance in the structures [64,65]. However, there are some intrinsic factors such as the hydrophilic nature of fibers, which can create obstacles in using natural fiber as reinforcements. To tackle these hindrances, researchers have been conducting experiments to minimize these effects through hybridizing and chemical treatments. Critical concerns while manufacturing the composites have also been studied, and recommendations are given for future research [66]. If researchers can optimize the use of natural fibers such as PALF and sisal fibers in the reinforcements of polymers for attaining specific properties, natural fibers can make a significant impact in the aerospace, automobile, and structural industry. For this to be achieved, more extensive and comprehensive studies are required.

ACKNOWLEDGMENT

The authors would like to express their gratitude to the Military Institute of Science and Technology (MIST), Dhaka for supporting this work.

REFERENCES

1. Saba N, Paridah MT, Abdan K, Ibrahim NA. Effect of oil palm nano filler on mechanical and morphological properties of kenaf reinforced epoxy composites. *Constr Build Mater.* 2016 Oct;123:15–26.
2. Saba N, Jawaid M, Sultan MTH, Alothman OY. Green biocomposites for structural applications. In: Jawaid M, Salit MS, Alothman OY, editors. *Green Biocomposites* [Internet]. Cham: Springer International Publishing; 2017 [cited 2021 Apr 13]. pp. 1–27. (Green Energy and Technology). Available from: http://link.springer.com/10.1007/978-3-319-49382-4_1.
3. Hasan M, Hoque ME, Mir SS, Saba N, Sapuan SM. Manufacturing of coir fibre-reinforced polymer composites by hot compression technique. In: Salit MS, Jawaid M, Yusoff NB, Hoque ME, editors. *Manufacturing of Natural Fibre Reinforced Polymer Composites* [Internet]. Cham: Springer International Publishing; 2015 [cited 2021 Apr 23]. pp. 309–30. Available from: http://link.springer.com/10.1007/978-3-319-07944-8_15.
4. Leão AL, Cherian BM, Narine S, Souza SF, Sain M, Thomas S. The use of pineapple leaf fibers (PALFs) as reinforcements in composites. In: *Biofiber Reinforcements in Composite Materials* [Internet]. Elsevier; 2015 [cited 2021 Apr 13]. pp. 211–35. Available from: https://linkinghub.elsevier.com/retrieve/pii/B9781782421221500071.
5. Saba N, Jawaid M, Alothman OY, Paridah MT. A review on dynamic mechanical properties of natural fibre reinforced polymer composites. *Constr Build Mater.* 2016 Mar;106:149–59.
6. Senthil Kumar K, Siva I, Jeyaraj P, Winowlin Jappes JT, Amico SC, Rajini N. Synergy of fiber length and content on free vibration and damping behavior of natural fiber reinforced polyester composite beams. *Mater Des.* 2014 Apr;56:379–86.
7. Vimal R, Subramanian KHH, Ashwin C, Logeswaran V, Ramesh M.. Comparisonal study of succinylation and phthalicylation of jute fibres: Study of mechanical properties of modified fibre reinforced epoxy composites. *Mater Today: Proceedings* 2015; 2(4–5): 2918–2927.
8. Ramesh M, Palanikumar K, Reddy KH. Influence of fiber orientation and fiber content on properties of sisal-jute-glass fiber-reinforced polyester composites. *J Appl Polym Sci.* 2016;133(6):42968.
9. Qin Y. 3 – A brief description of textile fibers. In: Qin Y, editor. *Medical Textile Materials* [Internet]. Woodhead Publishing; 2016 [cited 2021 Aug 2]. pp. 23–42. (Woodhead Publishing Series in Textiles). Available from: https://www.sciencedirect.com/science/article/pii/B9780081006184000030.
10. Ramesh M, Palanikumar K, Reddy KH. Plant fibre based bio-composites: Sustainable and renewable green materials. *Renew Sustain Energy Rev.* 2017 Nov 1;79:558–84.

11. Djafari Petroudy SR. Physical and mechanical properties of natural fibers. In: *Advanced High Strength Natural Fibre Composites in Construction* [Internet]. Elsevier; 2017 [cited 2021 Apr 14]. pp. 59–83. Available from: https://linkinghub.elsevier.com/retrieve/pii/B9780081004111000030.

12. Arib RMN, Sapuan SM, Ahmad MMHM, Paridah MT, Zaman HMDK. Mechanical properties of pineapple leaf fibre reinforced polypropylene composites. *Mater Des.* 2006 Jan;27(5):391–6.

13. Yan L, Kasal B,.Huang L A review of recent research on the use of cellulosic fibres, their fibre fabric reinforced cementitious, geo-polymer and polymer composites in civil engineering. *Compos Part B.* 2016;(92):94–132.

14. Ansell MP, Mwaikambo LY. The structure of cotton and other plant fibres. In: *Handbook of Textile Fibre Structure* [Internet]. Elsevier; 2009 [cited 2021 Apr 14]. pp. 62–94. Available from: https://linkinghub.elsevier.com/retrieve/pii/B9781845697303500020.

15. Osorio L, Trujillo E, Van Vuure AW, Verpoest I. Morphological aspects and mechanical properties of single bamboo fibers and flexural characterization of bamboo/epoxy composites. *J Reinf Plast Compos.* 2011 Mar;30(5):396–408.

16. Jacquemin F. The hygroscopic behavior of plant fibers: A review. *Front Chem.* 2014;1:43.

17. Mishra S, Misra M, Tripathy SS, Nayak SK, Mohanty AK. Potentiality of pineapple leaf fibre as reinforcement in PALF-polyester composite: Surface modification and mechanical performance. *J Reinf Plast Compos.* 2001 Mar;20(4):321–34.

18. Lopattananon N, Panawarangkul K, Sahakaro K, Ellis B. Performance of pineapple leaf fiber–natural rubber composites: The effect of fiber surface treatments. *J Appl Polym Sci.* 2006 Oct 15;102(2):1974–84.

19. Chand N, Tiwary RK, Rohatgi PK. Bibliography Resource structure properties of natural cellulosic fibres? An annotated bibliography. *J Mater Sci.* 1988 Feb;23(2):381–7.

20. Tran AV. Chemical analysis and pulping study of pineapple crown leaves. *Ind Crops Prod.* 2006 Jul;24(1):66–74.

21. Asim M, Abdan K, Jawaid M, Nasir M, Dashtizadeh Z, Ishak MR, et al. A review on pineapple leaves fibre and its composites. *Int J Polym Sci.* 2015;2015:1–16.

22. Lobo MG, Siddiq M. Overview of pineapple production, postharvest physiology, processing and nutrition. *Handbook of Pineapple Technology: Production, Postharvest Science,Processing and Nutrition.* 2016 Dec. pp. 1–15

23. Mishra S, Mohanty AK, Drzal LT, Misra M, Hinrichsen G. A review on pineapple leaf fibers, sisal fibers and their biocomposites. *Macromol Mater Eng.* 2004 Nov 19;289(11):955–74.

24. Bengtsson M, Gatenholm P, Oksman K. The effect of crosslinking on the properties of polyethylene/wood flour composites. *Compos Sci Technol.* 2005 Aug;65(10):1468–79.

25. Banik S, Nag D, Debnath S. Utilization of pineapple leaf agro-waste for extraction of fibre and the residual biomass for vermicomposting. *Indian J Fibre Text Res.* 2011;6:172–177.

26. Todkar SS, Patil SA. Review on mechanical properties evaluation of pineapple leaf fibre (PALF) reinforced polymer composites. *Compos Part B Eng.* 2019 Oct;174:106927.

27. Mohamed AR, Sapuan SM, Shahjahan M, Khalina A. Effects of simple abrasive combing and pretreatments on the properties of pineapple leaf fibers (Palf) and Palf-vinyl ester composite adhesion. *Polym Plast Technol Eng.* 2010 Aug 17;49(10):972–8.

28. Wan Nadirah WO, Jawaid M, Al Masri AA, Abdul Khalil HPS, Suhaily SS, Mohamed AR. Cell wall morphology, chemical and thermal analysis of cultivated pineapple leaf fibres for industrial applications. *J Polym Environ.* 2012 Jun;20(2):404–11.

29. Luo H, Hanagud S. Dynamics of delaminated beams. *Int J Solids Struct.* 2000 Mar;37(10):1501–19.

30. Lee S, Park T, Voyiadjis GZ. Vibration analysis of multi-delaminated beams. *Compos Part B Eng.* 2003 Oct;34(7):647–59.

31. Della CN, Shu D. Free vibration analysis of composite beams with overlapping delaminations. *Eur J Mech-A Solids.* 2005 May;24(3):491–503.
32. Pothan LA, Oommen Z, Thomas S. Dynamic mechanical analysis of banana fiber reinforced polyester composites. *Compos Sci Technol.* 2003 Feb;63(2):283–93.
33. Romanzini D, Lavoratti A, Ornaghi HL, Amico SC, Zattera AJ. Influence of fiber content on the mechanical and dynamic mechanical properties of glass/ramie polymer composites. *Mater Des.* 2013 May;47:9–15.
34. Paul V, Kanny K, Redhi GG. Mechanical, thermal and morphological properties of a bio-based composite derived from banana plant source. *Compos Part Appl Sci Manuf.* 2015 Jan;68:90–100.
35. Jawaid M, Abdul Khalil HPS, Alattas OS. Woven hybrid biocomposites: Dynamic mechanical and thermal properties. *Compos Part Appl Sci Manuf.* 2012 Feb;43(2):288–93.
36. Mohanty S, Nayak SK. Interfacial, dynamic mechanical, and thermal fiber reinforced behavior of MAPE treated sisal fiber reinforced HDPE composites. *J Appl Polym Sci.* 2006 Nov 15;102(4):3306–15.
37. Zhang Z, Wang P, Wu J. Dynamic mechanical properties of EVA polymer-modified cement paste at early age. *Phys Procedia.* 2012;25:305–10.
38. de Medeiros ES, Agnelli JAM, Joseph K, de Carvalho LH, Mattoso LHC. Mechanical properties of phenolic composites reinforced with jute/cotton hybrid fabrics. *Polym Compos.* 2005 Feb;26(1):1–11.
39. Hameed N, Sreekumar PA, Francis B, Yang W, Thomas S. Morphology, dynamic mechanical and thermal studies on poly(styrene-co-acrylonitrile) modified epoxy resin/glass fibre composites. *Compos Part Appl Sci Manuf.* 2007 Dec;38(12):2422–32.
40. Geethamma VG, Kalaprasad G, Groeninckx G, Thomas S. Dynamic mechanical behavior of short coir fiber reinforced natural rubber composites. *Compos Part Appl Sci Manuf.* 2005 Nov;36(11):1499–506.
41. Le Guen M-J, Newman RH, Fernyhough A, Staiger MP. Tailoring the vibration damping behaviour of flax fibre-reinforced epoxy composite laminates via polyol additions. *Compos Part Appl Sci Manuf.* 2014 Dec;67:37–43.
42. Flynn J, Amiri A, Ulven C. Hybridized carbon and flax fiber composites for tailored performance. *Mater Des.* 2016 Jul;102:21–9.
43. Sumesh KR. Effect of TiO_2 nano-filler in mechanical and free vibration damping behavior of hybrid natural fiber composites. *J Braz Soc Mech Sci Eng.* 2020;42:1–12.
44. Senthilkumar K, Saba N, Chandrasekar M, Jawaid M, Rajini N, Alothman OY, et al. Evaluation of mechanical and free vibration properties of the pineapple leaf fibre reinforced polyester composites. *Constr Build Mater.* 2019 Jan;195:423–31.
45. George J, Bhagawan SS, Thomas S. Thermogravimetric and dynamic mechanical thermal analysis of pineapple fibre reinforced polyethylene composites. *J Therm Anal.* 1996 Oct;47(4):1121–40.
46. Devi LU, Bhagawan SS, Thomas S. Dynamic mechanical properties of pineapple leaf fiber polyester composites. *Polym Compos.* 2011;32(11):1741–50.
47. Devi LU, Bhagawan SS, Thomas S. Dynamic mechanical analysis of pineapple leaf/glass hybrid fiber reinforced polyester composites. *Polym Compos.* 2010 Jun;31(6):956–65.
48. Aji IS, Zainuddin ES, Khalina A, Sapuan SM. Optimizing processing parameters for hybridized kenaf/PALF reinforced HDPE composite. *Key Eng Mater.* 2011 Feb;471–472:674–9.
49. Asim M, Jawaid M, Paridah MT, Saba N, Nasir M, Shahroze RM. Dynamic and thermo-mechanical properties of hybridized kenaf/PALF reinforced phenolic composites. *Polym Compos.* 2019 Oct;40(10):3814–22.
50. Murugan R, Ramesh R, Padmanabhan K. Investigation on static and dynamic mechanical properties of epoxy based woven fabric glass/carbon hybrid composite laminates. *Procedia Eng.* 2014;97:459–68.

51. Zhang K, Wang F, Liang W, Wang Z, Duan Z, Yang B. Thermal and mechanical properties of bamboo fiber reinforced epoxy composites. *Polymers*. 2018;10:608.
52. Doddi PRV, Chanamala R, Dora SP. Dynamic mechanical properties of epoxy based PALF/basalt hybrid composite laminates. *Mater Res Express*. 2019 Sep 4;6(10):105343.
53. Luz FSD, Monteiro SN, Tommasini FJ. Evaluation of dynamic mechanical properties of PALF and coir fiber reinforcing epoxy composites. *Mater Res*. 2018;21(suppl 1):e20171108.
54. Siakeng R, Jawaid M, Ariffin H, Sapuan SM. Mechanical, dynamic, and thermomechanical properties of coir/pineapple leaf fiber reinforced polylactic acid hybrid biocomposites. *Polym Compos*. 2019;40(5):2000–11.
55. Asim M, Jawaid M, Nasir M, Saba N. Effect of fiber loadings and treatment on dynamic mechanical, thermal and flammability properties of pineapple leaf fiber and Kenaf phenolic composites. *J Renew Mater*. 2018 Jun 1;6(4):383–93.
56. Ramesh M. Kenaf (Hibiscus cannabinus L.) fibre based bio-materials: A review on processing and properties. *Prog Mater Sci*. 2016;78:92.
57. Ho M. Critical factors on manufacturing processes of natural fibre composites. *Compos Part B*. 2012;4. pp. 3549–3562
58. Li Y, Mai Y-W, Ye L. Sisal fibre and its composites: A review of recent developments. *Compos Sci Technol*. 2000;60:2037–55.
59. Senthilkumar K, Saba N, Rajini N, Chandrasekar M, Jawaid M, Siengchin S, et al. Mechanical properties evaluation of sisal fibre reinforced polymer composites: A review. *Constr Build Mater*. 2018 Jun;174:713–29.
60. Alvarez VA, Ruscekaite RA, Vazquez A. Mechanical properties and water absorption behavior of composites made from a biodegradable matrix and alkaline-treated sisal fibers. *Journal of Composite Materials*. 37(17):1575–1588
61. Ali ZAAA. Sisal natural fiber reinforcement influenced with experimental and numerical investigation onto vibration and mechanical properties of composite plate. *Int J Energy Environ*. 2016 Nov 1;7(6):497–509.
62. Chandra R, Singh SP, Gupta K. Damping studies in fiber-reinforced composites – A review. Compos Struct. 1999;46:41–51.
63. Liu W, Misra M, Askeland P, Drzal LT, Mohanty AK. 'Green' composites from soy based plastic and pineapple leaf fiber: Fabrication and properties evaluation. *Polymer*. 2005 Mar;46(8):2710–21.
64. Arifur Rahman M, Parvin F, Hasan M, Hoque ME. Introduction to manufacturing of natural fibre-reinforced polymer composites. In: Salit MS, Jawaid M, Yusoff NB, Hoque ME, editors. *Manufacturing of Natural Fibre Reinforced Polymer Composites* [Internet]. Cham: Springer International Publishing; 2015 [cited 2021 Apr 17]. pp. 17–43. Available from: http://link.springer.com/10.1007/978-3-319-07944-8_2.
65. Pruncu CI, Gürgen S, Hoque ME, editors. *Fiber-Reinforced Polymers: Processes and Applications*. New York: Nova Science Publishers; 2021. 1 p. (Polymer science and technology).
66. Ketabchi MR, Hoque ME, Khalid Siddiqui M. Critical concerns on manufacturing processes of natural fibre reinforced polymer composites. In: Salit MS, Jawaid M, Yusoff NB, Hoque ME, editors. *Manufacturing of Natural Fibre Reinforced Polymer Composites* [Internet]. Cham: Springer International Publishing; 2015 [cited 2021 Apr 23]. pp. 125–38. Available from: http://link.springer.com/10.1007/978-3-319-07944-8_6.

5 Influence of Fiber Length and Content on the Free Vibration and Damping Characteristics of Nanofiller-Added Natural Fiber-Based Polymer Composites

G. Rajeshkumar and S. Arvindh Seshadri
PSG Institute of Technology and Applied Research

CONTENTS

5.1 INTRODUCTION

Environmental awareness around the world has resulted in the increased usage of natural fiber-based polymeric composites (NFPCs) in numerous applications. Natural fibers offer numerous benefits such as biodegradability, lightweight, less energy consumption, considerable mechanical properties, low cost, and availablity in abundance (Rajeshkumar, et al. 2021a; Rajeshkumar, et al. 2021c). In particular, the degradability of these fibers assists the recyclability characteristics of the composites, which is the need of the hour (Vigneshwaran et al. 2020). The NFPCs are classified as partially biodegradable and fully degradable composites. The partially biodegradable composites are prepared by reinforcing the natural fibers into petroleum-based resins

DOI: 10.1201/9781003173625-6

such as polyester and epoxy, whereas the fully degradable composites are obtained by reinforcing the natural fibers into the biodegradable resins such as starch, polylactic acid, and polybutylene succinate (Gholampour and Ozbakkaloglu 2020).

With the aim of enhancing the performance of fiber-reinforced composites, a secondary reinforcement is used, which results in the production of hybrid composites. The secondary reinforcement may be either a nanofiller or another natural fiber. Among these, better properties are witnessed when nanofillers are used (Rajeshkumar et al. 2021b). Such hybrid composites are fabricated with the help of different manufacturing techniques such as compression or injection molding, resin transfer molding, and pultrusion, and these methods influence the properties of the composites.

In recent days, the NFPCs are widely used in automobile and aerospace applications (Rajeshkumar et al. 2021d) and the vibrations frequently occur in the automobile and aerospace components due to dynamic loads. Therefore, to avoid the failure of such components high-performance composites are prepared by reinforcing natural fibers along with the nanofillers into the polymers. The existence of hemicellulose in the cellulosic fibers helps to improve the free vibration characteristics (FVCs) of the NFPCs (Vigneshwaran et al. 2020). Apart from this, the stiffness of the composites plays a vital role in improving the natural frequency (NF) of composites. By taking the above points into account, this chapter reviews the FVC of composites added with natural fiber and nanoclay.

5.2   NATURAL FIBERS

Natural fiber polymer composites are made up of two main components: a polymer matrix and a natural fiber reinforcement. Although the polymer matrices are lightweight in nature, they do not possess the mechanical properties that will be necessary to withstand the loads in structural applications. As a result, the polymers are reinforced with natural fibers with excellent strength and stiffness to obtain composite materials (Rajeshkumar et al. 2021e). Such natural fibers are obtained from three major sources, namely plant, animal, and mineral, and are illustrated in Figure 5.1 (Madhu et al. 2019; Rajeshkumar et al. 2021a).

In polymer composites, plant fibers are used predominantly as reinforcement, when compared to animal and mineral fibers. This is due to their widespread availability and ease of extraction. Plant fibers primarily consist of three components: cellulose, hemicellulose, and lignin. Furthermore, the presence of wax, pectin, etc., is also observed in small amounts (Nagarajan et al. 2021). Cellulose dictates the resultant mechanical properties of the plant fibers and it is because of the presence of hydrogen bonds in cellulose and alignment of microfibrils in fiber direction (Thyavihalli et al. 2019). On the other hand, properties such as moisture absorption, thermal degradation, and biodegradation behavior of plant fibers are influenced by hemicellulose. Lignin, which is amorphous and hydrophobic, is instrumental in providing additional structural stability to the plant fibers (Ramamoorthy et al. 2015).

The inclusion of natural fibers into polymer matrices enhances their performance and effectively decreases the composites' cost. In addition, natural fibers possess unique properties such as high strength and modulus, low weight, easy availability,

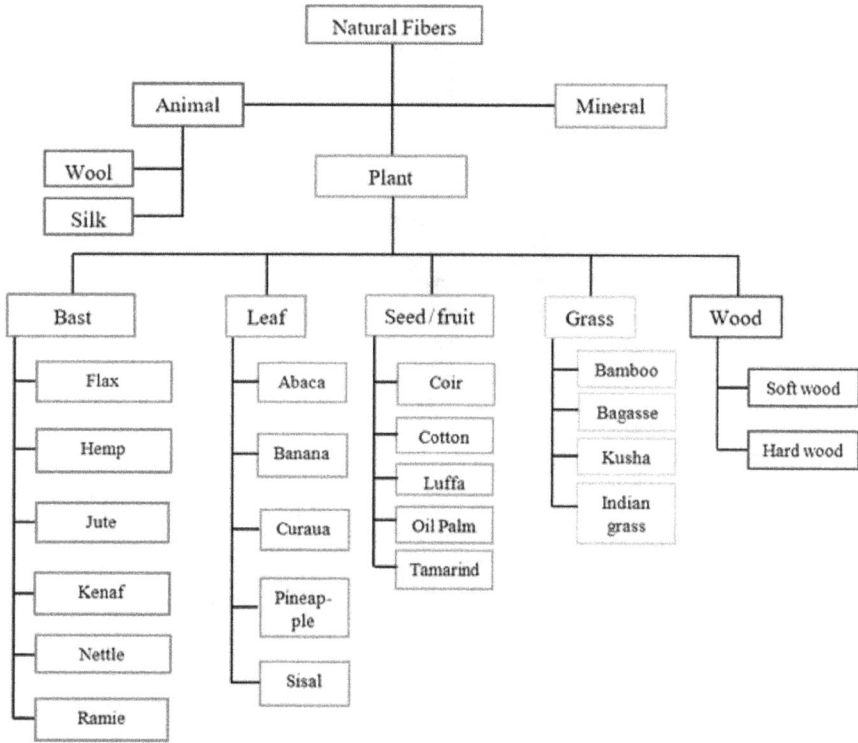

FIGURE 5.1 Classification of natural fibers.

biodegradability, renewability and good thermal insulation, and acoustic characteristics. Furthermore, the utilization of natural fibers leads to lower greenhouse gas emissions, minimal health, and other environmental issues (Rajeshkumar et al. 2021f).

5.3 NANOFILLERS

Nanomaterials are those materials whose dimensions are in the order of nanometer (10^{-9}m) scale. Nanomaterials are majorly divided into three categories depending on the number of their dimensions that are in the nanometer range. They are nanoparticles, nanolayers, and nanotubes. Nanoparticles and nanotubes play a vital role as filler materials in polymer matrix composites by enhancing their physical and mechanical properties (Saba, Tahir, and Jawaid 2014). This is attributed to their unique size in the nanometer range, which gives them the ability to produce large specific surface areas (more than 1,000 m²/g), combined with other peculiar characteristics (Njuguna, Pielichowski, and Alcock 2007). The addition of nanofillers results in the improvement of specific properties in composites such as mechanical, thermal, optical, fire-retardant, or electrical properties. The nanofillers are added in the range of 1–10 wt.% to the polymer matrices (Marquis, Guillaume, and Chivas-Joly 2011). The different nanofillers that can be added to polymer matrix composites

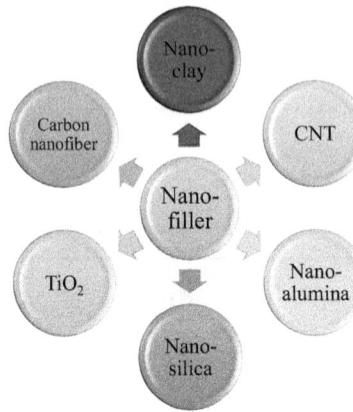

FIGURE 5.2   Different types of nanofillers.

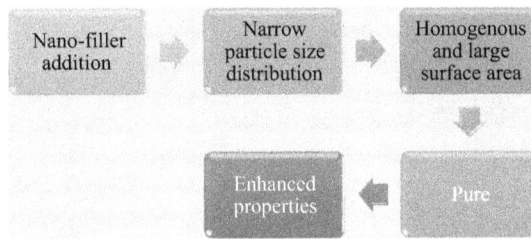

FIGURE 5.3   Modifications due to the addition of nanofillers in composites.

are presented in Figure 5.2. Additionally, Figure 5.3 represents the modifications that occur in composites due to the inclusion of nanofillers.

5.4   VIBRATION

The vibration characteristics of a component such as NF, damping ratio (DR), and mode shapes play a vital role in applications such as aerospace, automotive, and machine components. (Krishnasamy Senthilkumar et al. 2017). The presence of vibration in the structures will result in loud noises, premature wear, fatigue failure, and dangerous operating conditions. If the NF of vibrations caused by the dynamic loading becomes equal to the NF of the component, then resonance will occur, eventually leading to catastrophic failure of the component (Haldar, Singh, and Prince 2011). As a result, it is critical to explore the free vibration and damping characteristics of NFPCs.

The vibration and damping properties of NFPCs are influenced by various factors, namely fiber length, fiber content, fiber-matrix bonding, Young's modulus, density, fiber orientation, loading direction, and chemical treatment of fibers (Akoussan et al. 2016; Ni et al. 2015; Chung 2003). Energy during vibration is dissipated through shear deformation in polymers. On the contrary, there are different energy dissipation mechanisms in the case of composites. These include friction generated by slip

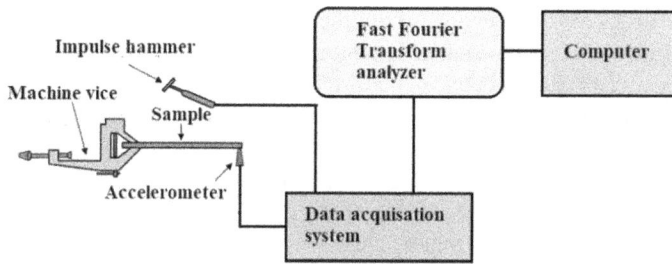

FIGURE 5.4 Experimental setup of impulse hammer test.

at fiber-matrix interface, viscoelastic nature of matrix and fibers, delamination produced at damaged locations, visco-plastic and thermoelastic damping, and energy dissipation at cracks (Chandra, Singh, and Gupta 1999; Chauhan, Karmarkar, and Aggarwal 2009).

A vibration experiment is conducted to evaluate the NF and DR of composites. The experimental vibrational analysis is conducted using an impulse hammer test. This is depicted in Figure 5.4. NF and DR are generally associated with the first three modes, namely Mode 1 (bending), Mode 2 (twisting), and Mode 3. (second bending), which are examined. Composites, in the form of cantilever, are subjected to a constant force at one end using a piezoelectric impulse hammer. The response of the specimen caused by excitation is measured using an accelerometer that is fixed to the other end of the specimen. The primary measurement required in a vibration test is the frequency response function, which is the ratio of output response to input excitation force. This is obtained by utilizing a Fast Fourier Transform (FFT) analyzer, which converts time-domain signal received from the DEWE data acquisition system into the frequency response function. There are peaks observed in the frequency response function and this denotes the NF of samples (Senthil Kumar et al. 2014; Rajesh and Pitchaimani 2016).

5.5 FREE VIBRATION AND DAMPING CHARACTERISTICS OF NANOFILLER-ADDED NATURAL FIBER-BASED POLYMER COMPOSITES

The FVCs of nanoclay-added epoxy/*Phoenix* sp. composites were studied, in which fiber length (10–40 mm), fiber volume fraction (10%–50%), and nanoclay content (0–7 wt.%) were varied to examine their effect (Kumar, Hariharan, and Saravanakumar 2021). It was evident from the results that NF of the composites increased as longer fibers were used, irrespective of the fiber content. When fiber length is enlarged, contact area between matrix and reinforcement becomes greater, and hence, NF of the composite increases. Similarly, NF also increased upon fiber addition up to 40%, after which agglomeration of fibers ensued in a reduction in the NF. NF of composite samples incorporated with fibers of 30 mm in length and 40 vol.% was found to be the maximum (30.44 Hz). The addition of nanoclay (organically modified montmorillonite (MMT)) led to an augmentation in the NF of the composites up to 5 wt.%, beyond which there was a decline in the NF because of nanoclay agglomeration. The enhancement in the NF with the inclusion of nanoclay is due to improved bonding between matrix and fiber, caused by the increased dispersion of nanoclay and improved stiffness of the samples. A maximum of 42.65 Hz NF was obtained when 5 wt.% of MMT was incorporated into the composites.

Ramakrishnan et al. (2021) analyzed the influence of nanoclay (Cloisite 20A) on the vibration behavior of jute reinforced epoxy composites. The fiber length and fiber content were kept constant (30 mm and 20 wt.%), while the nanoclay content was varied (0–7 wt.%). The vibration analysis was conducted using an impact hammer and an FFT analyzer was utilized to obtain the frequency response function. The NF of epoxy/jute composites increased up to a 5 wt.% nanoclay loading. The incorporation of nanoclay leads to better fiber-matrix bonding and increased stiffness of composites, due to which NF increases. When 7 wt.% of nanoclay was added, NF reduced due to agglomeration. A maximum NF of 59.95 Hz was obtained when 5 wt.% nanoclay was incorporated, which was greater than 1 wt.%, 3 wt.%, and 7 wt.% added epoxy/jute composites by 21.85%, 14%, and 9%, respectively.

Rajini et al. (2013) conducted a vibration analysis to examine the influence of nanoclay loading on the NF of coconut sheath/polyester composites. The fiber content was kept constant at 48 wt.% and nanoclay content was varied from 1 to 5 wt.%. With the addition of nanoclay, the fundamental NF increased up to 3 wt.% of nanoclay, following which it reduced. The raise in NF is ascribed to the increase in the value of Young's modulus, while at higher concentrations of nanoclay agglomeration and poor fiber-matrix bonding reduces the modulus and thus, NF decreases. Similar trend was also seen in 2nd and 3rd modes as well. However, nanoclay-added composites showed better vibration and damping properties compared to the composites without nanoclay.

Sumesh and Kanthavel (2020b) examined the influence of fiber quantity and nanofiller (nanoalumina) loading on the vibrational performance of sisal/coir reinforced epoxy hybrid composites. The fiber loading was varied between 0 and 20 wt.% and nanoalumina content was between 0 and 3 wt.%. The fibers had a fixed length of 5 mm. The maximum Mode 1 frequency of 26 Hz was obtained when the composites contained 20 wt.% sisal, 15 wt.% coir, and 1 wt.% nanoalumina, whereas the maximum Mode 2 and Mode 3 frequency of 257 and 537 Hz, respectively, were observed at 15 wt.% sisal, 10 wt.% coir, and 3 wt.% nanoalumina. The addition of nanoalumina led to a larger area of contact with fiber/matrix and matrix/filler interface, thus improving the stiffness and vibrational properties of the composites (Tang and Yan 2020). Furthermore, the maximum damping factor values of 0.0759, 0.0178, and 0.0087, for Mode 1, Mode 2, and Mode 3, respectively, were obtained at 20 wt.% sisal, 15 wt.% coir, and 1 wt.% nanoalumina. Damping is improved by energy dissipation improvement at large strain areas and friction between fiber and matrix (Chandra, Singh, and Gupta 1999). Energy dissipation improvement is caused by shear deformation and good interfacial bonding between matrix and fiber. The addition of 20 wt.% sisal and 15 wt.% coir ensures good fiber-matrix bonding, providing good dispersion of force with enhanced damping characteristics.

Arulmurugan and Venkateshwaran (2016) examined the effect of fiber content and nanoclay addition on the vibrational properties of polyester/jute composites.

Vibration tests were conducted with composites having different amounts of fiber (5–30 wt.%) and nanoclay (1–7 wt.%). The incorporation of jute fibers led to an augmentation in the NF of the composites. In nanoclay filled composite, adding 5 wt.% of jute led to the increase of NF from 28 to 29.3 Hz. Similarly, higher jute content of 10, 15, 20, 25, and 30 wt.% resulted in the increase of NF to 31.21, 35, 40, 39.62, 41, and 55.31 Hz, respectively. In addition, DR was found to increase with both nanoclay and jute fiber addition.

The free vibration behavior of jute/banana hybrid polyester composites upon nanoclay addition was examined (Rajesh, Jeyaraj, and Rajini 2016). The composites were manufactured using compression molding. The vibration analysis was carried out using Fixed-Fixed and Fixed-Free boundary conditions. The NF and DR from the initial three modes were measured from the vibrational analysis. Nanoclay loading up to 2 wt.% leads to a rise in the NF of the samples. The addition of nanoclay raises the modulus of the composites and hence, NF increases. When nanoclay is added, it enables the polymer to withstand greater value of stress and allows a minimal stress transfer at the fiber-matrix interface. This results in a lower damping value. When larger amount of nanoclay is added, the bonding among the fiber and matrix enhances, which leads to better frictional resistance and mechanical interlocking between matrix and reinforcement. Due to this, composites with higher nanoclay content provide better damping properties (Rajini et al. 2013).

Sumesh, Kanthavel, and Vivek (2019) analyzed the influence of nanoalumina addition (0–3 wt.%) on the vibration performance of epoxy-based banana, coir, and sisal hybrid composites. The fiber quantity was maintained at 35 wt.% and the wt.% of nanoalumina was varied. Sisal/banana composites with 3 wt.% nanoalumina exhibited maximum NF values of 29, 425, and 687 Hz for Mode 1, Mode 2, and Mode 3, respectively, when compared to other specimens. This was because of the enhanced properties of both banana and sisal fibers and enhanced fiber-matrix that was formed. Additionally, the inclusion of nanoalumina improves the vibration performance by decreasing the voids situated in the fiber-matrix surface and hence, enhancing the stiffness of the composites (Hsieh, Huang, and Shen 2017; Khan et al. 2011). Similarly, the damping factor was also found to increase due to nanoalumina addition. The inclusion of nanoalumina provides good friction, enhances the energy dissipation, and hence, better damping factors are obtained. It was also pointed out fiber content plays a pivotal role in the enhancement of damping characteristics (Chandra, Singh, and Gupta 1999). The maximum damping factors were observed for sisal/coir and sisal/banana composites as the low diameter of sisal fibers results in a thicker interface and better damping properties (Mukherjee and Satyanarayana 1984).

An experimental work to analyze the FVCs of nanoclay-added basalt fiber-based epoxy composites was carried out (Bulut et al. 2020). The quantity of nanoclay was varied from 0 to 3 wt.%. It was noticed that the NF of the composites increased with nanoclay addition. The maximum NF value (52.54 Hz) was obtained when 2 wt.% of nanoclay was added. Further addition of nanoclay led to a decrease in NF from 52.54 to 51.81 Hz. When nanoclay is added, higher load transfer occurs between filler and matrix, and hence, the dynamic elastic modulus of the composites increases, and this results in higher NF. At higher concentrations, exfoliation of nanoclay causes a decrease in load transfer, interfacial stress, and stiffness between the matrix and fiber. Therefore, NF decreases. Similarly, DR also increases with nanoclay addition up to 2 wt.%, beyond which it decreases. The improvement in damping property is because of the strong interfacial adhesion between nanoclay, fiber, and matrix, which leads to efficient stress transfer between fiber and matrix.

Sumesh and Kanthavel (2020a) produced three different kinds of epoxy-based hybrid composites: sisal/banana (SB), pineapple/sisal (PS), and banana/pineapple (BP), and studied their FVCs. To enhance the properties, different contents (1–4 wt.%) of TiO_2 nanofiller were added to the composites. The natural frequencies of the composites were found to increase up to 3 wt.% of nanofiller loading. At 4 wt.% of TiO_2, the natural frequencies reduced owing to the weaker interface that was formed due to agglomeration. The maximum natural frequencies were obtained at 3 wt.% of TiO_2 because of the superior surface contact with resin/hybrid natural fibers and filler/resin interface, which enhanced the stiffness of the composites and thus, improved the NF. Additionally, it was pointed out that fiber length had no impact on NF of the composites (Senthil Kumar et al. 2014). Similar to NF, the DR of the composites also increased up to 3 wt.% of TiO_2 addition and reduced when 4 wt.% of TiO_2 was added. The authors concluded that pineapple-based hybrid composites exhibited better damping rate due to the higher interface thickness and lower diameter of pineapple fibers (Saravana Bavan and Mohan Kumar 2010). Lower diameters lead to higher aspect ratios, which eventually result in better contact area and stiffer composites (Senthilkumar et al. 2019).

5.6 CONCLUSIONS AND FUTURE PERSPECTIVES

Government policies and awareness of the environment are driving scientists, academia, and businesses to concentrate on the development of environmentally benign materials. Natural fibers were found to be a viable solution for developing composites owing to their merits such as being lightweight, eco-friendly, high strength-to-weight ratio, non-toxicity, and so on. Natural fiber composites are employed in numerous industrial applications and the continuous development in this field has resulted in the advancement of high-performance hybrid composites. In particular, the composites consisting of natural fiber and nanofiller offer better vibration characteristics when compared to single fiber-reinforced composites. This improvement is caused by the increased interfacial bonding between the matrix and reinforcements and the stiffness of the composites. However, only a few pieces of literature detailed the FVCs of such hybrid composites. A huge research gap is available in this area, wherein the hybrid composites with improved vibration characteristics can be developed by combining different natural fibers and nanofillers.

REFERENCES

Akoussan, K, H. Boudaoud, D. El Mostafa, Y. Koutsawa, and E. Carrera. 2016. "Sensitivity Analysis of the Damping Properties of Viscoelastic Composite Structures According to the Layers Thicknesses." *Composite Structures* 149. Elsevier Ltd: 11–25. doi:10.1016/j.compstruct.2016.03.061.

Arulmurugan, S., and N. Venkateshwaran. 2016. "Vibration Analysis of Nanoclay Filled Natural Fiber Composites." *Polymers and Polymer Composites* 24 (7): 507–16. doi:10.1177/096739111602400709.

Bulut, M, Ö. Y. Bozkurt, A. Erkliğ, H. Yaykaşlı, and Ö. Özbek. 2020. "Mechanical and Dynamic Properties of Basalt Fiber-Reinforced Composites with Nanoclay Particles." *Arabian Journal for Science and Engineering* 45 (2). Springer Berlin Heidelberg: 1017–33. doi:10.1007/s13369-019-04226-6.

Chandra, R., S. P. Singh, and K. Gupta. 1999. "Damping Studies in Fiber-Reinforced Composites - A Review." *Composite Structures* 46 (1): 41–51. doi:10.1016/S0263-8223(99)00041-0.

Chauhan, S., A. Karmarkar, and P. Aggarwal. 2009. "Damping Behavior of Wood Filled Polypropylene Composites." *Journal of Applied Polymer Science* 114 (4): 2421–26. doi:10.1002/app.30718.

Chung, D. D. L. 2003. "Structural Composite Materials Tailored for Damping." *Journal of Alloys and Compounds* 355 (1–2): 216–23. doi:10.1016/S0925-8388(03)00233-0.

Gholampour, A., and T. Ozbakkaloglu. 2020. A Review of Natural Fiber Composites: Properties, Modification and Processing Techniques, Characterization, Applications. *Journal of Materials Science* 55. Springer US. doi:10.1007/s10853-019-03990-y.

Haldar, A. K., S. Singh, and P. Prince. 2011. "Vibration Characteristics of Thermoplastic Composite." In *AIP Conference Proceedings*, vol. 1414, pp. 211–214. doi:10.1063/1.3669958.

Hsieh, T. H., Y. S. Huang, and M. Y. Shen. 2017. "Dynamic Properties of Carbon Aerogel/Epoxy Nanocomposite and Carbon Fiber-Reinforced Composite Beams." *Journal of Reinforced Plastics and Composites* 36 (23): 1745–55. doi:10.1177/0731684417728585.

Khan, S. U., C. Y. Li, N. A. Siddiqui, and J. K. Kim. 2011. "Vibration Damping Characteristics of Carbon Fiber-Reinforced Composites Containing Multi-Walled Carbon Nanotubes." *Composites Science and Technology* 71 (12). Elsevier Ltd: 1486–94. doi:10.1016/j.compscitech.2011.03.022.

Kumar, G. R, V. Hariharan, and S. S. Saravanakumar. 2021. " Enhancing the Free Vibration Characteristics of Epoxy Polymers Using Sustainable Phoenix Sp. Fibers and Nano-Clay for Machine Tool Applications." *Journal of Natural Fibers* 18 (4). Taylor & Francis: 531–38. doi:10.1080/15440478.2019.1636740.

Madhu, P., M. R. Sanjay, P. Senthamaraikannan, S. Pradeep, S. S. Saravanakumar, and B. Yogesha. 2019. "A Review on Synthesis and Characterization of Commercially Available Natural Fibers: Part-I." *Journal of Natural Fibers* 16 (8). Taylor & Francis: 1132–44. doi:10.1080/15440478.2018.1453433.

Marquis, D., E. Guillaume, and C. Chivas-Joly. 2011. "Properties of Nanofillers in Polymer." In *Nanocomposites and Polymers with Analytical Methods*, edited by J. Cuppoletti, p. 261. InTech. doi:10.5772/21694.

Mukherjee, P. S., and K. G. Satyanarayana. 1984. "Structure and Properties of Some Vegetable Fibres - Part 1 Sisal Fibre." *Journal of Materials Science* 19 (12): 3925–34. doi:10.1007/BF00980755.

Nagarajan, K. J., N. R. Ramanujam, M. R. Sanjay, S. Siengchin, B. Surya Rajan, K. Sathick Basha, P. Madhu, and G. R. Raghav. 2021. "A Comprehensive Review on Cellulose Nanocrystals and Cellulose Nanofibers: Pretreatment, Preparation, and Characterization." *Polymer Composites* 42 (4): 1588–630. doi:10.1002/pc.25929.

Ni, N., Y. Wen, D. He, X. Yi, T. Zhang, and Y. Xu. 2015. "High Damping and High Stiffness CFRP Composites with Aramid Non-Woven Fabric Interlayers." *Composites Science and Technology* 117. Elsevier Ltd: 92–9. doi:10.1016/j.compscitech.2015.06.002.

Njuguna, J., K. Pielichowski, and J. R. Alcock. 2007. "Epoxy-Based Fibre Reinforced Nanocomposites." *Advanced Engineering Materials* 9 (10): 835–47. doi:10.1002/adem.200700118.

Rajesh, M., and J. Pitchaimani. 2016. "Dynamic Mechanical Analysis and Free Vibration Behavior of Intra-Ply Woven Natural Fiber Hybrid Polymer Composite." *Journal of Reinforced Plastics and Composites* 35 (3): 228–42. doi:10.1177/0731684415611973.

Rajesh, M, P. Jeyaraj, and N. Rajini. 2016. "Mechanical, Dynamic Mechanical and Vibration Behavior of Nanoclay Dispersed Natural Fiber Hybrid Intra-Ply Woven Fabric Composite." In *Nanoclay Reinforced Polymer Composites*, pp. 281–96. doi:10.1007/978-981-10-0950-1_12.

Rajeshkumar, G., K. Naveen Kumar, M. Aravind, S. Seshadri Arvindh, S. Santhosh, and T. K. Gowtham Keerthi. 2021a. "A Comprehensive Review on Mechanical Properties of Natural Cellulosic Fiber Reinforced PLA Composites." In *Materials, Design, and Manufacturing for Sustainable Environment*, edited by S. Mohan, S. Shankar, and G. Rajeshkumar, pp. 227–37. Springer, Singapore. doi:10.1007/978-981-15-9809-8_19.

Rajeshkumar, G., S. Arvindh Seshadri, G.L. Devnani, M.R. Sanjay, S. Siengchin, J. Prakash Maran, N. A. Al-Dhabi, et al. 2021b. "Environment Friendly, Renewable and Sustainable Poly Lactic Acid (PLA) Based Natural Fiber Reinforced Composites – A Comprehensive Review." *Journal of Cleaner Production* 310 (January). Elsevier Ltd: 127483. doi:10.1016/j.jclepro.2021.127483.

Rajeshkumar, G., V. Hariharan, G. L. Devnani, J. Prakash Maran, M. R. Sanjay, S. Siengchin, N. A. Al-Dhabi, and K. Ponmurugan. 2021c. "Cellulose Fiber from Date Palm Petioles as Potential Reinforcement for Polymer Composites: Physicochemical and Structural Properties." *Polymer Composites*, May, pc.26106. doi:10.1002/pc.26106.

Rajeshkumar, G., A. Seshadri, K. R. Sumesh, and K. C. Nagaraja. 2021d. "Influence of Phoenix Sp. Fiber Content on the Viscoelastic Properties of Polymer Composites." In *Materials, Design, and Manufacturing for Sustainable Environment*, edited by S. Mohan, S. Shankar, and G. Rajeshkumar, pp. 131–39. Springer, Singapore. doi:10.1007/978-981-15-9809-8_10.

Rajeshkumar, G., S. Arvindh Seshadri, M. Bilal Mohammed, K. Srijith, B. Brahatheesh Vikram, and A. Sailesh. 2021e. "Eco-Friendly Wood Fibre Composites with High Bonding Strength and Water Resistance." In *Eco-Friendly Adhesives for Wood and Natural Fiber Composites*, edited by M. Jawaid, T. Ahmed Khan, M. Nasir, and M. Asim, pp. 105–122. Composites Science and Technology. Springer, Singapore. doi:10.1007/978-981-33-4749-6_5.

Rajeshkumar, G., S. Arvindh Seshadri, S. Ramakrishnan, M. R. Sanjay, S. Siengchin, and K. C. Nagaraja. 2021f. "A Comprehensive Review on Natural Fiber/Nano-Clay Reinforced Hybrid Polymeric Composites: Materials and Technologies." *Polymer Composites*. doi:10.1002/pc.26110.

Rajini, N., Jt Winowlin Jappes, S. Rajakarunakaran, and P. Jeyaraj. 2013. "Dynamic Mechanical Analysis and Free Vibration Behavior in Chemical Modifications of Coconut Sheath/Nano-Clay Reinforced Hybrid Polyester Composite." *Journal of Composite Materials* 47 (24): 3105–21. doi:10.1177/0021998312462618.

Ramakrishnan, S., K. Krishnamurthy, G. Rajeshkumar, and M. Asim. 2021. "Dynamic Mechanical Properties and Free Vibration Characteristics of Surface Modified Jute Fiber/Nano-Clay Reinforced Epoxy Composites." *Journal of Polymers and the Environment* 29 (4). Springer US: 1076–88. doi:10.1007/s10924-020-01945-y.

Ramamoorthy, S. K., M. Skrifvars, and A. Persson. 2015. "A Review of Natural Fibers Used in Biocomposites: Plant, Animal and Regenerated Cellulose Fibers." *Polymer Reviews* 55 (1): 107–62. doi:10.1080/15583724.2014.971124.

Saba, N, P. Md Tahir, and M. Jawaid. 2014. "A Review on Potentiality of Nano Filler/Natural Fiber Filled Polymer Hybrid Composites." *Polymers* 6 (8): 2247–73. doi:10.3390/polym6082247.

Saravana Bavan, D., and G. C. Mohan Kumar. 2010. "Potential Use of Natural Fiber Composite Materials in India." *Journal of Reinforced Plastics and Composites* 29 (24): 3600–13. doi:10.1177/0731684410381151.

Senthil Kumar, K., I. Siva, P. Jeyaraj, J. T. Winowlin Jappes, S. C. Amico, and N. Rajini. 2014. "Synergy of Fiber Length and Content on Free Vibration and Damping Behavior of Natural Fiber Reinforced Polyester Composite Beams." *Materials and Design* 56. Elsevier Ltd: 379–86. doi:10.1016/j.matdes.2013.11.039.

Senthilkumar, K., I. Siva, M. T. H. Sultan, N. Rajini, S. Siengchin, M. Jawaid, and A. Hamdan. 2017. "Static and Dynamic Properties of Sisal Fiber Polyester Composites - Effect of Interlaminar Fiber Orientation." *BioResources* 12 (4): 7819–33. doi:10.15376/biores.12.4.7819-7833.

Senthilkumar, K., N. Saba, M. Chandrasekar, M. Jawaid, N. Rajini, O. Y. Alothman, and S. Siengchin. 2019. "Evaluation of Mechanical and Free Vibration Properties of the Pineapple Leaf Fibre Reinforced Polyester Composites." *Construction and Building Materials* 195. Elsevier Ltd: 423–31. doi:10.1016/j.conbuildmat.2018.11.081.

Sumesh, K. R., and K. Kanthavel. 2020a. "Effect of TiO_2 Nano-Filler in Mechanical and Free Vibration Damping Behavior of Hybrid Natural Fiber Composites." *Journal of the Brazilian Society of Mechanical Sciences and Engineering* 42 (4). Springer, Berlin Heidelberg: 211. doi:10.1007/s40430-020-02308-3.

Sumesh, K. R., and K. Kanthavel. 2020b. "Synergy of Fiber Content, Al_2O_3 Nanopowder, NaOH Treatment and Compression Pressure on Free Vibration and Damping Behavior of Natural Hybrid-Based Epoxy Composites." *Polymer Bulletin* 77 (3): 1581–1604. doi:10.1007/s00289-019-02823-x.

Sumesh, K R, K Kanthavel, and S Vivek. 2019. "Mechanical/Thermal/Vibrational Properties of Sisal, Banana and Coir Hybrid Natural Composites by the Addition of Bio Synthesized Aluminium Oxide Nano Powder." *Materials Research Express* 6 (4): 045318. doi:10.1088/2053-1591/aaff1a.

Tang, X, and X. Yan. 2020. "A Review on the Damping Properties of Fiber Reinforced Polymer Composites." *Journal of Industrial Textiles* 49 (6): 693–721. doi:10.1177/1528083718795914.

Thyavihalli, G, Y. Gowda, S. Mavinkere Rangappa, J. Parameswaranpillai, and S. Siengchin. 2019. "Natural Fibers as Sustainable and Renewable Resource for Development of Eco-Friendly Composites: A Comprehensive Review." *Frontiers in Materials* 6 (September): 1–14. doi:10.3389/fmats.2019.00226.

Vigneshwaran, S., R. Sundarakannan, K. M. John, R. Deepak Joel Johnson, K. Arun Prasath, S. Ajith, V. Arumugaprabu, and M. Uthayakumar. 2020. "Recent Advancement in the Natural Fiber Polymer Composites: A Comprehensive Review." *Journal of Cleaner Production* 277. doi:10.1016/j.jclepro.2020.124109.

6 Free Vibration and Damping Characteristics of Completely Biodegradable Polymer-Based Composites

Vinyas Mahesh
National Institute of Technology, Silchar, Assam, India

Vishwas Mahesh
Siddaganga Institute of Technology, Tumkur, India
Indian Institute of Science, Bangalore, India

Subashchandra Kattimani and Vinayak Kallannavar
National Institute of Technology Karnataka, Surathkal, India

Dineshkumar Harursampath
Indian Institute of Science, Bangalore, India

CONTENTS

DOI: 10.1201/9781003173625-7

6.1 INTRODUCTION

With the rising concerns of global warming, environmental pollution and tedious material disposal and recycling, a shout-out for biodegradable materials is increasing day by day. Meanwhile, for various structural applications, the convergence of structural stability, economic benefits, manufacturing feasibility together with biodegradability is a challenging issue. However, with the rapid developments in material science, several natural fibers such as jute, hemp, ramie, coir, pineapple, and so on are found to be replacements for many of the synthetic fibers. Similarly, biopolymers such as cellulose, starch, proteins, and so on have found to replace petroleum-based polymers quite effectively and exhibit a superior life cycle than petroleum-based polymers. However, this study mainly focuses on natural fibers such as Jute and Rubber and Polylactic acid (PLA), biopolymer-based matrix. Therefore, the literature review is restricted to the concerned areas only. Many kinds of research on PLA suggest that it releases an insignificant amount of hazardous gases while recycling. Through hydrolysis, it can be readily converted into water and carbon dioxide by the action of microorganisms, which is worthy for developing various agricultural products. In addition, at elevated temperatures, it can be biodegraded in months, while it takes some years to degrade naturally. PLA can be synthesized by polymerization or ring-opening polymerization of Lactic acid and Lactide, respectively.

On the other hand, natural fibers can be extracted from various sources out of which plant and animal sources are prominent. In the case of the plant source, it can be extracted from the stalk, fruits, seeds, leaves, and bast of the plant. The main pros of natural fibers are their abundant availability, economic, and high recyclability. However, the drawback of poor wettability makes them fall far behind synthetic fibers. Further, the strength exhibited by the natural fiber depends on the cellulose content it possesses. The previous research noticed that jute displays high cellulose content, which makes it more apt for structural applications. Several works have been reported on the material characterization and free vibration characteristics of natural fiber composites. For a better understanding of the readers, they have been encapsulated here. Senthil et al. [1] evaluated the frequency response of pineapple leaf composite reinforced in the polyester matrix. Kumar et al. [2] probed the effect of the stacking sequence of coconut sheath and sisal-based hybrid composites on its natural frequencies.

Further, Rajini et al. [3] assessed the effect of chemical modifications on the free vibration characteristics of coconut sheath and nanoclay reinforced composites. Rajesh et al. [4] investigated the frequency response of banana/sisal-based composite beam. Uthayakumar et al. [5] experimentally investigated the influence of red mud on the free vibration response of banana-based composites. Rajesh et al. [6] studied the effect of dispersing nanoclay in the natural fibers on the overall vibrations of the hybrid intra-ply has woven fabric composites. Idicula et al. [7] investigated the dynamic mechanical behavior of randomly oriented hybrid composites constituting banana and sisal natural fibers. Kumar et al. [8] studied the frequency response of coconut sheath/banana fiber hybrid composites with various layering patterns. Etaati et al. [9] conducted an experimental investigation to know the vibration and damping behavior of short hemp fiber-based thermoplastic composite. El Mahi et al. [10]

studied the damping characteristics of unidirectional fiber composites and orthotropic laminates. In this study, both experimental and numerical results were obtained and evaluated. Chandradass et al. [11], through their work, suggested that a higher percentage of nanoclay improves the damping characteristics of glass fiber-reinforced vinyl ester composites. Rizal et al. [12] combined the experimental and numerical studies to assess the frequencies of Jute-reinforced composites. Ramakrishnan et al. [13] evaluated the influence of surface modification on the overall damping and free vibration characteristics of jute/clay-based epoxy composites. Roy et al. [14] demonstrated the effect of integrating jute and nanoclay on the overall material properties, including the dynamic properties of rubber composites. Chandra et al. [15] presented the different damping mechanisms associated with fiber-reinforced composites in their review article. Similarly, Saba et al. [16] reviewed the dynamic mechanical properties of natural fiber-reinforced composites.

Meanwhile, Ruksakulpiwat et al. [17] studied the effect of molding techniques on the mechanical properties of PLA/Jute/natural rubber. Ejaz et al. [18] probed on the biodegradability of Jute-reinforced PLA composites. Ma and Joo [19] highlighted the structural properties of surface-treated jute/PLA composites in their research work. Similar work was reported by Zafar et al. [20], focusing much on the effect of surface treatments on microstructural changes. Reinforcing various forms of natural fiber, including jute Gunti et al. [21] compared the biodegradability and mechanical properties of such natural composites.

From the exhaustive literature review carried out, it was revealed that very little works have been carried out on assessing the vibration and damping characteristics of jute/rubber-reinforced PLA-based sandwich composites. More particularly, the experimental evaluation of frequency response of such composites is available in scarce. To this end, this research attempts to fill the gap and come up with benchmark results that facilitate further research and development in this stream.

6.2 EXPERIMENTAL DETAILS

6.2.1 MATERIALS AND METHODS

This research makes use of PLA procured from the Vexma manufacturers, Baroda with L-100H, food-safe, biodegradable grade. Further, the jute fibers and rubber crumb have been procured from the local vendor of Haryana and Manjunatha Traders of Baikampady, India, respectively. The pictorial view of the various materials used in this work is shown in Figure 6.1. The constituents of the different configurations of the completely biodegradable sandwich jute/rubber (CBS-JR) composite are shown in Table 6.1.

6.2.2 COMPOSITE FABRICATION

The rubber crumb in the form of thin layers adheres to the woven jute layer through the binding gum. The resulted sandwich layer of the Jute and Rubber crumb is mixed with the PLA matrix using the compression molding technique. The molding temperature of 185°C and the pressure of 13.7 MPa were maintained for 6 hours. Further, the specimens of dimension 220 mm × 30 mm × 3 mm were cut to perform the vibration testing.

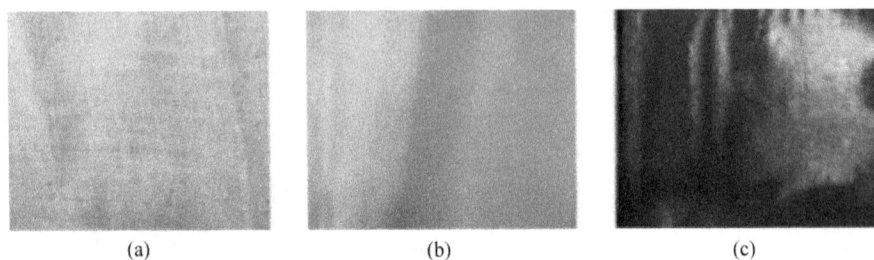

(a)   (b)   (c)

FIGURE 6.1   Schematic of (a) Jute, (b) Rubber, (c) sandwich composites prepared.

TABLE 6.1
The Percentage Weight of Jute, Rubber, and PLA in Different wt.%

Sl. No.	PLA (wt.%)	Jute (wt.%)	Rubber (wt.%)	Representation
1	80	20	0	Jute
2	78	20	2	J5R
3	77	20	3	J10R
4	75	20	5	J15R
5	Nil	10	90	JRJ
6	Nil	15	85	JRJRJ

6.2.3   EXPERIMENTAL SETUP

The natural frequencies and the corresponding damping parameters such as quality factor and loss factor for different compositions of the CBS-JR beam were studied experimentally using the setup shown in Figure 6.2. The setup consists of an accelerometer, impact hammer, and NI 9234 data acquisition (DAQ) unit. The impact hammer was used to provide initial excitation, and the corresponding acceleration and force signal were recorded using the accelerometer. Further, using the NI DAQ unit and LABVIEW programming, the signals were processed and plotted to obtain the frequency response.

6.3   RESULTS AND DISCUSSION

In this section, the frequency response and damping parameters of different CBS-JR beams are evaluated and discussed. As shown in Figure 6.3a and b, two different boundary conditions, clamped-free (cantilever; CFFF) and clamped-clamped (fixed; CFCF), are incorporated for this study. The initial excitation is provided using the impact hammer and the corresponding force signal is recorded. Also, the accelerometer captures the vibration amplitude through which the frequency response function plots are obtained after post-processing by the DAQ system.

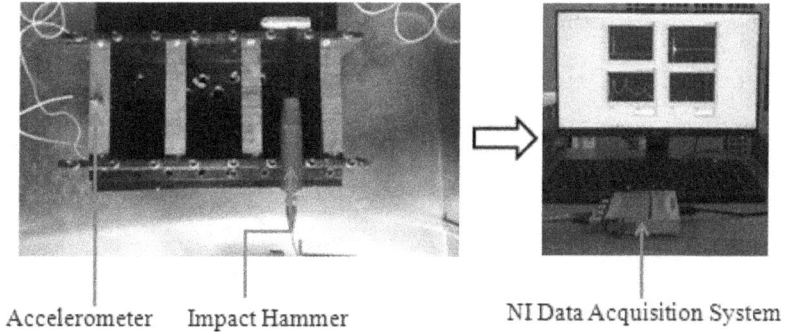

Accelerometer   Impact Hammer   NI Data Acquisition System

FIGURE 6.2   Experimental setup.

(a) cantilever   (b) fixed

FIGURE 6.3   Mechanical boundary constraints enforced on CBS-JR beam.

TABLE 6.2

Effect of Boundary Conditions and CBS-JR Beam Configuration on the Fundamental Frequency (in Hz)

| Boundary Conditions | Configuration~ | | | | | |
	Jute	J5R	J10R	J15R	JRJ	JRJRJ
CFFF	22.0193	23.49605	27.00945	23.91585	9.01965	11.0192
CFCF	146.4981	151.4943	164.99995	149.0038	46.9978	51.52365

From Table 6.2, it can be noticed that among all the selected configurations of Jute and rubber, a higher fundamental frequency has been reported for J10R configuration. As compared to a normal jute-based CBS-JR beam, the frequency improves up to 10% rubber. However, the stiffness of the beam then drastically reduces with further enhancement in the % of rubber. On the other hand, the composite JRJ beam displays the lowest natural frequency due to the absence of the PLA matrix. In addition, even though a lesser percentage of rubber is present in JRJ composites, its frequency remains lesser than the normal jute/PLA beam.

TABLE 6.3

Effect of Boundary Conditions and CBS-JR Beam Configuration on the Quality Factor

Boundary Condition	Configuration					
	Jute	J5R	J10R	J15R	JRJ	JRJRJ
CFCF	9.03919	7.1	7.00	7.58475	2.7744	3.66429
CFFF	27.61477	35.71893	26.55283	39.74641	3.80134	7.50261

TABLE 6.4

Effect of Boundary Conditions and CBS-JR Beam Configuration on the Loss Factor

Boundary Conditions	Configuration					
	Jute	J5R	J10R	J15R	JRJ	JRJRJ
CFCF	0.11063	0.1408	0.14274	0.1318	0.36043	0.2729
CFFF	0.03603	0.02799	0.03766	0.02519	0.263	0.13328

This emphasizes that the PLA matrix can provide more structural stiffness than mere adhesion of the rubber crumb with the jute layers. Meanwhile, as opposed to the cantilever condition, the higher frequency was noticed for fixed conditions, which is obvious due to the higher stiffness offered to the structure when it is clamped on both ends.

Parallelly, the damping characteristics of CBS-JR beams are also investigated. The parameters such as quality factor and loss factor are experimentally obtained. It is worthy to mention at this point that higher and lower values of quality and loss factors, respectively, refer to the lesser damping capabilities. From Tables 6.3 and 6.4, it can be noticed that even though lower frequency is witnessed for JRJ configuration, it exhibits an enhanced damping capability. This can be attributed to the greater thickness of rubber present in this configuration.

In addition, for the configurations involving PLA (jute, J5R, J10R, J15R), a significant damping characteristic was shown by the J10R configuration. The percentage increase or decrease in the damping parameters such as damping factor and loss factor compared with the pure jute/PLA CBS-JR beam is shown in Tables 6.5 and 6.6, respectively. As discussed previously, JRJ configuration shows a greater shift in the quality and loss factors that implies high damping. Therefore, from the experimental results, it can be recommended that when the optimum balance between frequency and damping is needed, it is preferred to use the J10R configuration. Meanwhile, when only damping is the prime importance, then JRJ configuration is best suited.

TABLE 6.5

Percentage Change in the Quality Factor Compared with Jute/PLA CBS-JR Beam

Boundary Conditions	Configuration				
	J5R	J10R	J15R	JRJ	JRJRJ
CFFF	−22.5	−21.45	−16.09	−69.30	−59.46
CFCF	39.44	3.66	55.168	−85.16	−70.71

TABLE 6.6

Percentage Change in the Loss Factor Compared with Jute/PLA CBS-JR Beam

Boundary Conditions	Configuration				
	J5R	J10R	J15R	JRJ	JRJRJ
CFFF	27.67	27.27	19.13	225.8	146.67
CFCF	−28.28	−3.510	−35.46	573.84	241.48

6.4 CONCLUSIONS

In this work, the free vibration response and damping behavior of a completely bio-degradable sandwich beam are studied experimentally. The matrix is considered to be food grade, biodegradable PLA, and reinforcements are Jute and Rubber crumb. The experimental setup is made of an impact hammer, accelerometer, and DAQ system. The contribution of reinforcements (jute and rubber) and PLA matrix on the frequency response is probed by considering different percentage weights of reinforcements. Also, the beam with and without the PLA matrix is also considered for the study to assess the influence of PLA on the natural frequency, quality factor, and loss factor. The results show that the CBS-JR beam with a 10% weight of the rubber crumb shows excellent frequency and damping response compared to other configurations. Meanwhile, the superior damping performance of the JRJ configuration without the PLA matrix is witnessed. The authors believe that the results presented in this research will pave the way for further progress in the field of natural composites.

ACKNOWLEDGMENT

The financial support by The Royal Society, London through Newton International Fellowship (NIF\R1\212432) is sincerely acknowledged by the author Vinyas Mahesh.

The financial support by Science and Engineering Research Board (SERB) through Teachers Associateship for Research Excellence (TAR/2021/000016) is sincerely acknowledged by the author Vishwas Mahesh.

Subhaschandra Kattimani acknowledges the Department of Science and Technology (DST), Goverment of India for the funding EEQ/2017/000744.

REFERENCES

1. Senthilkumar K, Saba N, Chandrasekar M, Jawaid M, Rajini N, Alothman OY, Siengchin S. Evaluation of mechanical and free vibration properties of the pineapple leaf fibre reinforced polyester composites. *Construction and Building Materials.* 2019; 195: 423–431.

2. Kumar KS, Siva I, Rajini N, Jeyaraj P, Jappes JW. Tensile, impact, and vibration properties of coconut sheath/sisal hybrid composites: effect of stacking sequence. *Journal of Reinforced Plastics and Composites.* 2014 Oct; 33(19): 1802–1812.

3. Rajini N, Jappes JW, Rajakarunakaran S, Jeyaraj P. Dynamic mechanical analysis and free vibration behavior in chemical modifications of coconut sheath/nano-clay reinforced hybrid polyester composite. *Journal of Composite Materials.* 2013; 47(24): 3105–3121.

4. Rajesh M, Pitchaimani J, Rajini NJ. Free vibration characteristics of banana/sisal natural fibers reinforced hybrid polymer composite beam. *Procedia Engineering.* 2016; 144: 1055–1059.

5. Uthayakumar M, Manikandan V, Rajini N, Jeyaraj P. Influence of redmud on the mechanical, damping and chemical resistance properties of banana/polyester hybrid composites. *Materials & Design.* 2014; 64: 270–279.

6. Rajesh M, Jeyaraj P, Rajini N. Mechanical, dynamic mechanical and vibration behavior of nanoclay dispersed natural fiber hybrid intra-ply woven fabric composite. In: Mohammad Jawaid, ou el Kacem Qaiss, chid Bouhfid (Eds). *Nanoclay Reinforced Polymer Composites* 2016: 281–296. Springer, Singapore.

7. Idicula M, Malhotra SK, Joseph K, Thomas S. Dynamic mechanical analysis of randomly oriented intimately mixed short banana/sisal hybrid fibre reinforced polyester composites. *Composites Science and Technology.* 2005; 65(7–8): 1077–1087.

8. Kumar KS, Siva I, Rajini N, Jappes JW, Amico SC. Layering pattern effects on vibrational behavior of coconut sheath/banana fiber hybrid composites. *Materials & Design.* 2016; 90: 795–803.

9. Etaati A, Mehdizadeh SA, Wang H, Pather S. Vibration damping characteristics of short hemp fibre thermoplastic composites. *Journal of Reinforced Plastics and Composites.* 2014; 33(4): 330–341.

10. El Mahi A, Assarar M, Sefrani Y, Berthelot JM. Damping analysis of orthotropic composite materials and laminates. *Composites Part B: Engineering.* 2008; 39(7–8): 1069–1076.

11. Chandradass J, Kumar MR, Velmurugan R. Effect of nanoclay addition on vibration properties of glass fibre reinforced vinyl ester composites. *Materials Letters.* 2007; 61(22): 4385–4388.

12. Rizal M, Mubarak AZ, Razali A, Asyraf M. Free vibration characteristics of jute fibre reinforced composite for the determination of material properties: numerical and experimental studies. *AIP Conference Proceedings.* 2019; 2187(1): 050020.

13. Ramakrishnan S, Krishnamurthy K, Rajeshkumar G, Asim M. Dynamic mechanical properties and free vibration characteristics of surface modified jute fiber/nano-clay reinforced epoxy composites. *Journal of Polymers and the Environment.* 2021; 29(4): 1076–1088.

14. Roy K, Debnath SC, Das A, Heinrich G, Potiyaraj P. Exploring the synergistic effect of short jute fiber and nanoclay on the mechanical, dynamic mechanical and thermal properties of natural rubber composites. *Polymer Testing.* 2018; 67: 487–493.

15. Chandra R, Singh SP, Gupta K. Damping studies in fiber-reinforced composites–a review. *Composite Structures.* 1999; 46(1): 41–51.

16. Saba N, Jawaid M, Alothman OY, Paridah MT. A review on dynamic mechanical properties of natural fibre reinforced polymer composites. *Construction and Building Materials*. 2016; 106: 149–159.

17. Ruksakulpiwat Y, Tonimit P, Kluengsamrong J. Mechanical properties of PLA-jute composites by using natural rubber and epoxidized natural rubber as impact modifiers: effect of molding technique. Chapter 6: Green chemistry, buildings, and constructions. In *Clean Technology*, 2010: 310–313. CRC Press, Houston, TX.

18. Ejaz M, Azad MM, Shah AU, Afaq SK, Song JI. Mechanical and biodegradable properties of jute/flax reinforced PLA composites. *Fibers and Polymers*. 2020; 21(11): 2635–2641.

19. Ma H, Joo CW. Structure and mechanical properties of jute—polylactic acid biodegradable composites. *Journal of Composite Materials*. 2011; 45(14): 1451–1460.

20. Zafar MT, Maiti SN, Ghosh AK. Effect of surface treatments of jute fibers on the microstructural and mechanical responses of poly (lactic acid)/jute fiber biocomposites. *RSC Advances*. 2016; 6(77): 73373–73382.

21. Gunti R, Ratna Prasad AV, Gupta AV. Mechanical and degradation properties of natural fiber-reinforced PLA composites: jute, sisal, and elephant grass. *Polymer Composites*. 2018; 39(4): 1125–1136.

7 Effect of Organic Nanofillers on the Free Vibration and Damping Characteristics of Polymer-Based Nanocomposites

Vinyas Mahesh
National Institute of Technology Silchar
and
City, University of London

Vishwas Mahesh
Siddaganga Institute of Technology
and
Indian Institute of Science, Bangalore

Sriram Mukunda
Nitte Meenakshi Institute of Technology, Bangalore

Arjun Siddharth
Indian Institute of Science, Bangalore

Athul S Joseph
Indian Institute of Science, Bangalore
and
Katholieke Universitiet Leuven

Dineshkumar Harursampath
Indian Institute of Science, Bangalore

DOI: 10.1201/9781003173625-8

CONTENTS

7.1   INTRODUCTION

The engineering domain is witnessing a tremendous shift in the use of nanocomposites. Almost all the prominent fields, including aerospace, medical, marine, defense, sports, and transportation,havealready been exposed to nanocomposites extremely. Many researchers have suggested that by reinforcingsuitable nanofillers based on the application,the mechanical properties of the overall composite structure can be significantly enhanced. Among the various manufacturing processes that can be incorporated to develop nanocomposite structures, additive manufacturing stands tall due to its flexibility, accuracy, and less material wastage. Fused deposition modeling (FDM) technique has grasped the attention of many pioneers due to the availability of low-cost machines and ease of technique. In this process,the filament is made to pass through the heated nozzle, which melts and facilitates easy deposition in the form of layers and as per the required shape.The mechanical properties of the 3D-printed structures can be altered through several process parameters such as nozzle temperature, printing speed, raster angle, infill density, layer thickness, etc. Similarly, the 3D printing of the final product/structure may significantly impact its structural properties, which can be owed to its printing parameters.

More often, the composite structures are subjected to dynamic loads, and it is desired that they withstand such loads and perform effectively. However, proper knowledge of such structures' frequency response is very much needed to avoid resonance and the resulting failure. The dissipation of unwanted energy during the service life of any composite structure is prominent, and it depends on its damping characteristics. Many researchers focused on assessing the natural frequencies and damping factors of composite structures made of different forms of nanocomposites which are of close relevance to the present work. Few have been encapsulated and discussed here.Kannan et al. [1] attempted to evaluate the dynamic properties of carbon fiber-reinforced Polyethylene terephthalate glycol (PETG) composites prepared usingan additive manufacturing routine. They found an increase in the natural frequencies by 17% when PETG is reinforced with carbon fibers. Mansour et al. [2] incorporated the fused filament fabrication approach and developed PETG/carbon fibercomposites. Their assessment performed modal analysis of 3D-printed specimens and concluded that pure PETG exhibited better damping than the PETG/carbon fibercomposites. Mayandi et al. [3] investigated the effect of reinforcing wood fibers in the polylactic acid (PLA) matrix on the frequency response of the 3D architecture composite beam. Their numerical studies reveal that pure PLA has better stiffness compared to wood-reinforced PLA.Wang et al. [4] examined the natural frequencies and evaluated the damping response of 3D-printed Kagome lattice embedded with viscoelastic material. Karami Khorramabadi [5] numerically estimated the influence of reinforcing clay on the free vibration response of epoxy matrix-based nanocompositesusing first-order shear deformation theory. In this study,the author considered both uniformly distributed and functionally graded clay distributions. The natural frequencies found a sudden decrement once the weight percentage of nanoclay exceeded 5%. Agarwal et al. [6] demonstrated the influence of adding boron nitride nanotube (BNNT) and carbon nanotube(CNT) fillers on the damping characteristics of polymer composites. The better damping properties were exhibited by CNT when compared with BNNT. Chandradass et al. [7] considered free vibration analysis of glass fiber-reinforced polyester-reinforced with organically modified montmorillonite (OMMT)nanoclay. DeValve and Pitchumani [8] considered the cases of stationary and rotating beams and assessed the influence of reinforcing CNTs on the damping performance of fiber-reinforced composites. They observed an increase of 130% and 150% in the damping characteristics of stationary and rotating beams,respectively. Ilangovan et al. [9] examined the free vibration response of nanoparticles reinforced basalt/epoxy and glass/epoxy composites through experimental,numerical, and analytical routines. Their study concluded that the damping performance of basalt/epoxy nanocomposites was superior to that of glass/epoxy nanocomposites. Joy et al. [10] probed on the free vibration,and damping characteristics of multiwalled CNT (MWCNT)-reinforced epoxy nanocomposites with carboxyl-terminated butadiene acrylonitrile (CTBN) and diglycidyl ether of bisphenol A as an adduct. They witnessed higher damping compared to neat epoxy. For free and forced vibration studies, Khan et al. [11] examined the damping properties of MWCNT polymer composites. They noticed an increase in the damping factors with higher percentage of CNTs in the composite. Khashaba [12] investigated thefree vibration and damping behavior of glass fiber-reinforced epoxy composites with MWCNTs. The natural frequencies tend to decrease with the enhancement in the free length of the beam. Li et al. [13] developed raphemenanoplatelets-reinforced polyetherimide composites usingsolution-processing technique. The experimental results revealed that superior damping characteristics could be obtained by using rapheme as a nanofiller. Malakooti et al. [14] attempted to develop zinc oxide nanowires on the surface of carbon fibers and investigated the damping properties of the overall nanocomposites. Compared to conventional carbon fiber composites,higher damping was witnessed for carbon fibers with nanowire of ZnO. Mansour et al. [15] carried out modal experiments to understand the influence of CTBN rubber in epoxy-based composites. They noticed that adding CTBN fillers enhanced the damping behavior but reduced the natural frequencies. Mansour et al. [16] assessed the influence of adding silica nanoparticles on the frequency response of hybrid composites made of aramid and carbon fibers. It was seen from their study that reinforcing silica nanoparticles would improve the damping behavior of these hybrid composites. Mohammed et al. [17] investigated the influence of adding MWCNT and Al_2O_3 nanoparticles on the free vibration of epoxy composites. The natural frequencies were noticed to improve with the addition of these nanofillers,but the damping ratio decreased drastically. Pistor et al. [18] demonstrated the variation of the dynamic mechanical properties with various fractions ofepoxycyclohexyl–POSS in an epoxy matrix. Lin et al. [19] assessed the contribution of MWCNTs toward the damping performance of polyaniline composite sensors. Rafiee et al. [20] investigated the usage of MWCNT in passive damping composite structures by assessing its damping characteristics. Tehrani et al. [21] proposed a hybrid composite reinforced with carbon fiberand CNTs developed through the graphitic structure by design technique for different engineering structures and assessed its damping behavior. Tsongas et al. [22] developed machine mounts based on acrylonitrile-butadiene rubber/MWCNT composites and experimentally evaluated its vibration isolation characteristics.

From the exhaustive literature survey, it was noticed that nanofillers play a prominent role in deciding the frequency and damping behavior of composite structures. In addition,to the best of the author's knowledge, it was evident that no work has been reported on experimentally evaluating the vibration and damping behavior of OMMT nanoclay-reinforced PLA composites prepared through additive manufacturing. This work makes the first attempt toward this end.

7.2 EXPERIMENTAL DETAILS

7.2.1 MATERIALS AND METHODS

In the present study,nanocomposites made of PLA matrix reinforced with OMMT nanoclay areconsidered for evaluation. The virgin PLA pellets and OMMT nanoclay were procured fromVexma manufacturers, Baroda and BYK Additives and Instruments (Cloisite SE3000), India, respectively. PLA pellets belong to L-100H, food-safe andbiodegradable grade. Further, the size of OMMT nanoclay was less than 10 μm and had a bulk density of 0.45g/cm³. The different compositions of PLA and OMMT nanoclay used in the current study are encapsulated in Table 7.1.

7.2.2 COMPOSITE FABRICATION

Initially,the different compositions of PLA and OMMT nanoclay mentioned in Table 7.1 are mechanically mixed. Later, using a twinscrew extruder with a hopper arrangement, the pellets were compounded. The compounded pellets were pre-dried at 65°C–70°C for about 300 minutes and extruded in filament form. During this

TABLE 7.1
Compositional Data of the Composites

Specification	PLA Content (wt.%)	OMMT Nanoclay Content (wt.%)
Pure PLA	100%	–
PLA + 1% OMMT	99%	1%
PLA + 3% OMMT	97%	3%
PLA + 5% OMMT	95%	5%

(a)   (b)

FIGURE 7.1 Raw materials (a) PLA pellets and (b) OMMT nanoclay used in the study.

process,the torque and temperature of the extruder were maintained at optimum values. Figure 7.1a and b shows the PLA pellets and OMMT nanoclay materials, respectively. The schematic of the extrusion line is shown in Figure 7.2.

7.2.3 3D Printing of PLA/OMMT Nanocomposite Beam

Using a single screw extruder, the 3D printable filaments with an average diameter of 1.5mm were obtained. Further, using FDM printer fitted with a brass nozzle,the specimens of dimension 220 mm × 30 mm × 3 mm to fit the vibration setup were-printed. The nozzle and print bed temperatures of 230°C and 75°C, respectively, were adopted. In addition, the printing speed of 55 mm/s and raster angle of 0° were used in this study.

7.2.4 Experimental Setup

The natural frequencies and the corresponding damping parameters such as quality factor and loss factor for different compositions of the PLA/OMMT nanocomposite beam were studied experimentally using the setup as shown in Figure 7.3. The setup consists of an accelerometer, impact hammer, and NI 9234 data acquisition

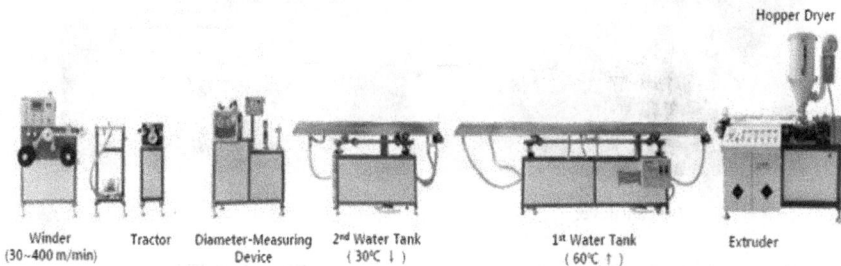

| Winder | Tractor | Diameter-Measuring | 2ⁿᵈ Water Tank | 1ˢᵗ Water Tank | Extruder |
| (30~400 m/min) | | Device | (30°C ↓) | (60°C ↑) | |

FIGURE 7.2 Setup for the extrusion of the monofilaments.

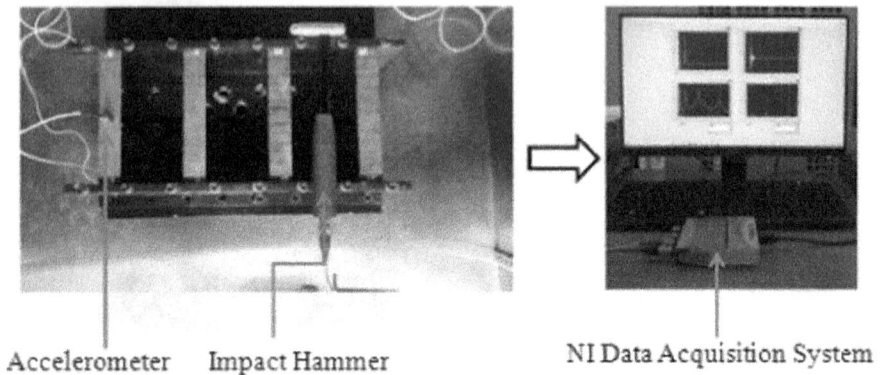

Accelerometer   Impact Hammer   NI Data Acquisition System

FIGURE 7.3   Experimental setup.

(DAQ) unit. The impact hammer was used to provide initial excitation,and the corresponding acceleration and force signal were recorded usingthe accelerometer. Further, using the NI DAQ unit and LABVIEW programming, the signals were processed and plotted to obtain the frequency response.

7.3   RESULTS AND DISCUSSION

The experimental results of the frequency response and damping characteristics of the PLA/OMMT nanocomposite beam are presented in this section. The results of fundamental natural frequency, quality factor, and loss factors are also discussed. The evaluation is carried out for two boundary conditions, cantilever (CFFF) and fixed (CFCF), as shown in Figure 7.4a and b, respectively. Table 7.2 shows the variation in the fundamental natural frequency of PLA/OMMT nanocomposite beam with different OMMT compositions. As opposed to pure PLA beam, a higher fundamental natural frequency is witnessed for the other compositions with OMMT nanoclay reinforced.It can be attributed to the internal material voids in pure PLA beam developed during 3D printing of the specimen,which reduces the stiffness of the overall structure. Meanwhile, with the addition of the OMMT nanoclay particles, these voids fill up, contributing to enhanced stiffness.

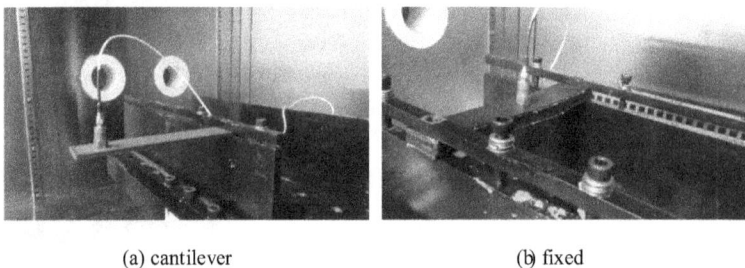

(a) cantilever   (b) fixed

FIGURE 7.4   Mechanical boundary constraints enforced on PLA/OMMT beam. (a) Cantilever. (b) Fixed.

TABLE 7.2

Effect of Boundary Conditions and PLA+OMMT Composition on the Fundamental Frequency (in Hz)

Boundary Conditions	Composition			
	PLA	PLA+ 1% OMMT	PLA+ 3% OMMT	PLA+ 5% OMMT
CFFF	12.54	14.24	17.88	15.68
CFCF	83.75	87.17	94.63	88.41

Further, it is worthy to note that among all the compositions selected highest natural frequency is noticed for PLA+ 3% OMMT nanoclay. Alongside, the natural frequency starts to reduce drastically for 5% addition of OMMT nanoclay. This is because the addition of nanoclay particles beyond 3% results in the matrix embrittlement,which leads to reduced frequency response. Meanwhile, as opposed to the cantilever condition,the higher frequency was noticed for fixed conditions, which is evident due to higher stiffness offered to the structure when it is clamped on both ends.

On the other hand,the damping parameters of PLA/OMMT nanocomposite beams are also studied and depicted in Tables 7.3 and 7.4 in terms of quality factor and loss factor, respectively. In order to consider a material to be of high damping, it should exhibit a higher value of loss factor but a lesser value of the quality factor. Therefore, from the experimental assessment made through Tables 7.5

TABLE 7.3

Effect of Boundary Conditions and PLA+OMMT Composition on the Quality Factor

Boundary Conditions	Composition			
	PLA	PLA+ 1% OMMT	PLA+ 3% OMMT	PLA+ 5% OMMT
CFFF	38.46	31.25	23.25	32.25
CFCF	7.353	6.622	4.762	5.524

TABLE 7.4

Effect of Boundary Conditions and PLA+OMMT Composition on the Loss Factor

Boundary Conditions	Composition			
	PLA	PLA+ 1% OMMT	PLA+ 3% OMMT	PLA+ 5% OMMT
CFFF	0.026	0.032	0.043	0.031
CFCF	0.136	0.151	0.210	0.167

TABLE 7.5
Percentage Change in the Quality Factor Compared with Pure PLA Beam

Boundary Conditions	Composition		
	PLA+ 1% OMMT	PLA+ 3% OMMT	PLA+ 5% OMMT
CFFF	−18.74	−39.54	−16.14
CFCF	−9.94	−35.23	−24.87

TABLE 7.6
Percentage Change in the Loss Factor Compared with Pure PLA Beam

Boundary Conditions	Composition		
	PLA+ 1% OMMT	PLA+ 3% OMMT	PLA+ 5% OMMT
CFFF	23.07	65.38	19.23
CFCF	11.02	54.41	22.79

and 7.6, it is clear that the addition of nanoclay particles improves the damping characteristics compared with pure PLA. In addition, PLA+ 3% OMMT nanocomposite exhibitssuperior damping characteristics. This holds good for both CFFF and
CFCF boundary conditions. However, for CFFF condition,the damping properties
exhibited by PLA+ 1% OMMT nanocomposite and PLA+ 5% OMMT nanocomposite are almost equivalent,even though PLA+ 1% OMMT nanocomposite hold an
upper hand. On the other hand, in the case of CFCF boundary condition, PLA+ 5%
OMMT nanocomposite shows a predominant damping effect and stands as the composition with the second-highest damping performance next to PLA+ 3% OMMT
nanocomposite.

Extending the assessment, the percentage change in the damping characteristics of
OMMT reinforced PLA nanocomposite as opposed to pure PLA beam is presented
in Tables 7.5 and 7.6 for better clarity to the readers. In Table 7.5, the negative sign
indicates an improvement in the damping capabilities as the quality factor is a measure that is inversely proportional to the effective damping. Table 7.6 shows that the
improvement in the damping characteristics of reinforcing 1% and 3% nanoclay is
significant when the beam is used in CFFF condition, whereas 5% nanoclay addition
provides better damping when constrained with CFCF condition. The reason may
be due to the variation in the nanoclay accumulation during 3D printing and matrix
embrittlement. In addition, all the inferences made in the previous section hold good
here as well.

7.4   CONCLUSIONS

The present research work deals with assessing the influence of adding nanofillers on
the frequency response of nanocomposites. To this end, OMMT nanoclay-reinforced

PLA nanocomposite is considered for evaluation. The specimens are prepared/printed usingthe FDM technique after compounding and extruding processes. The experimental setup to obtain the frequencies is made of an impact hammer, accelerometer, and DAQ system. The influence of OMMT nanoclay on the frequencies and damping parameters isstudied by varying its weight percentage by 1%,3%, and 5% in the PLA matrix. Also, the frequency response of pure PLA is recorded to make the comparison study effective. The results reveal that the fundamental natural frequencies increase with the addition of OMMT nanoclay particles with PLA+ 3% OMMT nanocomposite exhibiting superior frequency characteristics. However,a sharp decrease in the natural frequency is reported for PLA+ 5% OMMT nanocomposite due to reduced stiffness resulting from matrix embrittlement. The author believes that the results of this studymay be helpful for further research progress in the field of structural analysis of 3D-printed nanocomposites.

ACKNOWLEDGEMENTS

The financial support by The Royal Society, London through Newton International Fellowship (NIF\R1\212432) is sincerely acknowledged by the author Vinyas Mahesh.

The financial support by Science and Engineering Research Board (SERB) through Teachers Associateship for Research Excellence (TAR/2021/000016) is sincerely acknowledged by the author Vishwas Mahesh.

REFERENCES

1. Kannan, S., Ramamoorthy, M., Sudhagar, E. and Gunji, B., 2020. Mechanical characterization and vibrational analysis of 3D printed PETG and PETG reinforced with short carbon fiber. In *AIP Conference Proceedings* (Vol. 2270, No. 1, p. 030004). AIP Publishing LLC.
2. Mansour, M., Tsongas, K., Tzetzis, D. and Antoniadis, A., 2018. Mechanical and dynamic behavior of fused filament fabrication 3D printed polyethylene terephthalate glycol reinforced with carbon fibers. *Polymer-Plastics Technology and Engineering*, 57(16), pp.1715–1725.
3. Mayandi, K, Sethuramalingam, Ramanan, B., Ayrilimis, N., Rajini, N., Raju, R.P. and Rajkumar, M., 2020. Free vibrations and flexural strength analysis of 3D architected core sandwich polymer. *International Journal of Control and Automation*, 13(4), pp.1137–1151.
4. Wang, R., Shang, J., Li, X., Luo, Z. and Wu, W., 2018.Vibration and damping characteristics of 3D printed Kagome lattice with viscoelastic material filling. *Scientific Reports*, 8(1), pp.1–13.
5. Karami Khorramabadi, M., 2019.Free vibration of functionally graded epoxy/clay nanocomposite beams based on the first order shear deformation theory. *ADMT Journal*,12(2), pp.45–51.
6. Agrawal, R., Nieto, A., Chen, H., Mora, M. and Agarwal, A., 2013.Nanoscale damping characteristics of boron nitride nanotubes and carbon nanotubes reinforced polymer composites. *ACS Applied Materials &Interfaces*, 5(22), pp.12052–12057.
7. Chandradass, J., Kumar, M.R. and Velmurugan, R., 2007.Effect of nanoclay addition on vibration properties of glass fibre reinforced vinyl ester composites. *Materials Letters*, 61(22), pp.4385–4388.

8. DeValve, C. and Pitchumani, R., 2013.Experimental investigation of the damping enhancement in fiber-reinforced composites with carbon nanotubes. *Carbon, 63,* pp.71–83.

9. Ilangovan, S., Kumaran, S.S. and Naresh, K., 2020.Effect of nanoparticles loading on free vibration response of epoxy and filament winding basalt/epoxy and E-glass/epoxy composite tubes: Experimental, analytical and numerical investigations. *Materials Research Express, 7*(2), p.025007.

10. Joy, A., Varughese, S., Shanmugam, S. and Haridoss, P., 2019.Multiwalled carbon nanotube reinforced epoxy nanocomposites for vibration damping. *ACS Applied Nano Materials, 2*(2), pp.736–743.

11. Khan, S.U., Li, C.Y., Siddiqui, N.A. and Kim, J.K., 2011.Vibration damping characteristics of carbon fiber-reinforced composites containing multiwalled carbon nanotubes. *Composites Science and Technology, 71*(12), pp.1486–1494.

12. Khashaba, U.A., 2015.Toughness, flexural, damping and interfacial properties of hybridized GFRE composites with MWCNTs. *Composites Part A: Applied Science and Manufacturing, 68,* pp.164–176.

13. Li, B., Olson, E., Perugini, A. and Zhong, W.H., 2011.Simultaneous enhancements in damping and static dissipation capability of polyetherimide composites with organosilane surface modified graphene nanoplatelets. *Polymer, 52*(24), pp.5606–5614.

14. Malakooti, M.H., Hwang, H.S. and Sodano, H.A., 2015.Morphology-controlled ZnO nanowire arrays for tailored hybrid composites with high damping. *ACS Applied Materials &Interfaces, 7*(1), pp.332–339.

15. Mansour, G., Tsongas, K. and Tzetzis, D., 2016.Investigation of the dynamic mechanical properties of epoxy resins modified with elastomers. *Composites Part B: Engineering, 94,* pp.152–159.

16. Mansour, G., Tsongas, K. and Tzetzis, D., 2016.Modal testing of epoxy carbon–aramid fiber hybrid composites reinforced with silica nanoparticles. *Journal of Reinforced Plastics and Composites, 35*(19), pp.1401–1410.

17. Mohammed, S.M., Gamil, M. and Mohammed, S.S., 2017.Mechanical and dynamic behaviour of Epoxy/Mwcnts and Epoxy/Al_2O_3nanocomposites. *Nano Science & Nano Technology, 11*(2), p.121.

18. Pistor, V., Ornaghi, F.G., Ornaghi, H.L. and Zattera, A.J., 2012.Dynamic mechanical characterization of epoxy/epoxycyclohexyl–POSS nanocomposites. *Materials Science and Engineering: A, 532,* pp.339–345.

19. Lin, W., Rotenberg, Y., Ward, K.P., Fekrmandi, H. and Levy, C., 2017.Polyaniline/multiwalled carbon nanotube composites for structural vibration damping and strain sensing. *Journal of Materials Research, 32*(1), pp.73–83.

20. Rafiee, M., Nitzsche, F. and Labrosse, M.R., 2018.Effect of functionalization of carbon nanotubes on vibration and damping characteristics of epoxy nanocomposites. *Polymer Testing, 69,* pp.385–395.

21. Tehrani, M., Safdari, M., Boroujeni, A.Y., Razavi, Z., Case, S.W., Dahmen, K., Garmestani, H. and Al-Haik, M.S., 2013.Hybrid carbon fiber/carbon nanotube composites for structural damping applications. *Nanotechnology, 24*(15), p.155704.

22. Tsongas, K., Tzetzis, D. and Mansour, G., 2017.Mechanical and vibration isolation behaviour of acrylonitrile-butadiene rubber/multiwalled carbon nanotube composite machine mounts. *Plastics, Rubber and Composites, 46*(10), pp.458–468.

8 Influence of Fiber Treatment on the Damping Performance of Plant Fiber Composites

Md Zillur Rahman
Ahsanullah University of Science and Technology

CONTENTS

8.1 INTRODUCTION

The study on plant fiber-based polymer composites is widely growing due to their ease of manufacturing, low cost, high stiffness to weight ratio, high impact resistance, and high energy dissipative characteristics. Damping, mass, and stiffness determine the basic dynamic behavior of a composite structure. Stiffness along with mass store the energy, while damping dissipates the mechanical energy, typically by converting mechanical energy into other types of energy (Cremer, Heckl, and Petersson 2005, Khan et al. 2011, Rainieri and Fabbrocino 2014) such as heat, which happens inside materials as a result of deformations imposed on them (Khan et al. 2011, De Silva 2006). Plant fiber composites (PFCs) are frequently used in applications (such as automobile parts) where noise and vibration are a major concern, and a substantial amount of damping is needed. However, damping estimation is more challenging than that of mechanical properties (e.g., stiffness and strength) owing to a poor signal-to-noise ratio (Furtado et al. 2014). Damping in PFCs relies not only

DOI: 10.1201/9781003173625-9

on the viscoelastic behavior of fibers and polymers but also on fiber–matrix interfaces and material defects (e.g., cracks, voids, air inclusion, and moisture absorption), fiber porosity, delamination, and damaged fiber. The dynamic responses
(natural frequency, damping, and mode shape) of the structure are directly associated with the weight of the material, stiffness, damping of constituents, and geometry
(Rahman 2020).

Plant fibers are porous in nature and viscoelastic, and viscoelastic materials can
inherently dissipate energy during mechanical deformation by transforming vibration energy to heat energy. However, composites' damping performance is affected
by the nature of the fiber–matrix interfaces, and the interface can be weak, ideal,
or strong (Chandra, Singh, and Gupta 1999). Various physical and chemical treatments of plant fibers may be performed to enhance the interfacial bonding strength,
greatly influencing the composites' damping performance. Physical modification is
used to alter the structure and surface properties of plant fibers without varying their
chemical composition, allowing for improved adhesion capability between fibers and
matrices. In contrast, the chemical treatment reduces the hydrophilic nature of plant
fibers while increasing the adhesion behavior of the fiber and matrix. In this chapter, various fiber treatments and their effects on damping are discussed, along with
damping characterization techniques, damping measurement methods, and damping
mechanisms. Finally, some applications are mentioned, which are made considering
the significantly high damping performance of plant fibers.

8.2   VIBRATION MEASUREMENT TECHNIQUES

Three vibration-damping measuring techniques are commonly used in the case of
fiber-based composite materials, including dynamic mechanical analyzer (DMA),
impact hammer, and mechanical shaker. DMA measures the viscoelastic properties
(stiffness and damping) of a material as a function of time, frequency, temperature,
stress, and strain. It also characterizes the dynamic responses (storage and loss moduli, and damping) of a material over a range of frequencies at a fixed temperature or
over a range of temperatures at a fixed frequency. Typical DMA curves are illustrated
in Figure 8.1.

The excitation of the composite structure can also be performed either by a vibration generator or shaker or by using some form of transient input, such as a hammer blow over at any frequency range. Notably, shaker and impact tests have no
difference theoretically when pure forces are exerted to the structure except for
interacting between the structure and applied forces, and responses are estimated
using the massless transducer, which does not have any effect on the structure. In
general, there is no effect of shaker and response transducer on the structure during the vibration test. This is due to the structure suspension, mounted transducer
mass, the shaker/stinger arrangement's possible stiffening effects, and other factors (Avitabile 2017). The effect can be significant if the structure is lightweight.
However, an impact test is unaffected by these problems, and it can overcome some
of them using a noncontacting response transducer (e.g., laser doppler velocimeter)
to estimate the excitation forces and corresponding various responses. A typical frequency response curve under impact hammer testing is shown in Figure 8.2, which

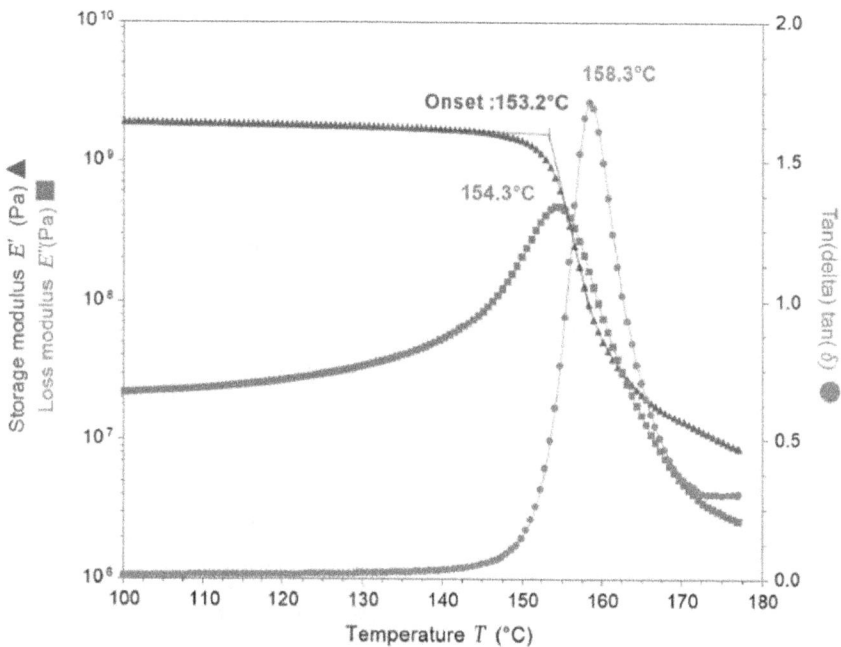

FIGURE 8.1 Typical DMA curves (TA Instruments 2019). (Open access.)

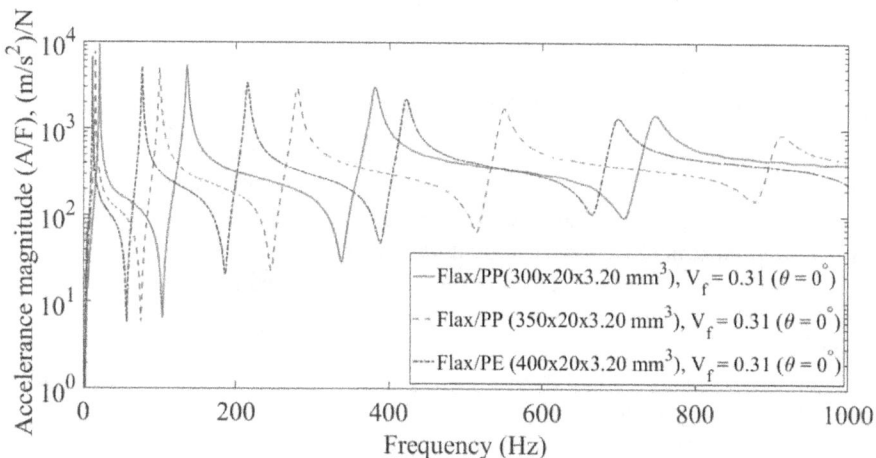

FIGURE 8.2 Frequency responses of composite beams of varying lengths from an impulse hammer test (Rahman 2020). (Reused with permission, license number 5118911455853.).

can be used to extract modal parameters (natural frequency, damping, and mode shapes) following the methods, as presented in Section 8.4. Note that the impact hammer technique is typically used to estimate the vibration behavior of a small-scale structure, whereas the mechanical shaker test is preferred for a relatively larger structure.

The DMA can give vibration behavior of a material at various temperatures and low frequencies, while other testing methods (e.g., impact hammer and mechanical shaker) may be applied to obtain vibration characteristics at higher frequencies, indicating the probability of estimating loss factors of PFCs at a wide range of frequencies. In addition, when comparing damping measures quantitively from various testing techniques, they are likely to be the same (Rahman 2017, Rueppel et al. 2017), although the external effects such as clamping effects and air resistance are required to quantify and understand the limitations of each testing procedure (Rueppel et al. 2017).

8.3   MEASURES OF DAMPING

There is a variety of definitions and ways of measuring vibration damping. However, for light damping, the various definitions are related to each other (Alan and Vikram 1992) as given in Equation 8.1:

$$\eta = \frac{\psi}{2\pi} = Q^{-1} = \frac{\delta}{\pi} = \tan\phi \approx \phi = \frac{E''}{E'} = 2\zeta = \frac{\Delta W}{2\pi W} = \frac{\lambda\alpha}{\pi} \tag{8.1}$$

where η is the loss factor (or damping), ψ is the specific damping capacity, Q is the quality factor, δ is the logarithmic decrement, ϕ is the phase angle by which stress leads to strain in a cyclic motion, E'' is the loss modulus, E' is the storage modulus, ζ is the viscous damping ratio (or viscous damping factor), ΔW is the energy loss per cycle, W is the maximum elastic stored energy, λ is the wavelength of elastic waves, and α is the attenuation. In this chapter, results are presented in terms of loss factor (or damping) (η).

8.4   DAMPING MEASUREMENT METHODS

After measuring the frequency response of a composite structure, the next step in analyzing the measurement is the extraction of the modal parameters such as natural frequency and loss factor. A wide variety of techniques has been developed over the past decade to extract the modal parameters. The most widely-used classification for these techniques is mainly based on frequency response and time response methods (Maia and Silva 1997). These are listed below.

I. Frequency Response Methods
- Peak-picking method
- Circle-fit method
- Inverse frequency response function (FRF) method
- Maximum quadrature component method
- Dobson's method
- Goyder's method
- Rational fraction polynomials

II. Time Response Methods
- Complex exponential method
- Least-squares complex exponential method

- Polyreference complex exponential method
- Ibrahim time-domain method
- Random decrement method
- Eigensystem realization algorithm method

The peak-picking method (also known as the half-power bandwidth method) is a commonly used method for estimating the loss factor of plant fiber-based composites. However, this approach overestimates damping since it assumes that the response around each natural frequency is dominated by a single-mode only. Nevertheless, off-resonant modes make a large contribution to the overall response at any natural frequency (Rahman, Jayaraman, and Mace 2017). Considering the mode number, modal overlap, and loss factor, the loss factor measurement errors for a multiple-degree-of-freedom system can be up to 20% or more (Wang, Jin, and Zhang 2012). In contrast, the single-degree-of-freedom circle-fit method (Ewins 2000) greatly increases the accuracy of loss factor measurement as opposed to the peak-picking method, as it takes into account the frequency range near the natural frequency. A detailed overview of this method can be found in Rahman (2017).

8.5 DAMPING OF POLYMER MATERIALS

Polymers also act like viscoelastic materials and exhibit high loss factors than conventional metals (Chung 2001). Viscoelastic materials are inherently capable of dissipating vibration energy as heat energy under dynamic loadings (Jones 2001). The intrinsic damping of composite constituents is the primary source of damping of PFCs. The viscoelastic nature of polymers contributes substantially to the damping of polymeric composites. Thermoplastics have higher energy dissipation capacity relative to thermosets, while the latter are usually selected for their better mechanical properties, strong chemical bond formation capability with various substates/surfaces, and excellent dimensional stability. An increase in viscoelastic polymers improves PFCs damping, but this has a detrimental effect on the mechanical properties (i.e., stiffness and strength) (Rahman, Mace, and Jayaraman 2016, Rahman, Jayaraman, and Mace 2017, 2018). Viscoelastic materials can retain and dissipate mechanical energy at the same time because of the existence of their long molecular chains (Richards and Lenzi 1984). Restoration of molecular chains results in damping following deformation (Darabi 2013). The loss factor of polymers is highly dependent on temperature and frequency because of their direct interaction between temperature and molecular movement. Plant fibers can be embedded into the viscoelastic materials (e.g., polymers), which can offer beneficial properties including damping, stiffness, strength, thermal stability, durability, and so on over the chosen ranges of temperature and frequency (Nashif, Jones, and Henderson 1985).

8.6 DAMPING OF PLANT FIBERS

Plant fibers have a multi-scale structure, which is a distinct advantage as reinforcing agents in composites. Figure 8.3 depicts the multi-scale structure of flax fibers. These fibers are made up of yarns of elementary fibers embedded in a pectin matrix;

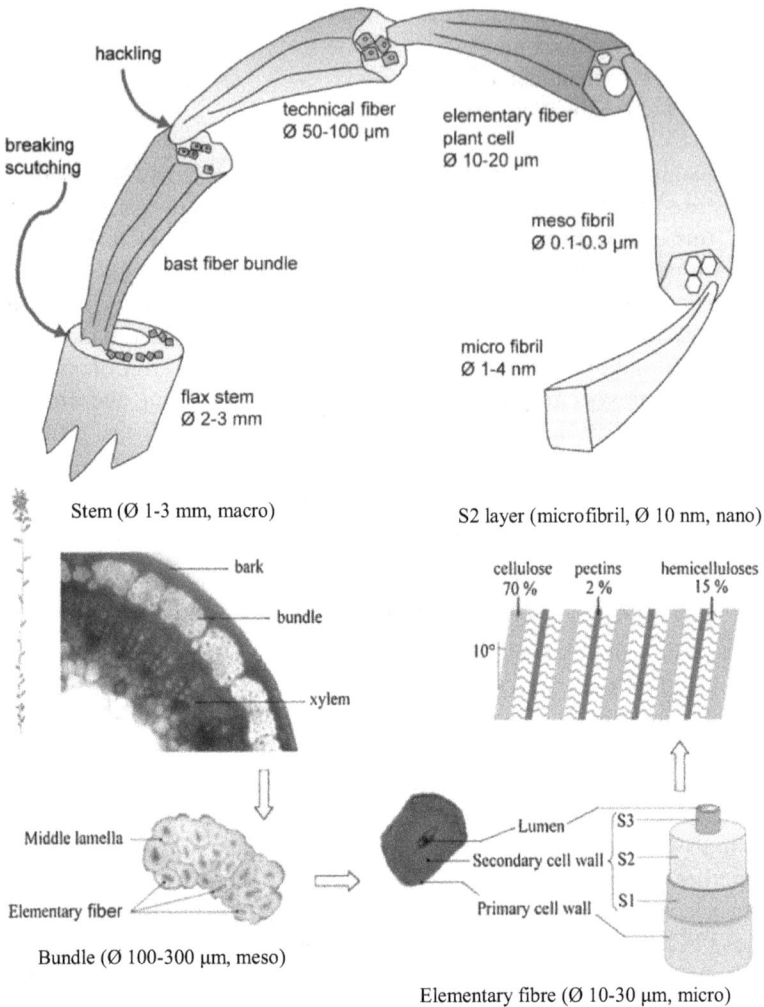

FIGURE 8.3 (a) Flax fiber from stem to microfibril (Bos, Müssig, and van den Oever 2006). (Reused with permission, license number 5118900258302.) (b) Multi-scale composite structure of flax (Charlet et al. 2007, 2010). (Reused with permission, license numbers 5118901126079 and 5118900811645, respectively.)

thereby, every elementary fiber makes a composite structure (see Figure 8.3b). In this structure, rigid cellulosic microfibrils are bonded together by hemicellulose and lignin matrix (Li, Luo, and Han 2010). The viscoelasticity and hierarchical structure of these fibers contribute to energy dissipation (Duc, Bourban, and Manson 2014b, Duc et al. 2014). Plant fiber typically consists of cellulose, hemicellulose, lignin, pectin, and wax in varying proportions (see Table 8.1). The quantities of plant fiber constituents differ due to the environmental conditions (soil, sun, rain, humidity, and temperature), maturity of the plant, and production process (Charlet et al. 2009). There may also have a lumen cavity with a diameter in the range of 4–65 μm (Nagaraja Ganesh

TABLE 8.1

Chemical Compositions of Plant Fibers (Ramesh, Palanikumar, and Reddy 2017)

Fiber	Cellulose (%)	Hemicellulose (%)	Lignin (%)	Pectin (%)	Wax (%)
Abaca	62.5	21	12	0.8	3
Alfa	45.4	38.5	38.5	–	2
Areca	57.35–58.21	13–15.42	23–24	–	0.12
Bagasse	37	21	22	10	–
Bamboo	34.5	20.5	26	–	–
Banana	62.5	12.5	7.5	4	–
Barley	31–45	27–38	14–19	–	2–7
Coir	456	0.3	45	4	–
Corn	38–40	28	7–21	–	3.6–7
Cotton	89	4	0.75	6	0.6
Curaua	73.6	5	7.5	–	–
Eucalyptus	41.7	32.56	25.4	8.2	0.22
Flax	72.5	14.5	2.5	0.9	–
Hemp	81	20	4	0.9	0.8
Henequen	60	28	8	–	0.5
Hibiscus	28	25	22.7	–	–
Isora	74	-	23	–	1.1
Jute	67	16	9	0.2	0.5
Kenaf	53.5	21	17	2	–
Phromium	67	30	11	–	–
Pineapple	80.5	17.5	8.3	4	–
Ramie	72	14	0.8	2	–
Rice husk	28–36	23–28	12–14	–	14–20
Sisal	60	11.5	8	1.2	–
Sorghum	27	25	11	–	–
Wheat	33–38	26–32	17–19	–	6.8

Source: Reused with permission, license number 5118911131993.

and Rekha 2020) for various plant fibers at the middle of plant fiber's multi-scale structure. In general, the cavity accounts for 3%–4% of an elementary fiber's cross-sectional area (Charlet et al. 2007). Plant fibers (such as flax) have high vibration-damping capability due to their complex microstructure and porous nature, including microfibrils incorporated into an amorphous matrix inside the elementary fibers and weakly bonded elementary fibers (Pil et al. 2016).

8.7 DAMPING MECHANISMS

The study on the mechanisms of vibration damping of polymers has been extensively performed by Povolo and Goyanes (1996), Song, Hourston, and Schafer (2001), and Sperling (1990). However, PFCs damping mechanisms differ from those of traditional materials (Petrone et al. 2012, Kumar, Siva, Jeyaraj, et al. 2014), for example, metals, alloys, and polymers. Energy dissipation in PFCs is caused by a variety of factors, including:

1. Viscoelastic behavior of fiber and/or matrix materials (Akash, Thyagaraj, and Sudev 2013, Pothan, Oommen, and Thomas 2003, Ornaghi et al. 2010, Hameed et al. 2007, Chandra, Singh, and Gupta 1999). This impact is significant in polymer matrix composites (Dong and Gauvin 1993).

2. Damping as a result of the interface. As the fibers are incorporated into the bulk matrix, the area adjacent to the fiber surface is termed as the interface that appears over the fiber length (Gibson, Hwang, and Kwak 1991, Chandra, Singh, and Gupta 2003). The nature of the interfacial adhesion (weak, ideal, or strong) influences the mechanical and damping performances of composites (Chandra, Singh, and Gupta 1999). The damping is strongly affected by the interface (Chaturvedi and Tzeng 1991, Hwang and Gibson 1993, Finegana and Gibson 1999). However, improved interfacial adhesion between fiber and matrix reduces damping (Cinquin et al. 1990, Chandra, Singh, and Gupta 1999).

3. Damping due to damaged fibers, matrix cracks, and delamination in unbonded regions (Júnior et al. 2012). As opposed to composites without slip, composites with interfacial slip display greater damping (Nelson and Hancock 1978, Greif and Hebert 1995).

   Energy loss due to friction arising from the slipping between the fiber–matrix interfaces (Sun, Wu, and Gibson 1987). Inter-cell wall (between the cell walls) and intra-cell wall (between hemicellulose/lignin matrix and cellulose microfibrils) frictions are two friction mechanisms that promote intrinsic energy dissipation in the plant fiber structure. Intra-yarn (between the fibers within the yarn) and inter-yarn (between the yarns) frictions also increase energy dissipation (Duc, Bourban, and Manson 2014a, b, Duc et al. 2014, Daoud et al. 2017).

4. Damping owing to the presence of porosity in the plant fiber (Kumar, Siva, Jeyaraj, et al. 2014).

5. Thermoplastic composites demonstrate high nonlinear damping at an increased amplitude of vibration or increased stress. Plasticity contributes more to damping than any other energy dissipating source owing to the high vibration amplitudes (Kenny and Marchetti 1995).

6. Damping energy is partly retained in the microstructure (void, crazes, shear band, and microcracks), and the rest is dissipated as heat (Chakraborty and Ratna 2020).

The fiber surface treatment, fiber content, fiber length, fiber layup, and fiber architectures (e.g., twill, triaxial, satin, and knitted fabric) affect the damping of PFCs as similar to the glass and carbon fiber composites. The presence of water in the fiber also influences its damping performance. However, the effects of fiber treatment on the damping of PFCs are discussed in the following section.

8.8   FIBER TREATMENT AND MODIFICATION

The interfacial bonding between fiber and matrix influences the modal parameters (i.e., natural frequency, damping, and mode shapes) of a composite. In general, the interfacial adhesion requires to be weak to attain the enhanced damping performance. The hydrophilic and hydrophobic behavior of fibers and matrices, respectively, often result in little interaction between them. This leads to weak interfacial adhesion increasing the damping of PFCs. However, the interfacial bonding strength can be improved by physical and chemical treatments (see Figure 8.4). Fiber treatments and modifications can be classified into three types — fiber pretreatment (e.g., mercerization, an alkaline treatment to fibrillate and cleanse fibers (partly removal of wax, oil, hemicellulose, lignin, and pectin)), surface coating modified by binding chemicals, and in situ compatibilization during processing based on practical applications (Li et al. 2020). Physical approaches such as plasma, corona, ultraviolet, fiber beating, or heat treatment are used to improve the fiber surface roughness and/or remove oils and waxes from the fiber surface, resulting in better mechanical interlocking between the fiber and matrix. In contrast, alkali, silane, acetylation, benzoylation, maleic anhydride-grafted polymers, acrylation and acrylonitrile grafting, isocyanate, permanganate, and other chemicals are used as a compatibilizer or binding agent between the fibers and matrices in chemical approaches (Bisanda 2000, Gholampour and Ozbakkaloglu 2020). Alkaline treatment improves fiber surface purity and fiber–matrix interfacial bonding strength, thereby eliminating fiber pull-out and contributing to the properties of the fiber (Ramesh, Palanikumar, and Reddy 2017).

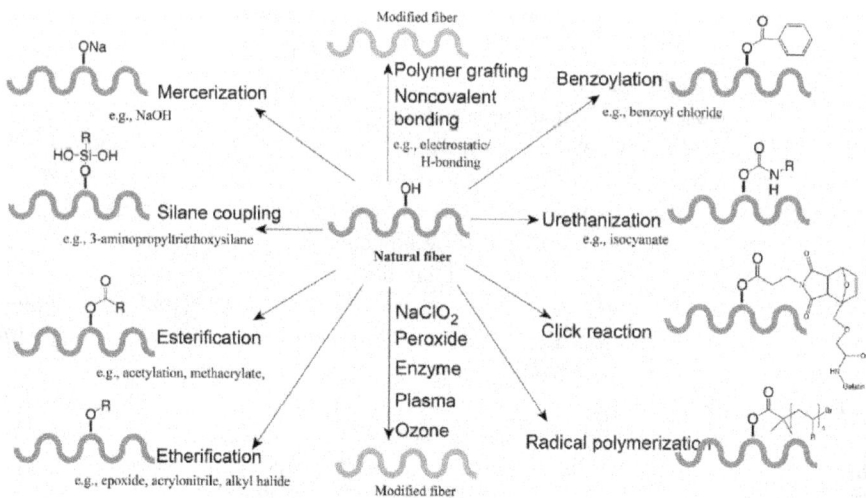

FIGURE 8.4   Typical methods for treating and modifying plant fibers (Li et al. 2020). (Reused with permission, license number 5118851303843.)

8.8.1 FIBER TREATMENT EFFECT ON DAMPING

The effect of the interface on PFCs shows similar results to that of synthetic fiber composites. Since the movement of molecular chains at the interface reduces, an enhancement in the interfacial adhesion between fiber and matrix reduces damping (Raditoiu et al. 2013, Saba et al. 2016). Chua (1987) also reported that unidirectional glass fiber/polyester composite specimens with weak interfacial adhesion between the fiber and matrix appear to dissipate additional energy than specimens with strong interfacial adhesion. Other researchers (Aziz and Ansell 2004a, b, Geethamma et al. 2005, Pothan, Oommen, and Thomas 2003, Dong and Gauvin 1993, Acha, Reboredo, and Marcovich 2007, Ashida, Noguchi, and Mashimo 1984, Silverajah et al. 2012, Paul et al. 2010, Felix and Gatenholm 1991, Pothan, Thomas, and Groeninckx 2006, Houshyar, Shanks, and Hodzic 2005, Iqbal et al. 2007, Edie et al. 1993, Li, Schlarb, and Evstatiev 2009, Mohanty and Nayak 2006, Pilla 2011, Akil and Mazuki 2011, Zhao et al. 2010, Shumao et al. 2010, Manoharan et al. 2014, Romanzini et al. 2013, Guo et al. 2006, He and Liu 2005, Karaduman et al. 2014) also identified a similar trend. The inclusion of compatibilizers (e.g., maleic anhydride-grafted polyethylene octane and maleic anhydride-grafted polypropylene (MAPP)) stimulates improved interfacial adhesion between the fiber and matrix, leading to reduced loss factor, as revealed by Etaati et al. (Etaati et al. 2013). Other researchers identified the same trend using an alkaline treatment on fiber's surface in their studies (Yan 2012, Mylsamy and Rajendran 2011a, Shukor et al. 2014, Singhal and Tiwari 2014), while studies (Mohanty, Verma, and Nayak 2006, Saw, Sarkhel, and Choudhury 2012) observed the same behavior following maleic anhydride-grafted polyethylene and furfuryl alcohol treatments, respectively.

The effects of fiber treatments on the damping of PFCs are presented in Table 8.2. The damping of untreated flax and linen fabric composites is greater than that of alkali-treated ones, with a 7.4% and 9.3% drop in loss factor of flax- and linen-epoxy composites, respectively. In contrast, the treated composite's natural frequency increases due to the increased stiffness of the composites (Yan 2012). A similar result is found in the cases of untreated and alkali-treated flax fiber/cement composites by Lai et al. (2019). Slipping between the fiber and matrix and two friction mechanisms (which occurs between elementary fibers and between adjacent fiber cell walls) are attributable to flax fiber damping. Alkali treatments partially remove wax, oil, pectin, hemicellulose, and lignin, which is the key cause of having higher damping of untreated flax fibers than alkali-modified flax fibers.

The removal of unnecessary materials by alkali treatment and then applying silane coating enhances the stiffness of hybrid composites (coconut sheath/ *Sansevieria cylindrica* fiber polyester). As a result, the composites exhibit a rise in natural frequency and reduction in damping (Bennet et al. 2015). The enhanced natural frequency of treated composite is attributed to the strengthened fiber/matrix adhesion, which is attained by exposing the more effective area of the fiber surface after removing impurities and waxy layers (Rajini et al. 2013). Vazquez et al. (1998) also found that epoxy coating on the fiber reduces damping. This infers to the decrease in the molecular mobility in the region next to the fiber surface because of chemical interactions that occurred between epoxy-amine and di-epoxy coating

TABLE 8.2

Effect of Fiber Treatments on the Damping of PFCs

Composites	Chemicals for Fiber Treatments	Measurement Techniques	Testing Conditions	Damping (Increase/ Decrease)	References
Doum fiber/PP	SEBS-g-MA[a]	DMA	1 Hz, 30°C–120°C	Increase	Essabir et al. (2013)
Banana fiber/Phenol formaldehyde	Alkali, silane, acetylation, formic acid, benzoylation, potassium permanganate	DMA	0.1, 20, and 50 Hz, 0°C–200°C	Increase[b], Decrease[c]	Indira, Jyotishkumar, and Thomas (2014)
Hybrid coconut sheath-sisal fiber/polyester	Alkali, tri-chloro vinyl silane	IH	Not stated	Increase	Kumar, Siva, Rajini, et al. (2014)
Jute fiber/PP	Alkali	DMA	1 Hz, 20°C–200°C	Decrease	Karaduman et al. (2014)
Kenaf and hemp fiber/polyester	Alkali	DMA	1 Hz, 30°C–180°C	Decrease	Aziz and Ansell (2004a)
Agave fiber/epoxy	Alkali	DMA	1 Hz, 30°C–135°C	Decrease	Mylsamy and Rajendran (2011b)
Flax fiber/cement	Alkali	DMA	0.1–20 Hz	Decrease	Lai et al. (2019)
Hybrid sisal-coir fiber/epoxy	Alkali	IH	0–750 Hz	Increase	Sumesh and Kanthavel (2020)
Hybrid banana-jute woven fabric/polyester	Alkali, benzoyl chloride, silane, potassium permanganate	DMA, IH	1 Hz, 0°C–200°C, 0–2,000 Hz	Increase, Decrease	Rajesh and Pritchaimani (2017)
Jute fiber/polyester	Dried	Shaker, IH	0–200 Hz	Decrease	Furtado et al. (2014)
Flax- and linen- fabric/epoxy	Alkali	IH	Not stated	Decrease	Yan (2012)
Flax fiber/epoxy	Glycerol, polyglycerol	DMA, IH	−25°C to 175°C, 0–10,000 Hz	Increase	Guen et al. (2014)
Hybrid *Sansevieria cylindrica*– coconut sheath/polyester	Alkali, silane	IH	Not stated	Decrease	Bennet et al. (2015)

(Continued)

TABLE 8.2 (*Continued*)
Effect of Fiber Treatments on the Damping of PFCs

Composites	Chemicals for Fiber Treatments	Measurement Techniques	Testing Conditions	Damping (Increase/Decrease)	References
Oil palm empty fruit bunch and jute fiber/epoxy	2-hydroxy ethyl acrylate, dicumyl peroxide	DMA	1Hz, −50°C to 150°C	Decrease	Jawaid et al. (2015)
Jute fiber/HDPE	Maleic anhydride-grafted polyethylene	DMA	10Hz, −120°C to 100°C	Decrease	Mohanty, Verma, and Nayak (2006b)
Oil palm fiber/LLDPE	Alkali	DMA	1, 10, and 20Hz, 100°C to 150°C	Increase	Shinoj et al. (2011)
Woven coconut fiber/polyester	Alkali, silane	DMA, IH	20Hz, 20°C–300°C, not stated	Increase	Rajini et al. (2013)
Jute fiber/vinyl ester	Alkali	DMA	1Hz, 30°C–210°C	Increase	Ray et al. (2002)
Kenaf fiber/polyester, Kenaf fiber/PP	Alkali	DMA	1Hz, −30°C to 100°C, 30°C–200°C	Decrease	Han et al. (2008)
Jute fiber/epoxy	Alkali, potassium permanganate, benzoyl chloride, maleic anhydride, silane, acetone, ethanol	DMA	1Hz, 0°C–200°C	Decrease	Singhal and Tiwari (2014b)
Bagasse fiber/polyester	Alkali, acrylic acid	DMA	30°C–170°C	Almost same	Vilay et al. (2008)
Coconut sheath fiber/epoxy	Sodium hydroxide, acetic acid	DMA	1Hz, 0°C–140°C	Decrease	Kumar, Duraibabu, and Subramanian (2014)

a   Styrene–(ethylene–butene)–styrene three-block copolymer grafted with maleic anhydride.
b   Damping increases after silane, acetylation, and benzoylation treatments.
c   Damping decreases after alkali, formic acid, and potassium permanganate treatments.

with the amine silane sizing. However, the mono-epoxy treatment of the fiber offers maximum damping, suggesting less crosslinking density of the interface and poor interfacial shear strength. The jute fiber and 2-hydroxy ethyl acrylate treated oil palm empty fruit bunch-reinforced epoxy hybrid composites also show less damping than untreated ones. Since the treatment causes improved adhesion, rigid and strong fiber–matrix interface induces a decrease in molecular mobility at the interface (Jawaid et al. 2015). Saha et al. (1999) also observed a similar trend using cyanoethylated jute/polyester composites. It is evident that untreated composites exhibit higher damping owing to the presence of weaker fiber–matrix interfaces where more energy dissipation occurred by internal frictional mechanisms. In addition to the fiber–matrix interface, energy dissipation also relies on the interphase region, frictional resistance, matrix cracks, and damaged fibers (Mohanty, Verma, and Nayak 2006). In another study, Mylsamy and Rajendran (2011b) concluded that composite with weak interfacial adhesion exhibit more energy dissipation relative to composites with a strong interfacial adhesion considering agave fiber/epoxy composites (untreated and alkaline treated). The damping (0.33 at 84.1°C) of coconut sheath fiber/epoxy composites (untreated) is better than that (0.27 at 82.4°C) of treated fiber composite. The decline in loss factor is attributable to the restriction of polymeric molecules' movement, an improvement in storage modulus, and a decrease in the viscoelastic lag between stress and strain. Consequently, matrix content is insufficient to adequately dissipate the vibration energy, thus reducing damping. The damping of structural composites improves as the temperature rises owing to matrix softening, which raises matrix damping (Chakraborty and Ratna 2020). However, high interfacial adhesion ensures that composite has a good load-carrying capability and an improved stress transfer (Kumar, Duraibabu, and Subramanian 2014). When interfacial adhesion is very strong, the interface conducts the entire load transfer functions and contributes little to damping (Chandra, Singh, and Gupta 1999).

Several studies (Shinoj et al. 2011, Essabir et al. 2013, Indira, Jyotishkumar, and Thomas 2014, Kumar, Siva, Rajini, et al. 2014, Rajesh and Pitchaimani 2017, Guen et al. 2014) reported that chemical treatments increase the damping of PFCs. Two studies (Shinoj et al. 2011, Essabir et al. 2013) conducted the treatments of oil palm fiber and doum fiber using alkali and styrene–(ethylene–butene)–styrene three-block copolymer grafted with maleic anhydride, respectively. In contrast, banana fiber was treated by silane, acetylation, and benzoylation (Indira, Jyotishkumar, and Thomas 2014). The addition of fibers restricts the polymer chain mobility and raises the storage modulus and viscoelastic lag between stress and strain; hence, composite damping increases (Essabir et al. 2013, Ray et al. 2002). Kumar, Siva, Rajini, et al. (2014) revealed that silane- and alkali-treated coconut sheath/sisal fiber-reinforced polyester hybrid composites produce higher natural frequency as compared to untreated composites. This is because treatments enhance the fiber–matrix adhesion, leading to improved stiffness and, as a result, increased natural frequency. The hybrid composites (alkali- and silane-treated) also dampen better than the untreated fiber composites. Although treatments typically reduce the damping, the porosity (i.e., absorb energy by cavitation mechanism under dynamic loading) of the coconut fiber may mitigate any potential loss in damping due to improved adhesion. Moreover, the damping of treated and untreated coconut sheath/sisal hybrid polyester composites outperforms the glass/epoxy, glass/vinyl ester, Kevlar/epoxy, and carbon/epoxy composites (Kumar, Siva, Rajini, et al. 2014). Rajesh and Pitchaimani (2017) also found that potassium permanganate and silane surface-treated composites (i.e., hybrid jute/banana woven fabric composite) have higher loss factor than that of untreated composites because of decreased crystalline behavior of oven fabric of plant fiber, contributing to the rise in free molecular mobility in the polymer chains and thus enhance the fiber–matrix interaction. Poor fiber–matrix adhesion in these chemically modified composites leads to more interaction between the fiber and matrix and thus increases energy dissipation. On the contrary, the damping of alkali- and benzoyl chloride-treated composites is minimized because of the better fiber–matrix interface bonding. The polyol-treated flax fiber composite has higher damping than untreated flax, aramid, and carbon fiber composites (Guen et al. 2014). Interestingly, the increase in damping with the increasing frequency of flax composites is because of the higher internal friction occurred by heterogeneous chemical compounds' molecular vibration. Amorphous compounds of flax fiber have a certain degree of movement when excited at molecular levels, which potentially can lead to dissipating energy. At 100 Hz, the damping of flax fiber composite materials improves by 25% and 10%, with polyglycerol and glycerol treatments, respectively. The damping in PFCs is derived from matrix and fiber, interphase, friction because of slipping in the unbound zones between the fiber and matrix, delamination, matrix cracks, and fiber breakage.

8.9 APPLICATIONS

PFCs have several benefits, including high stiffness and strength-to-weight ratios, high damping, low density, and biodegradability, making them appropriate in various applications. Potentially extensive applications can be seen in areas where substantial loss factor is necessary, for instance, automobiles, machinery components, aircraft interiors, sports, and recreation equipment. The advantages of lightweight PFCs with high damping are the extension of the PFC components' lifespan, declination of the audible noise and vibration, reduction of the dynamic loading effect on structural responses, and weight reduction.

PFC has superior damping property; some products are manufactured of plant fibers to take advantage of this property, as illustrated in Figures 8.5 and 8.6. Pil et al. (2016) stated that the high damping capability of plant fibers, for example, flax fibers can be hybridized with high stiffness and strength of carbon fiber to produce tennis rackets. Flax fibers with a volume fraction of 0.15 (of the total fiber content, the remaining is carbon fiber) enhance the loss factor of 22%, leading to reduced vibration and hence the possibility of muscle injury. High damping also results in improving the athlete's control over the athletic products. In addition, speaker kits are made by hybridizing flax (which absorbs energy) and glass (which provides controlled stiffness), contributing to improved dynamics and a more uniform sound across a broader spectrum and sound neutrality. Furthermore, flax/carbon hybrid composites are used to construct structural members of racing

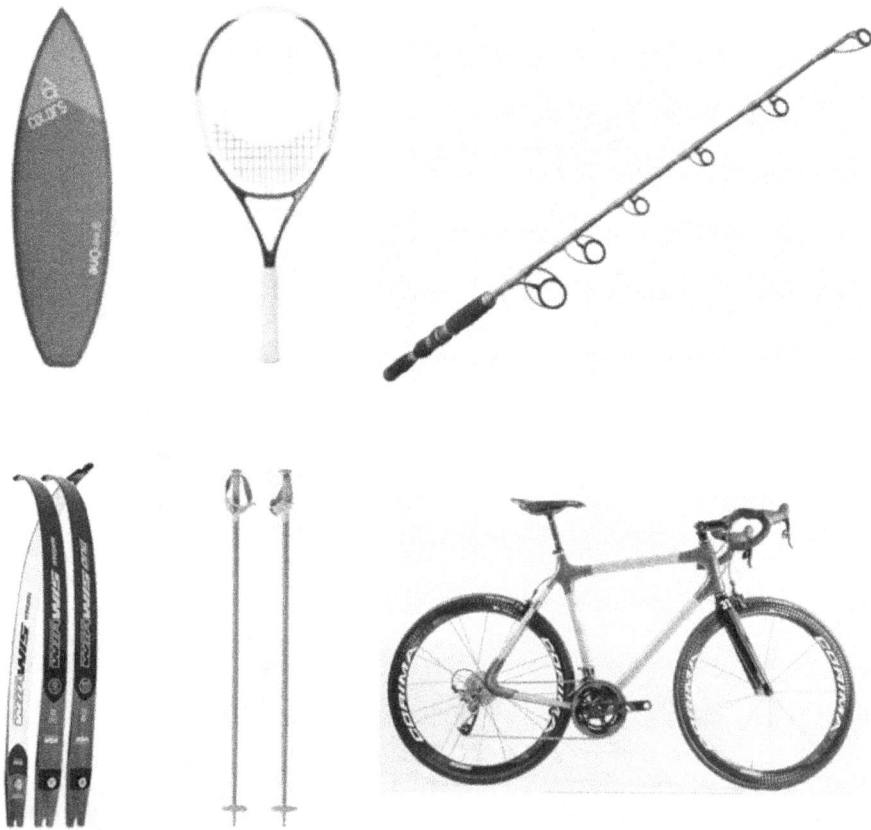

FIGURE 8.5 Applications of lightweight PFCs considering their high damping and stiffness: (a) Surfboard, (b) Tennis racket, (c) Fishing rod, (d) Archery, (e) Ski poles, and (f) Bike (Pil et al. 2016). (Reused with permission, license number 5118910413973.)

bicycles, offering an appropriate combination of mechanical and damping performances (Pil et al. 2016).

Phillips and Lessard (2012) have also applied flax prepregs composite to create top-plates for string musical instruments (e.g., violins, guitars, and ukuleles). PFCs with thermoset and thermoplastic matrices are extensively and increasingly applied in automobile parts by manufacturers for door panels, headliner panels, truck liners, noise-insulating panels, seat backs, dashboards, trays, and interior components. However, plant fibers such as flax, hemp, sisal, and kenaf fiber-reinforced thermoplastic composites have dominated the manufacturing of automobile components (Li et al. 2020, Furtado et al. 2014). For example, flax, hemp, and sisal fibers are employed to manufacture door cladding, floor panels, seat backs linings, and car disk brakes, while coconut fibers are for making seat bottoms, head restraints, and back cushions, cotton fibers are for soundproofing, as well as abaca and kenaf fibers are for the underfloor body and door inner panels, respectively.

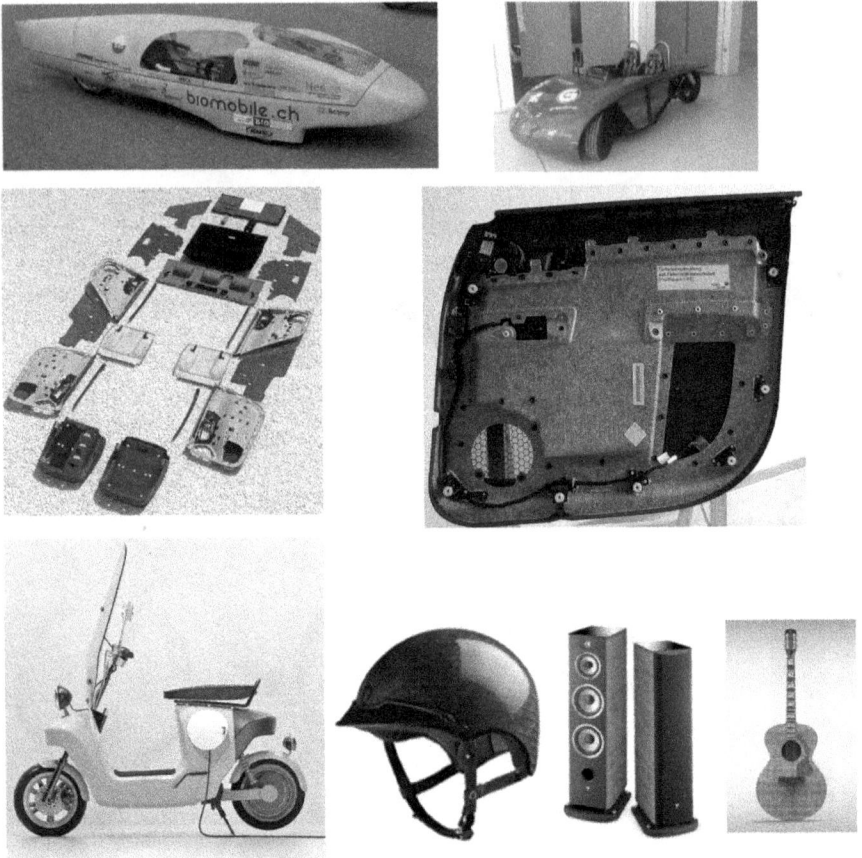

FIGURE 8.6 Applications of lightweight PFCs considering their high damping and stiffness: (a) Biomobile (Pil et al. 2016), (b) I-car racing (Pil et al. 2016), (c) Car parts, (d) Car door (Gahle 2007) (open access), (e) Electric scooter (Pil et al. 2016), (f) Helmet (Pil et al. 2016), (g) Speaker (Pil et al. 2016), and (h) Guitar (Pil et al. 2016). (Reused with permission, license number 5118910413973.)

8.10 CONCLUSIONS

Compared to metallic materials and synthetic fiber composites, PFCs generally show a higher damping capacity due to the viscoelasticity of the polymeric matrix and the viscoelasticity and hierarchical structure of the plant fiber. Most studies have found that fiber treatment reduces the damping due to the improved fiber–matrix adhesion, while some studies show enhancement in damping performance after treatments, where factors such as porous nature and amorphous compounds of the fiber ameliorate any potential loss in damping caused by improved interfacial bonding strength. Fiber surface treatment may degrade the fiber's texture because treated fiber undergoes twisting and becomes thinner, causing the treated fiber more brittle relative to the nontreated fiber. The treatment may also be toxic (e.g., isocyanates) and/

or expensive (e.g., silanes), negatively affecting the plant fiber's low cost, environmentally friendly image, and enhancements on interfacial bonding strength decrease vibration damping along with other properties such as impact strength. However, PFCs have the potential to be a viable alternative to synthetic fibers (glass and carbon fibers) composites for automobile and aerospace applications, especially where a weight reduction is needed while retaining high damping performance.

REFERENCES

Acha, Betiana A., María M. Reboredo, and Norma E. Marcovich. 2007. "Creep and dynamic mechanical behavior of PP–jute composites: effect of the interfacial adhesion." *Composites Part A: Applied Science and Manufacturing* 38 (6):1507–1516. https://doi. org/10.1016/j.compositesa.2007.01.003.

Akash, D. A., N. R. Thyagaraj, and L. J. Sudev. 2013. "Experimental study of dynamic behaviour of hybrid jute/sisal fibre reinforced polyester composites." *International Journal of Science and Engineering Applications* 2 (7):170–172.

Akil, Hazizan Md, and Adlan Akram Mohamad Mazuki. 2011. "Sustainable biocomposites based for construction applications." In *Handbook of Bioplastics and Biocomposites Engineering Applications*, edited by Srikanth Pilla, 285–316. John Wiley & Sons Inc.

Alan, Wolfenden, and K. Kinra Vikram. 1992. *M3D III: Mechanics and mechanisms of material damping*. ASTM.

Ashida, Michio, Toru Noguchi, and Satoshi Mashimo. 1984. "Dynamic moduli for short fiber-CR composites." *Journal of Applied Polymer Science* 29 (2):661–670. https://doi. org/10.1002/app.1984.070290222.

Avitabile, Peter. 2017. *Modal Testing: A Practitioner's Guide*. John Wiley & Sons.

Aziz, Sharifah H., and Martin P. Ansell. 2004a. "The effect of alkalization and fibre alignment on the mechanical and thermal properties of kenaf and hemp bast fibre composites: Part 1–polyester resin matrix." *Composites Science and Technology* 64 (9):1219–1230.

Aziz, Sharifah H., and Martin P. Ansell. 2004b. "The effect of alkalization and fibre alignment on the mechanical and thermal properties of kenaf and hemp bast fibre composites: Part 2–cashew nut shell liquid matrix." *Composites Science and Technology* 64 (9):1231–1238.

Bennet, C., N. Rajini, J. T. Winowlin Jappes, I. Siva, V. S. Sreenivasan, and S. C. Amico. 2015. "Effect of the stacking sequence on vibrational behavior of Sansevieria cylindrica/coconut sheath polyester hybrid composites." *Journal of Reinforced Plastics and Composites* 34 (4):293–306.

Bisanda, E. T. N. 2000. "The effect of alkali treatment on the adhesion characteristics of sisal fibres." *Applied Composite Materials* 7 (5–6):331–339.

Bos, Harriëtte L, Jörg Müssig, and Martien J. A. van den Oever. 2006. "Mechanical properties of short-flax-fibre reinforced compounds." *Composites Part A: Applied Science and Manufacturing* 37 (10):1591–1604.

Chakraborty, B. C., and D. Ratna. 2020. "Chapter 6 - Experimental techniques and instruments for vibration damping." In *Polymers for Vibration Damping Applications*, edited by B. C. Chakraborty and Debdatta Ratna, 281–325. Elsevier.

Chandra, R., S. P. Singh, and K. Gupta. 1999. "Damping studies in fiber-reinforced composites – a review." *Composites Structures* 46:41–51.

Chandra, R., S. P. Singh, and K. Gupta. 2003. "A study of damping in fiber-reinforced composites." *Journal of Sound and Vibration* 262 (3):475–496. https://doi.org/10.1016/ s0022-460x(03)00107-x.

Charlet, K., C. Baley, C. Morvan, J. P. Jernot, M. Gomina, and J. Bréard. 2007. "Characteristics of Hermès flax fibres as a function of their location in the stem and properties of the derived unidirectional composites." *Composites Part A: Applied Science and Manufacturing* 38 (8):1912–1921.

Charlet, K., J. P. Jernot, M. Gomina, J. Bréard, C. Morvan, and C. Baley. 2009. "Influence of an Agatha flax fibre location in a stem on its mechanical, chemical and morphological properties." *Composites Science and Technology* 69 (9):1399–1403. https://doi.org/10.1016/j.compscitech.2008.09.002.

Charlet, K., J. P. Jernot, S. Eve, M. Gomina, and J. Bréard. 2010. "Multi-scale morphological characterisation of flax: from the stem to the fibrils." *Carbohydrate Polymers* 82 (1):54–61.

Chaturvedi, Shive K., and Guang Yau Tzeng. 1991. "Micromechanical modeling of material damping in discontinuous fiber three-phase polymer composites." *Composites Engineering* 1 (1):49–60.

Chua, Ping Seng. 1987. "Dynamic mechanical analysis studies of the interphase." *Polymer Composites* 8 (5):308–313.

Chung, D. D. L. 2001. "Review: materials for vibration damping." *Journal of Materials Science* 36:5733–5737.

Cinquin, J., B. Chabert, J. Chauchard, E. Morel, and J. P. Trotignon. 1990. "Characterization of a thermoplastic (polyamide 66) reinforced with unidirectional glass fibres. Matrix additives and fibres surface treatment influence on the mechanical and viscoelastic properties." *Composites* 21 (2):141–147.

Cremer, Lothar, Manfred Heckl, and Björn A. T. Petersson. 2005. *Structure-Borne Sound: Structural Vibrations and Sound Radiation at Audio Frequencies*. Springer.

Daoud, Hajer, Abderrahim El Mahi, Jean-Luc Rebiere, Mohamed Taktak, and Mohamed Haddar. 2017. "Characterization of the vibrational behaviour of flax fibre reinforced composites with an interleaved natural viscoelastic layer." *Applied Acoustics* 128: 23–31.

Darabi, Babak. 2013. "Dissipation of vibration energy using viscoelastic granular materials." PhD Thesis, The University of Sheffield.

De Silva, Clarence W. 2006. *Vibration: Fundamentals and Practice*. CRC Press.

Dong, S., and R. Gauvin. 1993. "Application of dynamic mechanical analysis for the study of the interfacial region in carbon fiber/epoxy composite materials." *Polymer Composites* 14 (5):414–420.

Duc, F., P. E. Bourban, C. J. G. Plummer, and J. A. E. Manson. 2014. "Damping of thermoset and thermoplastic flax fibre composites." *Composites Part A: Applied Science and Manufacturing* 64:115–123. https://doi.org/10.1016/j.compositesa.2014.04.016.

Duc, F., P. E. Bourban, and J. A. E. Manson. 2014a. "Dynamic mechanical properties of epoxy/flax fibre composites." *Journal of Reinforced Plastics and Composites* 33 (17):1625–1633. https://doi.org/10.1177/0731684414539779.

Duc, F., P. E. Bourban, and J. A. E. Manson. 2014b. "The role of twist and crimp on the vibration behaviour of flax fibre composites." *Composites Science and Technology* 102:94–99. https://doi.org/10.1016/j.compscitech.2014.07.004.

Edie, D. D., J. M. Kennedy, R. J. Cano, and R. A. Ross. 1993. "Evaluating surface treatment effects on interfacial bond strength using dynamic mechanical analysis." *Composite Materials: Fatigue and Fracture* 4:419–429.

Essabir, H., A. Elkhaoulani, K. Benmoussa, R. Bouhfid, F. Z. Arrakhiz, and A. Qaiss. 2013. "Dynamic mechanical thermal behavior analysis of doum fibers reinforced polypropylene composites." *Materials & Design* 51:780–788.

Etaati, A., S. A. Mehdizadeh, H. Wang, and S. Pather. 2013. "Vibration damping characteristics of short hemp fibre thermoplastic composites." *Journal of Reinforced Plastics and Composites* 33 (4):330–341. https://doi.org/10.1177/0731684413512228.

Ewins, D. J. 2000. *Modal Testing: Theory, Practice and Application*. Research Studies Press Ltd.

Felix, Johan M., and Paul Gatenholm. 1991. "The nature of adhesion in composites of modified cellulose fibers and polypropylene." *Journal of Applied Polymer Science* 42 (3):609–620.

Finegana, Ioana C., and Ronald F. Gibson. 1999. "Recent research on enhancement of damping in polymer composites." *Composite Structures* 44 (2–3):89–98.

Furtado, Samuel C. R., A. L. Araújo, Arlindo Silva, Cristiano Alves, and A. M. R. Ribeiro. 2014. "Natural fibre-reinforced composite parts for automotive applications." *International Journal of Automotive Composites* 1 (1):18–38.

Gahle, Christian. 2007. "Interior carpeting of a cars door made by a biocomposite of hemp fibres and polyethylen." Accessed 30 March. https://commons.wikimedia.org/wiki/File:T%C3%BCrinnenverkleidung_Hanf-PP_nova.jpg.

Geethamma, V. G., G. Kalaprasad, Gabriël Groeninckx, and Sabu Thomas. 2005. "Dynamic mechanical behavior of short coir fiber reinforced natural rubber composites." *Composites Part A: Applied Science and Manufacturing* 36 (11):1499–1506.

Gholampour, Aliakbar, and Togay Ozbakkaloglu. 2020. "A review of natural fiber composites: properties, modification and processing techniques, characterization, applications." *Journal of Materials Science* 55 (3):829–892.

Gibson, R. F., S. J. Hwang, and H. Kwak. 1991. "Micromechanical modeling of damping in composites including interphase effects." Proceedings of 36th International SAMPE Symposium, San Diego, CA.

Greif, Robert, and Benjamin Hebert. 1995. "Experimental techniques for dynamic characterization of composite materials." *Journal of Engineering Materials and Technology* 117 (1):94–100. https://doi.org/10.1115/1.2804378.

Guen, M. J. L., R. H. Newman, A. Fernyhough, and M. P. Staiger. 2014. "Tailoring the vibration damping behaviour of flax fibre-reinforced epoxy composite laminates via polyol additions." *Composites Part A: Applied Science and Manufacturing* 67:37–43. https://doi.org/10.1016/j.compositesa.2014.08.018.

Guo, C., Y. Song, Q. Wang, and C. Shen. 2006. "Dynamic-mechanical analysis and SEM morphology of wood flour/polypropylene composites." *Journal of Forestry Research* 17 (4):315–318. https://doi.org/10.1007/s11676-006-0072-7.

Hameed, Nishar, P. A. Sreekumar, Bejoy Francis, Weimin Yang, and Sabu Thomas. 2007. "Morphology, dynamic mechanical and thermal studies on poly(styrene-co-acrylonitrile) modified epoxy resin/glass fibre composites." *Composites Part A: Applied Science and Manufacturing* 38 (12):2422–2432. https://doi.org/10.1016/j.compositesa.2007.08.009.

Han, Young Hee, Seong OK Han, Donghwan Cho, and Hyung-Il Kim. 2008. "Dynamic mechanical properties of natural fiber/polymer biocomposites: the effect of fiber treatment with electron beam." *Macromolecular Research* 16 (3):253–260.

He, L. H., and Y. L. Liu. 2005. "Damping behavior of fibrous composites with viscous interface under longitudinal shear loads." *Composites Science and Technology* 65 (6):855–860. https://doi.org/10.1016/j.compscitech.2004.09.003.

Houshyar, S., R. A. Shanks, and A. Hodzic. 2005. "The effect of fiber concentration on mechanical and thermal properties of fiber-reinforced polypropylene composites." *Journal of Applied Polymer Science* 96 (6):2260–2272.

Hwang, S. J., and R. F. Gibson. 1993. "Prediction of fiber-matrix interphase effects on damping of composites using a micromechanical strain energy/finite element approach." *Composites Engineering* 3 (10):975–984. https://doi.org/10.1016/0961-9526(93)90005-5.

Indira, K. N., P. Jyotishkumar, and Sabu Thomas. 2014. "Viscoelastic behaviour of untreated and chemically treated banana fiber/PF composites." *Fibers and Polymers* 15 (1):91–100.

Iqbal, Azhar, Lars Frormann, Anjum Saleem, and Muhammad Ishaq. 2007. "The effect of filler concentration on the electrical, thermal, and mechanical properties of carbon particle and carbon fiber-reinforced poly (styrene-co-acrylonitrile) composites." *Polymer Composites* 28 (2):186–197.

Jawaid, Mohammad, Othman Y. Alothman, Naheed Saba, Paridah Md Tahir, and H. P. S. Abdul Khalil. 2015. "Effect of fibers treatment on dynamic mechanical and thermal properties of epoxy hybrid composites." *Polymer Composites* 36 (9):1669–1674.

Jones, David I. G. 2001. *Handbook of Viscoelastic Vibration Damping.* John Wiley & Sons Ltd.

Júnior, José Humberto Santos Almeida, Heitor Luiz Ornaghi Júnior, Sandro Campos Amico, and Franco Dani Rico Amado. 2012. "Study of hybrid intralaminate curaua/glass composites." *Materials & Design* 42:111–117.

Karaduman, Y., M. M. A. Sayeed, Levent Onal, and A. Rawal. 2014. "Viscoelastic properties of surface modified jute fiber/polypropylene nonwoven composites." *Composites Part B: Engineering* 67:111–118.

Kenny, J. M., and M. Marchetti. 1995. "Elasto-plastic behavior of thermoplastic composite laminates under cyclic loading." *Composite Structures* 32 (1–4):375–382. https://doi.org/10.1016/0263-8223(95)00052-6.

Khan, Shafi Ullah, Chi Yin Li, Naveed A Siddiqui, and Jang-Kyo Kim. 2011. "Vibration damping characteristics of carbon fiber-reinforced composites containing multi-walled carbon nanotubes." *Composites Science and Technology* 71 (12):1486–1494.

Kumar, K. S., I. Siva, N. Rajini, P. Jeyaraj, and J. T. Winowlin Jappes. 2014. "Tensile, impact, and vibration properties of coconut sheath/sisal hybrid composites: effect of stacking sequence." *Journal of Reinforced Plastics and Composites* 33 (19):1802–1812.

Kumar, K. S., I. Siva, P. Jeyaraj, J. T. W. Jappes, S. C. Amico, and N. Rajini. 2014. "Synergy of fiber length and content on free vibration and damping behavior of natural fiber reinforced polyester composite beams." *Materials & Design* 56:379–386. https://doi.org/10.1016/j.matdes.2013.11.039.

Kumar, S. M. Suresh, D. Duraibabu, and K. Subramanian. 2014. "Studies on mechanical, thermal and dynamic mechanical properties of untreated (raw) and treated coconut sheath fiber reinforced epoxy composites." *Materials & Design* 59: 63–69.

Lai, Pengfei, Xudong Zhi, Shizhao Shen, Zheng Wang, and Ping Yu. 2019. "Strength and damping properties of cementitious composites incorporating original and alkali treated flax fibers." *Applied Sciences* 9 (10):2002.

Li, Mi, Yunqiao Pu, Valerie M. Thomas, Chang Geun Yoo, Soydan Ozcan, Yulin Deng, Kim Nelson, and Arthur J. Ragauskas. 2020. "Recent advancements of plant-based natural fiber–reinforced composites and their applications." *Composites Part B: Engineering* 200:108254. https://doi.org/10.1016/j.compositesb.2020.108254.

Li, Wenjing, Alois K. Schlarb, and Michael Evstatiev. 2009. "Study of PET/PP/TiO2 microfibrillar-structured composites. Part 1: Preparation, morphology, and dynamic mechanical analysis of fibrillized blends." *Journal of Applied Polymer Science* 113 (3):1471–1479. https://doi.org/10.1002/app.29993.

Li, Y., Y. Luo, and S. Han. 2010. "Multi-scale structures of natural fibers and their applications in making automobile parts." *Journal of Biobased Materials and Bioenergy* 4 (2):164–171.

Maia, N. M. M., and J. M. M. E. Silva. 1997. *Theoretical and Experimental Modal Analysis.* Research Studies Press Ltd.

Manoharan, Sembian, Bhimappa Suresha, Govindarajulu Ramadoss, and Basavaraj Bharath. 2014. "Effect of short fiber reinforcement on mechanical properties of hybrid phenolic composites." *Journal of Materials* 2014:1–9.

Mohanty, Smita, and Sanjay K. Nayak. 2006. "Interfacial, dynamic mechanical, and thermal fiber reinforced behavior of MAPE treated sisal fiber reinforced HDPE composites." *Journal of Applied Polymer Science* 102 (4):3306–3315. https://doi.org/10.1002/app.24799.

Mohanty, Smita, Sushil K. Verma, and Sanjay K. Nayak. 2006. "Dynamic mechanical and thermal properties of MAPE treated jute/HDPE composites." *Composites Science and Technology* 66 (3–4):538–547.

Mylsamy, K., and I. Rajendran. 2011a. "Influence of alkali treatment and fibre length on mechanical properties of short Agave fibre reinforced epoxy composites." *Materials & Design* 32 (8):4629–4640.

Mylsamy, K., and I. Rajendran. 2011b. "The mechanical properties, deformation and thermo-mechanical properties of alkali treated and untreated Agave continuous fibre reinforced epoxy composites." *Materials & Design* 32 (5):3076–3084.

Nagaraja Ganesh, B., and B. Rekha. 2020. "Intrinsic cellulosic fiber architecture and their effect on the mechanical properties of hybrid composites." *Archives of Civil and Mechanical Engineering* 20 (4):1–12.

Nashif, Ahid D., David I. G. Jones, and John P. Henderson. 1985. *Vibration Damping.* John Wiley & Sons Inc.

Nelson, D. J., and J. W. Hancock. 1978. "Interfacial slip and damping in fibre reinforced composites." *Journal of Materials Science* 13 (11):2429–2440.

Ornaghi, Heitor Luiz, Alexandre Sonaglio Bolner, Rudinei Fiorio, Ademir Jose Zattera, and Sandro Campos Amico. 2010. "Mechanical and dynamic mechanical analysis of hybrid composites molded by resin transfer molding." *Journal of Applied Polymer Science* 118 (2):887–896.

Paul, Sherely Annie, Christoph Sinturel, Kuruvilla Joseph, G. D. Mathew, Laly A. Pothan, and Sabu Thomas. 2010. "Dynamic mechanical analysis of novel composites from commingled polypropylene fiber and banana fiber." *Polymer Engineering & Science* 50 (2):384–395.

Petrone, G., S. Rao, S. De Rosa, B. R. Mace, and D. Bhattacharyya. 2012. "Vibration characteristics of fiber reinforced honeycomb panels: experimental study." International Conference on Noise and Vibration Engineering, Leuven, Belgium.

Phillips, Steven, and Larry Lessard. 2012. "Application of natural fiber composites to musical instrument top plates." *Journal of Composite Materials* 46 (2):145–154.

Pil, Lut, Farida Bensadoun, Julie Pariset, and Ignaas Verpoest. 2016. "Why are designers fascinated by flax and hemp fibre composites?" *Composites Part A: Applied Science and Manufacturing* 83:193–205. https://doi.org/10.1016/j.compositesa.2015.11.004.

Pilla, Srikanth. 2011. "Engineering applications of bioplastics and biocomposites—An overview." In *Handbook of Bioplastics and Biocomposites Engineering Applications*, edited by Srikanth Pilla, 1–15. John Wiley & Sons Inc.

Pothan, Laly A., Sabu Thomas, and G. Groeninckx. 2006. "The role of fibre/matrix interactions on the dynamic mechanical properties of chemically modified banana fibre/polyester composites." *Composites Part A: Applied Science and Manufacturing* 37 (9):1260–1269. https://doi.org/10.1016/j.compositesa.2005.09.001.

Pothan, Laly A., Zachariah Oommen, and Sabu Thomas. 2003. "Dynamic mechanical analysis of banana fiber reinforced polyester composites." *Composites Science and Technology* 63 (2):283–293. https://doi.org/10.1016/S0266-3538(02)00254-3.

Povolo, F., and S. N. Goyanes. 1996. "Amplitude dependent damping in vinyl polymers." *Le Journal de Physique IV* 6 (C8):579–582.

Raditoiu, Alina, Valentin Raditoiu, Raluca Gabor, Cristian-Andi Nicolae, Marius Ghiurea, and Gheorghe Hubca. 2013. "Thermo-mechanical behavior of cellulosic textiles coated with some colored silica hybrids." *University Politehnica of Bucharest Scientific Bulletin, Series A: Applied Mathematics and Physics* 75 (4):285–292.

Rahman, Md Zillur. 2017. "Static and dynamic characterisation of flax fibre-reinforced polypropylene composites." ResearchSpace@Auckland.

Rahman, Md Zillur. 2020. "Mechanical and damping performances of flax fibre composites–a review." *Composites Part C: Open Access*:100081.

Rahman, Md Zillur, Brian Richard Mace, and Krishnan Jayaraman. 2016. "Vibration damping of natural fibre-reinforced composite materials." *17th European Conference on Composite Material*, Munich.

Rahman, Md Zillur, Krishnan Jayaraman, and Brian Richard Mace. 2017. "Vibration damping of flax fibre-reinforced polypropylene composites." *Fibers and Polymers* 18 (11):2187–2195.

Rahman, Md Zillur, Krishnan Jayaraman, and Brian Richard Mace. 2018. "Influence of damping on the bending and twisting modes of flax fibre-reinforced polypropylene composite." *Fibers and Polymers* 19 (2):375–382.

Rainieri, Carlo, and Giovanni Fabbrocino. 2014. *Operational Modal Analysis of Civil Engineering Structures.* Springer.

Rajesh, M., and Jeyaraj Pitchaimani. 2017. "Mechanical characterization of natural fiber intra-ply fabric polymer composites: influence of chemical modifications." *Journal of Reinforced Plastics and Composites* 36 (22):1651–1664.

Rajini, N., J. T. Winowlin Jappes, S. Rajakarunakaran, and P. Jeyaraj. 2013. "Dynamic mechanical analysis and free vibration behavior in chemical modifications of coconut sheath/nano-clay reinforced hybrid polyester composite." *Journal of Composite Materials* 47 (24):3105–3121.

Ramesh, M., K. Palanikumar, and K. Hemachandra Reddy. 2017. "Plant fibre based biocomposites: sustainable and renewable green materials." *Renewable and Sustainable Energy Reviews* 79:558–584.

Ray, Dipa, B. K. Sarkar, S. Das, and A. K. Rana. 2002. "Dynamic mechanical and thermal analysis of vinylester-resin-matrix composites reinforced with untreated and alkali-treated jute fibres." *Composites Science and Technology* 62 (7–8):911–917.

Richards, E. J., and A. Lenzi. 1984. "On the prediction of impact noise. VII: The structural damping of machinery." *Journal of Sound and Vibration* 97 (4):549–586.

Romanzini, Daiane, Alessandra Lavoratti, Heitor L. Ornaghi, Sandro C. Amico, and Ademir J. Zattera. 2013. "Influence of fiber content on the mechanical and dynamic mechanical properties of glass/ramie polymer composites." *Materials & Design* 47: 9–15.

Rueppel, Marvin, Julien Rion, Clemens Dransfeld, Cyril Fischer, and Kunal Masania. 2017. "Damping of carbon fibre and flax fibre angle-ply composite laminates." *Composites Science and Technology* 146:1–9.

Saba, Naheed, Mohammad Jawaid, Othman Y. Alothman, and M. T. Paridah. 2016. "A review on dynamic mechanical properties of natural fibre reinforced polymer composites." *Construction and Building Materials* 106:149–159.

Saha, A. K., S. Das, D. Bhatta, and B. C. Mitra. 1999. "Study of jute fiber reinforced polyester composites by dynamic mechanical analysis." *Journal of Applied Polymer Science* 71 (9):1505–1513.

Saw, Sudhir Kumar, Gautam Sarkhel, and Arup Choudhury. 2012. "Preparation and characterization of chemically modified jute–coir hybrid fiber reinforced epoxy novolac composites." *Journal of Applied Polymer Science* 125 (4):3038–3049.

Shinoj, S., R. Visvanathan, S. Panigrahi, and N. Varadharaju. 2011. "Dynamic mechanical properties of oil palm fibre (OPF)-linear low density polyethylene (LLDPE) biocomposites and study of fibre–matrix interactions." *Biosystems Engineering* 109 (2):99–107.

Shukor, Faseha, Azman Hassan, Mahbub Hasan, Md Saiful Islam, and Munirah Mokhtar. 2014. "PLA/kenaf/APP biocomposites: effect of alkali treatment and ammonium polyphosphate (APP) on dynamic mechanical and morphological properties." *Polymer-Plastics Technology and Engineering* 53 (8):760–766.

Shumao, Li, Ren Jie, Yuan Hua, Yu Tao, and Yuan Weizhong. 2010. "Influence of ammonium polyphosphate on the flame retardancy and mechanical properties of ramie fiber-reinforced poly (lactic acid) biocomposites." *Polymer International* 59 (2):242–248.

Silverajah, V. S., Nor Azowa Ibrahim, Norhazlin Zainuddin, Wan Md Zin Wan Yunus, and Hazimah Abu Hassan. 2012. "Mechanical, thermal and morphological properties of poly (lactic acid)/epoxidized palm olein blend." *Molecules* 17 (10):11729–11747.

Singhal, Priya, and S. K. Tiwari. 2014. "Effect of various chemical treatments on the damping property of jute fibre reinforced composite." *International Journal of Advanced Mechanical Engineering* 4 (4):413–424.

Song, M., D. J. Hourston, and F.-U. Schafer. 2001. "Correlation between mechanical damping and interphase content in interpenetrating polymer networks." *Journal of Applied Polymer Science* 81 (10):2439–2442.

Sperling, L. H. 1990. "Sound and vibration damping with polymers." In *Sound and Vibration Damping with Polymers*, edited by Robert D. Corsaro and L. H. Sperling, 5–22. American Chemical Society.

Sumesh, K. R., and K. Kanthavel. 2020. "Synergy of fiber content, Al_2O_3 nanopowder, NaOH treatment and compression pressure on free vibration and damping behavior of natural hybrid-based epoxy composites." *Polymer Bulletin* 77 (3):1581–1604.

Sun, C. T., J. K. Wu, and R. F. Gibson. 1987. "Prediction of material damping of laminated polymer matrix composites." *Journal of Materials Science* 22 (3):1006–1012.

TA Instruments. 2019. "Dynamic Mechanical Analysis", accessed 20 March. https://tainstruments.com/wp-content/uploads/Boston-DMA-Training-2019.pdf.

Vazquez, A., M. Ambrustolo, S. M. Moschiar, M. M. Reboredo, and J. F. Gérard. 1998. "Interphase modification in unidirectional glass-fiber epoxy composites." *Composites Science and Technology* 58 (3–4):549–558.

Vilay, V., M. Mariatti, R. Mat Taib, and Mitsugu Todo. 2008. "Effect of fiber surface treatment and fiber loading on the properties of bagasse fiber–reinforced unsaturated polyester composites." *Composites Science and Technology* 68 (3–4):631–638.

Wang, Jin-Ting, Feng Jin, and Chu-Han Zhang. 2012. "Estimation error of the half-power bandwidth method in identifying damping for multi-DOF systems." *Soil Dynamics and Earthquake Engineering* 39:138–142.

Yan, Libo. 2012. "Effect of alkali treatment on vibration characteristics and mechanical properties of natural fabric reinforced composites." *Journal of Reinforced Plastics and Composites* 31 (13):887–896.

Zhao, Yong-Qing, Hoi-Yan Cheung, Kin-Tak Lau, Cai-Ling Xu, Dan-Dan Zhao, and Hu-Lin Li. 2010. "Silkworm silk/poly (lactic acid) biocomposites: dynamic mechanical, thermal and biodegradable properties." *Polymer Degradation and Stability* 95 (10):1978–1987.

9 Influence of Compatibilizer on Free Vibration and Damping Behavior of Polymer Biocomposites

Md Mahmudul Haque Milu
Jashore University of Science and Technology (JUST)

Anika Anjum
Bangladesh University of Engineering
and Technology (BUET)

Md Asadur Rahman and Md Enamul Hoque
Military Institute of Science and Technology (MIST)

CONTENTS

DOI: 10.1201/9781003173625-10

9.1 COMPATIBILIZERS

Compatibilizer is a general terminology used for combining two polymers in a physical mixture (blend). When some polymer materials are added to an immiscible blend, the new bonded material stabilizes its morphology. Compatibilizers are used to sustain interfacial adhesion between pairs of polymers that do not form a homogenous mixture when mixed. Compatibilizing or polymer blending is an appropriate path for the development of different belongings of polymeric materials. Properties of blended polymers are improved than the existing individual polymer after compatibilization [1]. This unification of different polymers usually occurs in different processing machines, for example, screw extruders [2]. The individual polymer in greater concentration forms an incessant phase. On the other hand, individual polymers with a lower concentration will disperse in the constant matrix. It is notified in Ref. [3] that the intermolecular linkage between the incessant and the dispersed phase is very fragile. As a result, the blend's mechanical performance is bad.

There are two types of compatibilizers. The purpose of the compatibilizer is both to enhance mechanical properties and it frequently yields properties that are not typically achievable in their single pure component. One type of compatibilizing agent is named graft or copolymers and the other one is reactive compatibilization. The copolymers are composed of two immiscible materials. The reducing dimension of phase-separated elements in the polymer blend is the reason behind the increased stability [4]. Another form of compatibilization is reactive compatibilization, which involves making copolymers in a solution to make immiscible polymer blends. Reactive compatibilization follows such strategies as trans reaction, ionically bonded structure, reactive formation of graft block, and mechanochemical blending that can result recombination and copolymerization [3].

9.2 IMPACTS OF COMPATIBILIZERS ON DISTINCTIVE MECHANICAL CHARACTERISTICS OF POLYMER BIOCOMPOSITES

Compatibilizers are used to enhance the mechanical characteristics of different polymer blends in polymer chemistry. The compatibilization process is used to reduce the interfacial tension of polymer particles, it can facilitate chain dispersion, can enhance between two polymeric phases, stabilizes morphologies against severe melt processing conditions, and promotes break-up of droplets during processing [5]. Besides these, viscoelasticity and glass transition temperature are two important factors that have an impact on the vibration and restraining characteristics of polymers. Chemical composition, storage stability is also affected by the hybrid compatibilizers.

9.3 IMPACT OF COMPATIBILIZER ON INTERFACIAL TENSION AND DISPERSION OF POLYMER BIOCOMPOSITES

After mixing two immiscible polymers, important interactions occur between the components' molecular fragments, resulting in bulky interfacial rigidity in the melt with little scattering of all the constituents collectively. The modification of the edge will result in anatomical transition, which relates to a reduction of coalescence and shrinkage in interfacial tension [6,7]. As a result, the modified interface is accompanied by a reduction of droplet dimension as well as a thinner droplet size circulation of the disseminated phase. The reducing scattered phase size following the extension of nanostructured fabric is evidence of the nanofillers' compatibilizing effect. In fact, scattered phase size decrease within inhomogeneous polymer mixes holding nanofillers is caused by a complicated procedure that occurs during compounding where the break-up mechanisms are restricted to consolidating [7]. Therefore, various issues are involved, particularly local flow parameters (residence time, temperature, and shear rate). Hence, the following factors leading to morphological changes are discussed in Refs. [8–10]:

  i. Interfacial energy reduction.
  ii. Hindrance of amalgamation due to the existence of a hard-compact blockade throughout the minor phase droplet.
  iii. The viscoelastic ratio varies when fillers are present in one of the two stages.
  iv. Inhibition of the droplet or matrix motion due to the formation of physical linkage of elements.
  v. The interaction of macromolecules with the particle surface during adsorption.
  vi. The enhancement of interfacial adhesion, though it is hard to evaluate directly, there is little proof in the literature [11].

Blends of LDPE/PA12 (low-density polyethylene – LDPE, polyamides – PA) containing 1–6 wt.% of montmorillonite (MMT) were prepared by Huitric et al. [12]. The droplet size was reduced. To investigate the reason behind the reduction of droplet size several calculations had been done. When polyethylene (PE) was the matrix (80PE/20PA), the MMT platelets were located at the boundary, and the dimension of the scattered stage was concentrated owing to coalescence hindrance for the solid-like blockade impact of MMT. On the other hand, the platelets were detected both in the PA matrix and at the interface when PA became the matrix (20PE/80PA). Here, the coalescence barrier effect and the break-up mechanism are the reason behind the size reduction.

Styrene–ethylene–butylene–styrene (SEBS) block copolymer compatibilized PE/PS blends in the range of 7% for blend configurations of 20/80 to 80/20. The Charpy impact strength is amplified by a factor of 4 in the best case according to the study reported in Ref. [13]. Morphological features showed a reduction in the dimension of the discrete elements and enhanced interfacial linkage amid both stages [13].

9.4 POLYMER BIOCOMPOSITES

A biocomposite is a fused material made up of a matrix and natural fiber reinforcement. Polymer biocomposites consist of two different polymers that show unique properties when bound together. The matrix of the biocomposite material is responsible for holding the structure of natural fibers from environmental degradation and it enhances different mechanical properties. Biofibers are derived from biological origins. Biofibers are the principal components of biocomposites. Some examples of biofiber origin are crops/fruits (cotton, flax, hemp, kenaf, jute, bamboo, and pineapple), recycled wood, waste paper, and regenerated cellulose fiber. [14–21].

9.4.1 VIBRATION AND DAMPING BEHAVIOR OF GLASS FIBRE REINFORCED POLYMER (GFRP) AND BASSALT FIBRE REINFORCED POLYMER (BFRP) LAMINATED BIOCOMPOSITES

Fiber-reinforced plastic is a complex material that is made of a polymer matrix covered with natural fibers. The fibers are usually carbon (in carbon fiber-reinforced polymer), glass (in fiberglass), aramid, or basalt [22,23]. Wood, paper, or asbestos have rarely been used in this case. These composites have noble static and dynamic properties and their weight is very low in environment. These composite materials are used commonly in structures because they have good strength, stiffness, and high damping characteristics.

To design the biocomposite materials, static and dynamic properties, storage modulus, and damping play a very significant role. Damping is related to the energy dissipation of compounds and storage modulus is associated with the rigidity of compounds [24]. It is proved that the damping effect can be improved through toughening of the polymer matrix. Accumulation of some nanoparticles with natural fibers enhances vibration, mechanical, and thermal characteristics of architectural composites [25]. The noteworthy reason for composite damping is viscoelastic feature of matrix found in the study reported in Ref. [26]. A collective numerical and laboratory-based investigation on the free vibration of composite GFRP plates has been conducted in the study reported in Refs. [27,28] and executed unrestricted vibration analysis of the coated fused beam with numerous periphery environments [29]. They also inspected the damping behavior compounds with several arrays. Woven textile-protected fused constituents are nonmiscible and huge resin-rich zones are shaped in the boundary from the interlocked warps and fills. The difference of damping is superior than the stiffness in the arrangement of high-performance fiber–polymer matrices [30].

The natural frequency of GFRP composites increases when treated with NaOH. The restraining feature of basalt/epoxy/Carbon nanotube (CNT) fusion was investigated at oceanic water absorption condition in [31] and showed that the vibration damping is amplified to 50%. To acquire improved vibration, a small number of procedures have been conducted in fiber clusters, especially for BFRP composites.

9.4.2 Vibration and Damping Behavior of Biocomposites Reinforced with Vegetable Fibers

When biopolymer matrix is used to bind reinforcement, this is called fiber-reinforced biocomposites. Natural fiber-reinforced composites (NFRCs) or biocomposites are widely used for decreasing the hazardous effects of synthetic polymer matrix on the environment. NFRC is getting extra consideration nowadays in the applications of automobile, aviation, packaging, and marine industries due to increasing strength and rigidity of the natural fiber in polymer material [32]. This compatibilization affects on damping nature of the biocomposites. Fiber-reinforced compound elements with higher stiffness values and strong damping characteristics are appropriate for an extensive range of anti-vibration operations [33]. The automobile and aerospace industries are working to improve vibration and noise reduction in secondary structural apparatuses to help passengers and drivers relax. Because of their inherent damping property, NFRC can provide noteworthy benefits over artificial fiber composites. Munde et al. reported the vibration studies in Ref. [34], by applying numerical and experimental approaches on natural fiber-reinforced thermoplastic and thermoset polymer matrix mixtures encouraging the scientists and manufacturers to investigate the vibrational response of NFRC.

Dynamic mechanical analysis can be used to measure the damping response of composites for low frequency and low strain up to 100 Hz only; however, experimental modal analysis (EMA) can be used for high frequency and amplitude ranging to 1,000 Hz [34]. The EMA fixes associated modal damping and the natural frequency as vital parameters affecting the composite structure's dynamic nature. Figure 9.1 depicts the diagram of a vibration assessment system. A fixed free boundary condition holds one end of the specimen and the soft tip impact hammer (PCB Piezoelectric

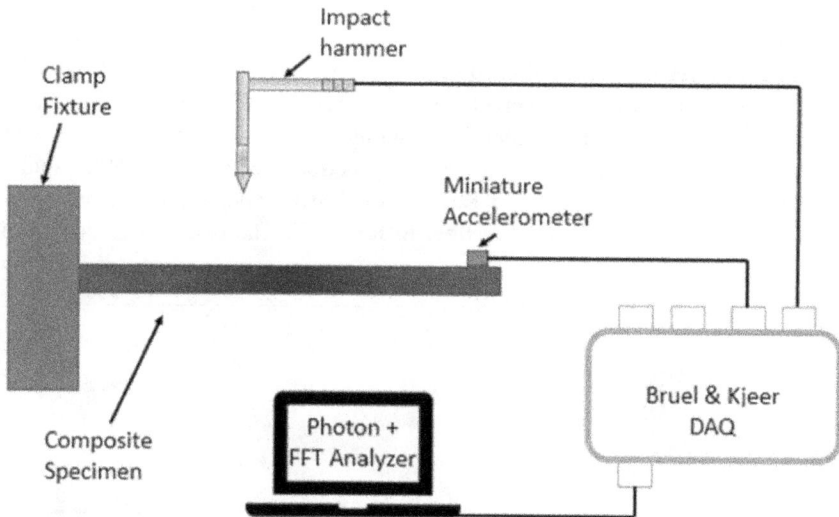

FIGURE 9.1 Experimental modal analysis (EMA) assessment system [34].

model 086C03) carried out excitation. Displacement of a signal is acquired by the lightweight scaled-down accelerometer (commonly, Bruel and Kjaer model 352B10). A fast Fourier transform is used in the data recording system (Bruel and Kjaer with Photon software) to obtain time-domain signal and frequency response signals. Separate stages are used to extract damping and natural frequency from the received signals. In Figure 9.2, the crest in the frequency response function (FRF) curve is shown which represents the natural frequency of compound beam.

The damping ratio of mixtures is achieved by three methods are as follows: (a) logarithmic decrement-based method, (b) circle fit method, and (c) half-power bandwidth. The logarithmic decrement-based method measures the damping by the relations (9.1) and (9.2) that equate acceleration amplitude with respect to time records.

$$\delta = \ln\left(\frac{Ao}{An}\right) \tag{9.1}$$

$$\xi = \frac{\delta}{\sqrt{4\pi^2 + \delta^2}} \tag{9.2}$$

The damping ratio is calculated using Equation 9.3:

$$\xi = \frac{\Delta\omega}{\omega_n} \tag{9.3}$$

Here, ξ = damping ratio, ω = bandwidth, and ω_n = natural frequency.

Senthil et al. [35] explored the impact of fiber weight and length rate on free vibration properties of PE compounds made of tiny sisal and small banana fibers. The natural frequency and fiber loading were found to be linearly interrelated in their study. Meantime, sisal and banana have the most noteworthy natural frequencies at smallest fiber sizes when 50% fiber stacking is used. The exploratory comes in [36] which supported the prior discoveries. Banana and cotton fiber composites are tested against dynamic loading at different fiber loadings in this experiment. With 40 wt.% fiber loading, the maximum natural frequency is documented. Bulut et al. studied the damping activity of 10-layered Kevlar (K)/glass(G) epoxy hybrid shields with various relative volume fractions of fiber in Ref. [37]. The researchers discovered

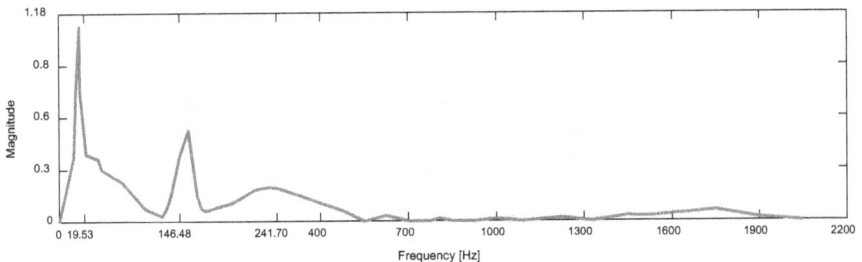

FIGURE 9.2 A typical FRF curve [34].

K10 composites have the highest NF and damping compared to G10 complexes, while hybrid composites are in the middle of the two. According to the findings of the reviewed reports, the lower the fiber length and the higher the loading is the higher the polymer natural frequency and the greater the chance of good damping. Arulmurugan and Venkateshwaran [38] created nanoclay-filled jute/polyester composites by changing the relative weight ratios of cross fillers (1%, 3%, 5%, and 7%) and fiber (5%, 10%, 15%, 20%, 25%, and 30%) to inspect their impacts on vibrational characteristics. As the filler percentage is increased up to 5%, the damping factor and natural frequencies are increased linearly. The impact of nanoclay expansion on vibration characteristics examined [39] for interlaced banana/jute half breed polyester and in the study reported in Ref. [40] for interlaced coconut sheath/PE composite also backs up the previous discovery. Natural frequency was highest with banana/jute composites up to 2% nanoclay addition, whereas it was highest with coconut sheath composites up to 3%.

9.4.3 VIBRATION AND DAMPING BEHAVIOR OF BIOCOMPOSITES REINFORCED WITH NONWOVEN MATS

Composite ingredients are usually used for better dynamic and static appearances. Sound vibration and damping absorption are two acute dynamic factors. Composite structures might improve vibration damping because composite materials can dissipate energy numerous orders of magnitude faster than conventional engineering materials [41–44], which is beneficial for components subjected to dynamic loading.

Natural fiber composites have greater acoustic and vibration restraining properties than artificial fiber composites [45,46]. When flax filaments were used as reinforcement in an epoxy network composite, the acoustic retention factor of the flax/epoxy compound was nearly 20% advanced than glass fiber composite at both the high and low frequency levels. This composite had a vibration-damping factor that was roughly 50% greater than the glass/epoxy composite [47]. Cotton filaments have been viably used for sound absorption by blending with other sorts of fibers and the diverse mixing proportions [48,49]. Natural fibers and polylactic acid (PLA) fibers were combined, laid up, carded, and needle-punched to nonwoven mats to explore the consequence of reinforcement fiber on the vibration-damping performance and sound absorption of biocomposites. Three categories of natural fibers: bamboo, cotton, and flax fibers were used for the production of composite laminate. These fiber biocomposites were subjected to vibration damping, three-point bending, and acoustic measurements.

In Ref. [50], the vibrational damping behavior of various samples was discovered. It is shown in Table 9.1 that cotton/PLA specimens have lower resonant frequencies than other composite material systems.

The structural damping ratios for all composite shields were calculated by means of 3 dB/half-power bandwidth technique in this analysis, which is displayed in Figure 9.3. It is worth mentioning that the damping characteristics of composite laminates are affected by both intrinsic parameters (fiber materials, viscoelastic nature of matrix, as well as fiber/matrix interfacial bonding) and extrinsic factors (applied frequency). The ratio of damping relates with the structure's ability to reduce

TABLE 9.1

First Three (Mode 3, Mode 2, and Mode 1) Resonant Frequencies of Natural Fiber Composite Laminates [50]

Specimen	Mode 3 (Hz)	Mode 2 (Hz)	Mode 1 (Hz)
Cotton/PLA	764	372	96
Bamboo/PLA	1299	646	228
Cotton/Bamboo/PLA	827	423	143
PLA/flax	1022	501	238
Hemp/Kenaf/PP	1117	556	131

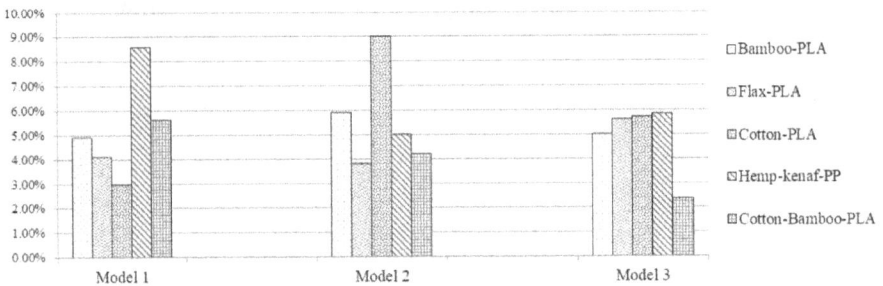

FIGURE 9.3 Damping ratios in first three modes of composite material [50].

objectionable vibrational and dynamic loads, subsequently, targeting to improve the functional life of a complex structure, it is preferable to use the design with the uppermost damping ratio.

Wave numbers were used to assess the acoustic features of PLA composite shields armored with cotton, flax, and bamboo. The velocity of sound in the air was schemed alongside the investigational data to determine the concurrent frequency for each material system. This frequency was then calculated at the intersection of the wave number-frequency and the velocity of sound. Greater coincidence frequency material systems produce more noise and have better acoustic performance.

Environment-friendly natural fiber composites have high efficiency for usage in acoustic absorption as a replacement for mock fiber/polymer composites. It also has ecological benefits. Cotton, flax, and bamboo fibers were combined with PLA to create nonwoven mats, which were then pressed into plastic panels at a high temperature. When it comes to three-point twisting, the fabricated composites were paralleled to a self-propelled profitmaking natural fiber composite panel; these three-point bendings are Charpy effect, damping efficiency, and sound absorption. The bamboo/PLA composite had 2.5 Gpa bending stiffness, which was substantially higher than other composites, demonstrating the high mechanical properties of bamboo fibers. Despite the usage of a brittle PLA matrix, the cotton/PLA composite demonstrated upper impact toughness. The acoustic output of the cotton/bamboo/PLA composite laminate was found to be the strongest, which is correlated to the excellence of natural fibers, bending stiffness, and density of the composites.

9.4.4 VIBRATION AND DAMPING BEHAVIOR OF BIOCOMPOSITE SANDWICHED WITH AUXETIC CORE

Biocomposite materials are very ecological, biodegradable, and environmentally impactful as mentioned earlier in previous sections. These materials can be used widely because of their recyclability feature. To improve the composite performance, Monti et al. [51] investigated the dynamic characteristics of a sophisticated eco-sandwich construction with skin that is made of flax fiber composite and the core is made of Balsa wood. Dynamic behavior of sandwich structure with a pyramidal truss core was studied in Ref. [52], and it was discovered that the insertion of the viscoelastic layers enhances the dynamic properties. A new method of sandwich structures with a honeycomb core that re-entrant was sketched and manufactured as reported in Ref. [53] using 3D-printed technology. Trials of free vibration were directed to compute the loss factor coefficient and the natural flexural frequency. The effect of flax fiber coating and relative densities of core on sandwich composite energy consumption was also investigated in this research.

Sandwich composite was formed using PLA and PLA flax fibers (PFFs). The dimension of the printed specimen is 270 mm × 25 mm. The auxetic cell's relative density may be computed using:

$$\frac{\rho}{\rho_s} = \frac{t\left(\frac{h}{l}+2\right)}{2\cos\theta\left(\frac{h}{l}+\sin\theta\right)}$$

(9.4)

Here, ρ = material's density; ρ_s = auxetic cell's density; l = original dimension of the inclined cell walls; θ = initial angle; b = thickness of the specimen; and t = thickness of the cell wall. The parameters are shown in Figure 9.4 (Table 9.2).

The dynamic appearances of sandwich fused configuration were experimented by the arrangement displayed in Figure 9.4. Loss factor for damping was determined by:

$$\eta_i = \frac{\Delta f_i}{f_i} = \frac{f_2 - f_1}{f_i}$$

(9.5)

Here, f_i = resonant frequency and η_i = damping loss factor.

The Young modulus of auxetic core and complex skin was calculated for each flexural mode by:

$$E = \frac{12\rho l^4 f_n^2}{e^2 c_n^2}$$

(9.6)

Here, ρ = composite fabric density; l = beam free length; e = beam thickness; f_n = reverberation rate of recurrence of nth flexural mode; and c_n = coefficient of nth mode of clamped free bar.

For each flexural mode, the equivalent stiffness of the sandwich beam is measured by:

(a)

(b)   (c)

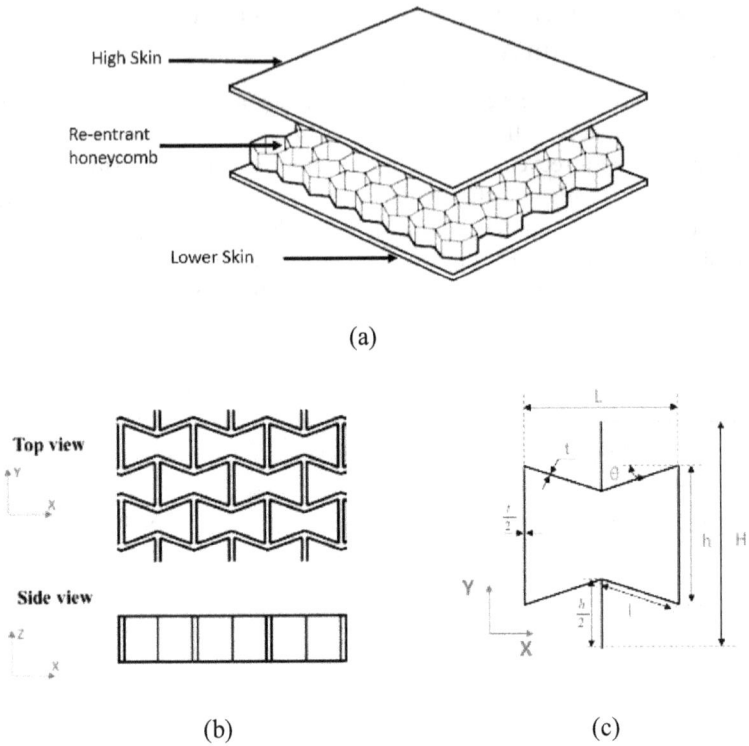

FIGURE 9.4 (a) Sandwich assembly with honeycomb cores that are re-entrant; (b) Sights of the auxetic core from upper and lateral position; and (c) auxetic structure single entity [53].

TABLE 9.2
Auxetic Core Design Specifications [53]

Cell No.	l (mm)	h (mm)	θ (°)	t (mm)	b (mm)	$\dfrac{\rho}{\rho_s}$ (%)
2 cells	6.7	8.5	-20.0	0.6	5.0	16.7
3 cells	4.4	5.7	-20.0	0.6	5.0	25.1

$$(EI)_{eq} = \left(\frac{m}{L}\right)(2\pi f)_n^2 \left(\frac{l^2}{(\beta_n l)^2}\right)^2 \tag{9.7}$$

Here, L=beam length; m=beam's mass; and $\beta_n l$=coefficient of n-th mode of clamped free beam. Vibration behavior of PLA and PFF composite skin is displayed in Figure 9.5, which represents loss factor and the Young modulus both as a function of frequency.

From the above graphical representations, we can see that as the frequency rises, the damping loss factor rises as well. The loss factor steadily upgrades in the range of

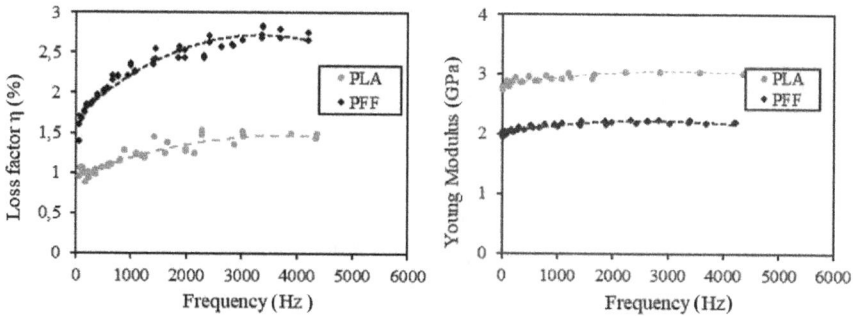

FIGURE 9.5 Advancement of the energetic characteristics (loss factor and Young's modulus) with the recurrence of the skins [53].

0–2,000 Hz. The damping figure is gradually increasing to a quasi-asymptotic value at this point. The damping factors of PFF fortification are significantly greater than PLA loss factor. As a result, the PFF bar is less hard than the PLA bar.

9.4.5 Vibration and Damping Behavior of Biocomposites Reinforced with Flax Fiber

Essassi et al. [53] investigated the damping activity of sandwich beams with PLA and PFF as superficial material and re-entrant honeycomb as core. For the two relative core densities, the flax fiber impact as composite brace was investigated. The identical stiffness and the damping coefficient of such columns were found out. Figure 9.6 depicts the influence of flax fiber on the sandwich's complex properties. Sandwich beams reinforced with flax fiber have a higher damping coefficient than those without, regardless of core relative density.

9.5 INVESTIGATING FREE VIBRATION AND DAMPING BEHAVIOR OF BIOCOMPOSITES VIA FINITE COMPONENT AND EXPLORATORY SYSTEMS

The vibration of biocomposites (natural material-based composites) due to compatibilization has been discussed in a very few works done recently. James et al. [54] investigated vibration damping along with the acoustic properties of biocomposites and found that normal strands can contribute a crucial part in giving an arrangement to the outflow of clamor predominantly in sandwich structures. With the help of a finite element model both experimentally and numerically, Garcia et al. [55] investigated the pulsation activity of GFRP composites fortified with nanofibers of nylon, they discovered the stiffness, fundamental frequencies, and damping of nanointerleaved compound coats of clamped-clamped beams. Comparable assessments were performed for examining the impact of pivotal squeezing stack on buckling. It was discovered that increasing the number of layers in a composite can improve its quality. As a result of the ongoing research and development on biocomposites, often found to be essential to examine the vibration characteristics of these NFCs.

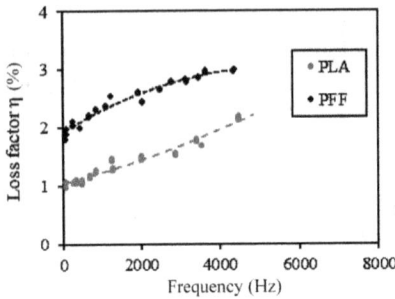

(a)

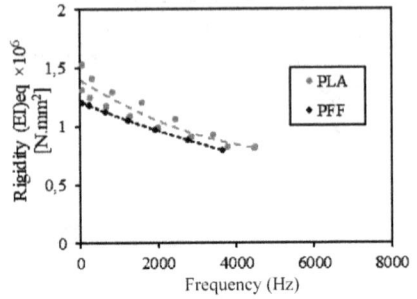

(b)

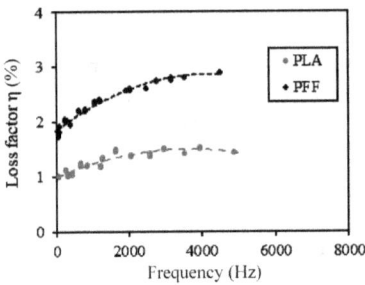

(c)

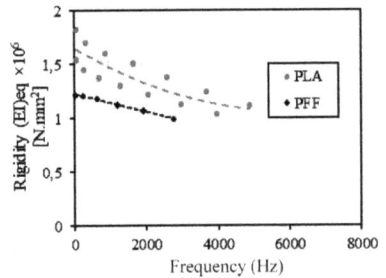

(d)

FIGURE 9.6   Dynamic properties of the sandwich as a function of frequency with (a) and (b) 16.7%; (c) and (d) 25.1% core relative density [53].

Ansys program was used to perform finite element analysis (FEA) [56]. FEA is a commonly used method for analyzing and solving a wide variety of technical and manufacturing problems. Ansys is finite element software that is one of the foremost requesting and widely used programs for limited component examination.

By experimenting with the Ansys software, several numerical findings were acquired for the shaking activities of Aloe Vera fiber-reinforced composite pillars. Concurring to the parametric studies on the Aloe Vera-fortified glass-epoxy combination, it is treated that the current analysis is composed of five layers, with Aloe Vera epoxy as the central layer. Aloe Vera fibers are oriented at 0°. Glass-epoxy layer is fixed correspondingly over and underneath the Aloe Vera epoxy. Figure 9.7 depicts the cover stacking arrangement. Figure 9.8 represents the composite pillar encompasses a cover heaping of [0°/90°/0°/90°/0°].

The hybridizing impact was examined through the analysis of the usual recurrence rate of the composite pillar by supplanting engineered fiber sheets. Three other materials were used to supplant the glass-epoxy layer: Kevlar epoxy, graphite epoxy, and boron epoxy. Here, the method for composite bar was inspected for an l/h proportion of 80 (Figure 9.9). It can be deduced from Figure 9.10 that Aloe Vera-fortified boron epoxy takes the most noteworthy common recurrence rate, while Aloe Vera-fortified glass-epoxy consumes the least common recurrence rate with Aloe Vera armored graphite and Aloe Vera strengthened Kevlar epoxy

| Glass-epoxy |
| Glass-epoxy |
| Aloe vera-epoxy |
| Glass-epoxy |
| Glass-epoxy |

FIGURE 9.7 Laminate stacking sequence [56].

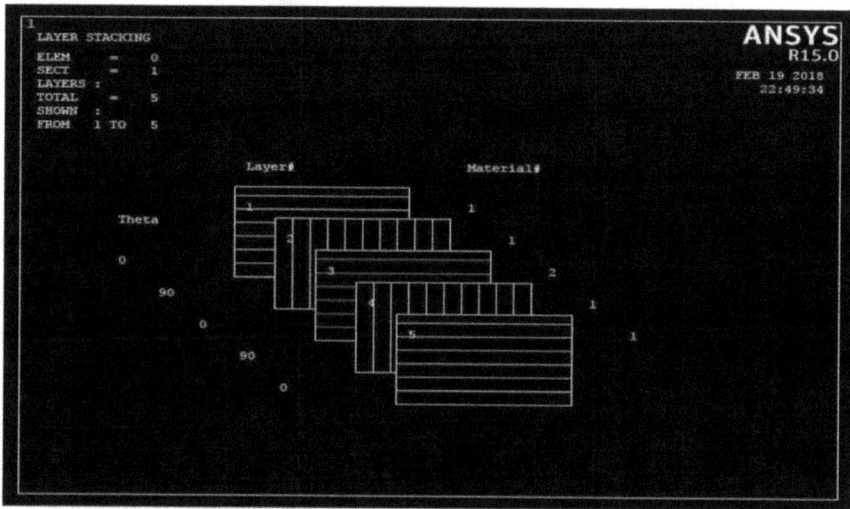

FIGURE 9.8 [0°/90°/0°/90°/0°] laminate sequence representation [56].

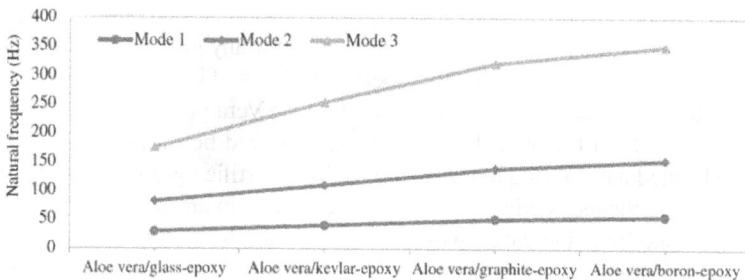

FIGURE 9.9 Relation of natural frequency versus combination of multiple materials [56].

existing between them [56]. Besides, a numerical comparison of the property – natural frequency is given in Table 9.3.

We can see from the above study that with the increment of number of layers, the normal frequency decreases and then saturates. Table 9.3 also shows that both

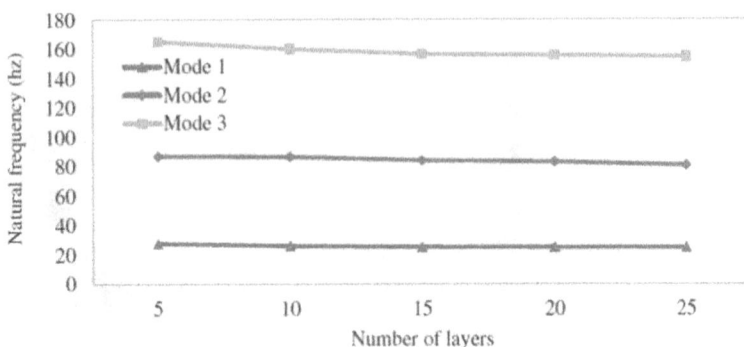

FIGURE 9.10   Relation between natural frequencies versus number of layers [56].

TABLE 9.3

Relationship between the Layer Number and Natural Frequency of Only Glass Epoxy and Aloe Vera Protected Glass Epoxy [56]

	Natural Frequency (Hz)	
Number of Layers	**Glass Epoxy**	**Aloe Vera Reinforced Glass Epoxy**
5	26.85	27.99
10	25.44	26.39
15	24.86	25.52
20	24.45	25.3
25	24.34	25.1

composites have comparable natural frequencies, implying that Aloe Vera fiber could be used to exchange artificial glass fiber within the biocomposite beam's core layer.

The above experimental analysis reveals that (a) for any given mode of frequency, the [0°/90°/0°/90°/0°] bundle has the uppermost and the [45°/45°/0°/45°/45°] bundle has the bottommost natural recurrence rate. (b) Aloe Vera protected glass-epoxy has the most reduced, on the other hand Aloe Vera covered boron epoxy that contains the most elevated natural frequency with Aloe Vera-fortified graphite–epoxy and the Aloe Vera strengthened Kevlar epoxy lying between them and (c) as the layers' number of composite sheet increases, natural frequency tends to diminish.

9.6 CONCLUSIONS

Immiscible blending of two different polymers enhances the tribological and mechanical properties of polymer biocomposites. Compatibilizing is used mainly to withstand interfacial linkage between the inhomogeneous polymers when mixed. Compatibilizers have lower interfacial stiffness, smoother chain dispersion, stabilizing melt processing, anti-vibration properties, etc. For this reason, researchers of manufacturing companies begin to use compatibilized polymers. Eventually, this chapter can be concluded as:

- The compatibilizers of polymer biocomposites have inferior vibrational and damping characteristics.
- Higher engagement of polymer fibers can upgrade the natural frequency of polymers.
- If any synthetic hybridization occurs, the natural frequency of the natural fibers degrades.
- Carbon in carbon fiber-reinforced polymers, glass in fiberglass, aramid, etc. are widely used in manufacturing structures for good strength and stiffness.
- The natural frequency of glass fiber-reinforced polymer composites enhances when it is treated with NaOH.
- NFRCs are hugely used in automobile and aerospace industries due to their inherent anti-vibrational properties. NFRCs have greater acoustic and vibrational characteristics than synthetic fiber composites.
- When bamboo, cotton, and flax fibers are mixed with PLA fibers; Bamboo/PLA composite has higher bending stiffness than others. Cotton/PLA has higher impact toughness and bamboo/cotton/PLA has the best acoustic properties.
- Sandwich beams reinforced with flax fiber have a higher damping coefficient than those without, regardless of core relative density.

REFERENCES

[1] R. Petrucci, L. Torre. (2017). *Modification of Polymer Properties.* pp. 23–46. William Andrew Applied Science Publisher. Retrieved from: https://www.sciencedirect.com/science/article/pii/B9780323443531000026.

[2] K. Cor, V. Duin, M. P. Christophe, J. Robert. (1998). Strategies for compatibilization of polymer blends. Vol. 23. Issue 4. pp. 707–757. *Progress in Polymer Science.* Retrieved from: https://orbi.uliege.be//bitstream/2268/4370/1/Koning_C_1998_Prog%20polym%20sc_%2023_4_707.pdf.

[3] V. Ambrogi, C. Carfagna, P. Cerruti, V. Marturano. (2017). *Modification of Polymer Properties.* pp. 87–108. William Andrew Applied Science Publisher. Retrieved from: https://www.sciencedirect.com/science/article/pii/B978032344353100004X.

[4] R. J. Roe. (1993). Use of Block Copolymer as Polymer Blend Compatibilizer. Retrieved from: https://apps.dtic.mil/dtic/tr/fulltext/u2/a260435.pdf.

[5] A. Ajji, L. A. Utracki. (1996). Interphase and compatibilization of polymer blends. Vol. 36. Issue 12. pp. 1574–1585. *Polymer Engineering and Science.* Retrieved from: https://onlinelibrary.wiley.com/doi/abs/10.1002/pen.10554.

[6] D. R. Paul, C. B. Bucknall. (2000). *Polymer Blends: Formulation and Performance.* Wiley. Retrieved from: https://www.wiley.com/en-us/Polymer+Blends%3A+Formulation+and+Performance%2C+Volumes+1+2%2C+Set-p-9780471248255.

[7] J. Vermant, S. Vandebril, C. Dewitte, P. Moldenaers. (2008). Particle-stabilized polymer blends. Vol. 47. pp. 835–839. *Rapid Communication.* Retrieved from: https://link.springer.com/article/10.1007/s00397-008-0285-0.

[8] S. Wu. (1987). Formation of dispersed phase in incompatible polymer interfacial and rheological effects. Vol. 27. pp. 335–343. *Polymer Engineering and Science.* Retrieved from: https://onlinelibrary.wiley.com/doi/abs/10.1002/pen.760270506

[9] G. Serpe, J. Jarrin, F. Dawans. (1990). Morphology-processing relationships in polyethylene-polyamide blends. Vol. 30. pp. 553–565. *Polymer Engineering Science*. Retrieved from: https://doi.org/10.1002/pen.760300908.

[10] F. Fenouillot, P. Cassagnau, J. C. Majesté. (2009). Uneven distribution of nanoparticles in immiscible fluids: Morphology development in polymer blends. Vol. 50. pp. 1333–1350. Retrieved from: https://hal.archives-ouvertes.fr/hal-00374439/.

[11] S. C. Agwuncha et al. (2016). *Design and Applications of Nanostructured Polymer Blends and Nanocomposite Systems: Immiscible Polymer Blends Stabilized with Nanophase*. Ch. 11. pp. 215–237. Retrieved from: https://doi.org/10.1016/B978-0-323-39408-6.00010-8.

[12] J. Huitrica, J. Ville, P. Médéric, M. Moan, T. Aubry. (2009). Rheological, morphological and structural properties of PE/PA/nanoclay ternary blends: Effect of clay weight fraction. Vol. 53. Issue 5. *Journal of Rheology*. Retrieved from: https://sor.scitation.org/doi/abs/10.1122/1.3153551

[13] T. Kallel, V. Massardier, M. Jaziri. (2002). Compatibilization of PE/PS and PE/PP blends. I. Effect of processing conditions and formulation. Vol. 90. Issue 9. pp. 2475–2484. *Journal of Applied Polymer Science*. Retrieved from: https://doi.org/10.1002/app.12873.

[14] Biocomposite. https://en.wikipedia.org/wiki/Biocomposite.

[15] K. F. Amin, S. A. Asrafuzzaman, Md Enamul Hoque. (2021). Bamboo/Bamboo Fiber Reinforced Concrete Composites and Their Applications in Modern Infrastructure. In: Jawaid M., Mavinkere Rangappa S., Siengchin S. (Eds), *Bamboo Fiber Composites. Composites Science and Technology*. Springer. Retrieved from: https://doi.org/10.1007/978-981-15-8489-3_15.

[16] A. B. Asha, A. Sharif, M. E. Hoque. (2017). Interface Interaction of Jute Fiber Reinforced PLA Biocomposites for Potential Applications. In: Jawaid M., Salit M., Alothman O. (Eds), *Green Biocomposites. Green Energy and Technology*. Springer. Retrieved from: https://doi.org/10.1007/978-3-319-49382-4_13

[17] M. R. Ketabchi, M. Khalid, C. T. Ratnam, S. Manickam, R. Walvekar, M. E. Hoque. (2016). Sonosynthesis of cellulose nanoparticles (CNP) from kenaf fiber: Effects of processing parameters. Vol. 17. Issue 9. pp. 1352–1358. *Fibers and Polymers*. Retrieved from: https://doi.org/10.1007/s12221-016-5813-4.

[18] F. M. Al-Oqla, S. M. Sapuan, T. Anwer, M. Jawaid, M. E. Hoque. (2015). eNatural fiber reinforced conductive polymer composites as functional materials: A review. Vol. 206, pp. 42–54. *Synthetic Metals*. Retrieved from: https://doi.org/10.1016/j.synthmet.2015.04.014.

[19] W. H. Haniffah, S. M. Sapuan, K. Abdan, M. Khalid, M. Hasan, M. E. Hoque. 2015. Kenaf fibre reinforced polypropylene composites: Effect of cyclic immersion on tensile properties. Vol. 2015. Article ID 872387, pp. 1–6. *International Journal of Polymer Science*. Retrieved from: https://doi.org/10.1155/2015/872387.

[20] M. Hasan, M. E. Hoque, S. S. Mir, N. Saba, S. M. Sapuan. (2015). Manufacturing of Coir Fiber Reinforced Polymer Composites by Hot Compression Technique. In: M. Jawaid, M. Sapuan Salit, Nukman Bin Yusoff, M. Enamul Hoque (Eds), *Manufacturing of Natural Fibre Reinforced Polymer Composites*. Springer-Verlag. Retrieved from: https://doi.org/10.1007/978-3-319-07944-8_15.

[21] M. Asim, K. Abdan, M. Jawaid, M. Nasir, Z. Dashtizadeh, M. R. Ishak, M. E. Hoque, Y. Deng. (2015). A review on pineapple leaves fibre and its composites. *International Journal of Polymer Science*. pp. 1–16. Retrieved from: https://doi.org/10.1155/2015/950567.

[22] M. C. Biswas, M. M. Lubna, Z. Mohammed, M. H. Ul Iqbal, M. E. Hoque. (2021). Graphene and Carbon Nanotube-Based Hybrid Nanocomposites: Preparation to Applications. In: Qaiss A. K., Bouhfid R., Jawaid M. (Eds), *Graphene and Nanoparticles Hybrid Nanocomposites. Composites Science and Technology*. Springer. Retrieved from: https://doi.org/10.1007/978-981-33-4988-9_3.

[23] K. Y. Tshai, A. B. Chai, I. Kong, M. E. Hoque, K. H. Tshai. (2014). Hybrid fibre poly-lactide acid composite with empty fruit bunch: Chopped glass strands. Vol. 2014. Article ID 987956, pp. 1–7. *Journal of Composites*. Retrieved from: https://doi.org/10.1155/2014/987956.

[24] Y. T. Wang, C. S. Wang, H. Y. Yin, L. L. Wang, H. F. Xie, R. S. Cheng. (2012). Carboxyl-terminated butadiene-acrylonitrile-toughened epoxy/carboxyl-modified carbon nanotube nanocomposites. Vol. 6. Issue 9. pp. 719–728. *eXPRESS Polymer Letters*. Retrieved from: https://doi.org/10.3144/expresspolymlett.2012.77.

[25] M. Biswal, S. Mohanty, S. K. Nayak. (2011). Mechanical, thermal and dynamic-mechanical behavior of banana fiber reinforced polypropylene nanocomposites. Vol. 32, Issue 8. pp. 1190–1201. Retrieved from: https://doi.org/10.1002/pc.21138.

[26] R. Chandraa, S. P. Singhb, K. Guptac. (2003). A study of damping in fiber-reinforced composites. Vol. 262. Issue 3. pp. 475–496. *Journal of Sound and Vibration*. Retrieved from: https://doi.org/10.1016/S0022-460X(03)00107-X.

[27] L. Sinha, D. Das, A. Nath, N. Shishir, K. Sahub. (2021). Experimental and numerical study on free vibration characteristics of laminated composite plate with/without cut-out. Vol. 256. *Composite Structure*. Retrieved from: https://doi.org/10.1016/j.compstruct.2020.113051.

[28] J. Alexander, H. A. Kumar, B. Augustine. (2020). Frequency response of composite laminates at various boundary conditions. pp. 11–15. *International Journal of Engineering Science Invention (IJESI)*. Retrieved from: http://www.ijesi.org/papers/AMAAS/C1004.pdf

[29] J. Tsai, Y. Chi. (2008). Effect of fiber array on damping behaviors of fiber composites. Vol. 39. pp. 1196–1204. *Composites: Part B*. Retrieved from: https://doi.org/10.1016/j.compositesb.2008.03.003.

[30] J. Alexander, B. S. M. Augustine. (2015). Free vibration and damping characteristics of GFRP and BFRP laminated composites at various boundary conditions. Vol. 8. Issue 12. *Indian Journal of Science and Technology*. Retrieved from: https://doi.org/10.17485/ijst/2015/v8i12/54208.

[31] M. M. Lubna, Z. Mohammed, M. C. Biswas, M. E. Hoque. (2021). Fiber-Reinforced Polymer Composites in Aviation. In: Iulian Pruncu, C., Gürgen, S., Hoque, M. E. (Eds), *Fiber-Reinforced Polymers: Processes and Applications*, pp. 177–210. Nova Science Publishers.

[32] M. T. Kima, K. Y. Rhee, I. Jung, S. J. Park, D. Huid. (2014). Influence of seawater absorption on the vibration damping characteristics and fracture behaviors of basalt/CNT/epoxy multiscale composites. Vol. 63. pp. 61–66. *Composites Part B: Engineering*. Retrieved from: https://doi.org/10.1016/j.compositesb.2014.03.010.

[33] C. Subramanian, S. B. Deshpande, S. Senthilvelan. (2012). Effect of reinforced fiber length on the damping performance of thermoplastic composites. Vol. 20. pp. 319–335. *Advance Composite Materials*. Retrieved from: https://doi.org/10.1163/092430410X550872.

[34] Y. S. Munde, R. B. Ingle, I. Siva. (2019). A comprehensive review on the vibration and damping characteristics of vegetable fiber-reinforced composites. *Journal of Reinforced Plastics and Composites*. Retrieved from: https://doi.org/10.1177/0731684419838340.

[35] K. Senthil, K. I. Siva, P. Jeyaraj, J. T. Jappes, S. C. Amico, N. Rajini. (2014). Synergy of fiber length and content on free vibration and damping behavior of natural fiber reinforced polyester composite beams. Vol. 56, pp. 379–386. *Material & Design*. Retrieved from: https://doi.org/10.1016/j.matdes.2013.11.039.

[36] T. P. Sathishkumar, J. Naveen, P. Navaneethakrishnan, S. Satheeshkumar, N. Rajini. (2016). Characterization of sisal/cotton fibre woven mat reinforced polymer hybrid composites. Vol. 47. Issue 4. *Journal of Industrial Textiles*. Retrieved from: https://doi.org/10.1177/1528083716648764.

[37] M. Bulut, A. Erkliğ, E. Yeter. (2015). Experimental investigation on influence of Kevlar fiber hybridization on tensile and damping response of Kevlar/glass/epoxy resin composite laminates. Vol. 50. Issue 14. *Journal of Composite Materials*. Retrieved from: https://doi.org/10.1177/0021998315597552.

[38] S. Arulmurugan, N. Venkateshwaran. (2016). Vibration analysis of nanoclay filled natural fiber composites. Vol. 24. Issue 7. *Polymers and Polymer Composites*. Retrieved from: https://doi.org/10.1177/096739111602400709.

[39] M. Rajesh, J. Pitchaimani. (2015). Dynamic mechanical analysis and free vibration behavior of intra-ply woven natural fiber hybrid polymer composite. Vol. 35. Issue 3. *Journal of Reinforced Plastics and Composites*. Retrieved from: https://doi.org/10.1177/0731684415611973.

[40] M. Rajesh, J. Pitchaimani, N. Rajini. (2016). Mechanical, Dynamic Mechanical and Vibration Behavior of Nanoclay Dispersed Natural Fiber Hybrid Intra-ply Woven Fabric Composite. In: Jawaid, M., Qaiss, A. K., Bouhfid, R. (Eds), *Nanoclay Reinforced Polymer Composites*. pp. 281–296. Retrieved from: https://doi.org/10.1007/978-981-10-0950-1_12.

[41] A. Treviso, B. V. Genechten, D. Mundo, M. Tournour. (2015). Damping in composite materials: Properties and models. Vol. 78. pp. 144–152. *Composites Part B: Engineering*. Retrieved from: https://doi.org/10.1016/j.compositesb.2015.03.081.

[42] R. F. Gibson. (2000). Modal vibration response measurements for characterization of composite materials and structures. Vol. 60. Issue 15. pp. 2769–2780. *Composites Science and Technology*. Retrieved from: https://doi.org/10.1016/S0266-3538(00)00092-0.

[43] S. Ashworth, J. Rongong, P. Wilson, J. Meredith. (2016). Mechanical and damping properties of resin transfer moulded jute-carbon hybrid composites. Vol. 105. pp. 60–66. *Composites Part B: Engineering*. Retrieved from: https://doi.org/10.1016/j.compositesb.2016.08.019.

[44] K. Wang, K. Okuno, M. Banu, B. I. Epureanu. (2017). Vibration-based identification of interphase properties in long fiber reinforced composites. Vol. 174. pp. 244–251. Retrieved from: https://doi.org/10.1016/j.compstruct.2017.04.018.

[45] J. Flynn, A. Amiri, C. Ulven. (2016). Hybridized carbon and flax fiber composites for tailored performance. Vol. 102. pp. 21–29. *Materials & Design*. Retrieved from: https://doi.org/10.1016/j.matdes.2016.03.164.

[46] M. Rueppel, J. Rion, C. Dransfeld, C. Fischer, K. Masania. (2017). Damping of carbon fibre and flax fibre angle-ply composite laminates. Vol. 146. pp. 1–9. *Composites Science and Technology*. Retrieved from: https://doi.org/10.1016/j.compscitech.2017.04.011

[47] S. Prabhakaran, V. Krishnaraj, M. S. Kumar, R. Zitoune. (2014). Sound and vibration damping properties of flax fiber reinforced composites. Vol. 97. pp. 573–581. *Procedia Engineering*. Retrieved from: https://doi.org/10.1016/j.proeng.2014.12.285.

[48] W. Zhu, V. Nandikolla, B. George. (2015). Effect of bulk density on the acoustic performance of thermally bonded nonwovens. Vol. 10. Issue 3. pp. 39–45. *Journal of Engineered Fibers and Fabrics*. Retrieved from: https://doi.org/10.1177/155892501501000316.

[49] M. Küçük, Y. Korkmaz. (2012). The effect of physical parameters on sound absorption properties of natural fiber mixed nonwoven composites. Vol. 82. Issue 20. pp. 2043–2053. *Textile Research Journal*. Retrieved from: https://doi.org/10.1177/0040517512441987.

[50] J. Zhang, A. A. Khatibi, E. Castanet, T. Baum, Z. Komeily-Nia, P. Vroman, X. Wang. (2019). Effect of natural fibre reinforcement on the sound and vibration damping properties of biocomposites compression moulded by nonwoven mats. Vol. 13. pp. 12–17. *Composites Communications*. Retrieved from: https://doi.org/10.1016/j.coco.2019.02.002

[51] L. Guillaumat, S. Terekhina, I. Derbali, A. Monti, A. E. Mahi, Z. Jendli. (2018). Flax Fibers Reinforced Thermoplastic Resin Based Biocomposites, a Future for Sustainable Composite Parts. Vol. 203. *MATEC Web of Conferences*. Retrieved from: https://doi.org/10.1051/matecconf/201820306019.

[52] J. Yang, J. Xiong, L. Ma, B. Wang, G. Zhang, L. Wu. (2013). Vibration and damping characteristics of hybrid carbon fiber composite pyramidal truss sandwich panels with viscoelastic layers. Vol. 106. pp. 570–580. *Composite Structures*. Retrieved from: https://doi.org/10.1016/j.compstruct.2013.07.015.

[53] K. Essassi, J. Rebiere, A. E. Mahi, M. Toure, M. B. Souf, A. Bouguecha, M. Haddar. (2019). Vibration Behaviour of a Biocomposite Sandwich with Auxetic Core. Vol. 283. *The 2nd Franco-Chinese Acoustic Conference (FCAC 2018)*. Retrieved from: DOI:10.1051/MATECCONF/201928309004.

[54] J. S. James, K. Hyung-Ick, A. Erik, S. Jonghwan. (2013). Sound and vibration damping characteristics in natural material based sandwich composite. Vol. 96. pp. 538–544. *Composite Structures*. Retrieved from: https://doi.org/10.1016/j.compstruct.2012.09.006.

[55] C. Garcia, J. Wilson, I. Trendafilova, L. Yang. (2017). Vibratory behaviour of glass fibre reinforced polymer (GFRP) interleaved with nylon nanofibers. Vol. 176. pp. 923–932. *Composite Structures*. Retrieved from: https://doi.org/10.1016/j.compstruct.2017.06.018.

[56] P. Thomas, M. P. Jenarthanan, V. M. Sreehari. (2018). Free vibration analysis of a composite reinforced with natural fibers employing finite element and experimental techniques. Vol. 17. Issue 5. pp. 688–699. *Journal of Natural Fibers*. Retrieved from: https://doi.org/10.1080/15440478.2018.1525466.

10 Characterization of Viscoelastic Properties of Biocomposites by Dynamic Mechanical Analysis
An Overview

Jyotishkumar Parameswaranpillai
Alliance University

Midhun Dominic C. D.
Sacred Heart College

Chandrasekar Muthukumar
Hindustan Institute of Technology & Science

Senthil Muthu Kumar Thiagamani
Kalasalingam Academy of Research and Education

Senthilkumar Krishnasamy
Francis Xavier Engineering College

CONTENTS

DOI: 10.1201/9781003173625-11

10.1   INTRODUCTION

Natural fibers such as coir, banana, cotton, flax, hemp, jute, and sisal are widely used with polymers to fabricate bio-based composites. Natural fibers have many advantages over traditional synthetic fibers (e.g., glass, carbon, and aramid fibers): low cost, lightweight, good strength, and biodegradability. On the other hand, they are hydrophilic, while most of the polymers used are hydrophobic. This caused poor compatibility between the natural fiber and polymer. The drawback can be overcome by the physical and chemical treatment of the natural fibers. Physical treatments such as ozonolysis, ultraviolet treatment, plasma treatment, etc., are widely accepted methods. Similarly, the chemical treatments such as alkalization, benzoylation, salination, and oxidation with potassium permanganate are the widely accepted methods. Recent technologies such as autoclave and ozonolysis, are employed for the surface modification of the natural fibers [1–3].

Polymer materials are viscoelastic, and they showed both elastic and viscous properties upon the application of mechanical stress and temperature [4]. The viscoelastic properties of the polymer composites provide information on the thermomechanical properties of the composites. The parameters such as storage modulus (E'), loss modulus (E''), and tan δ provide details on the material performance of the composites. Specifically, the E' gives information on composites' energy stored or stiffness during cyclic loading. The E'' provides information on the viscous response or energy dissipated during cyclic loading. The tan δ/damping factor provides information on the ratio of E''/E' [5]. Thus, the viscoelastic measurements provide information on the microscopic properties and molecular rearrangements [6]. A typical storage modulus profile consists of (a) glassy region, (b) glass transition region, and (c) rubbery region [7]. In the glassy region, the macromolecular chains are frozen and have no segmental mobility. The chain mobility increased in the glass transition region, resulting in a drop in modulus or a peak height in the E'' and tan δ profile [7,8]. In the rubbery region, the stress is dissipated, and hence the E' is not affected, while the E'' and tan δ are dropped.

The creep and stress relaxation of the polymers are also studied in the literature. By definition, creep is the deformation of a polymer under constant stress/load at a constant temperature. The creep of a polymer has three stages: (a) elastic deformation (Stage I), (b) viscoelastic deformation (Stage II), and (c) viscous deformation (Stage III). The type of polymer, the type of the fiber, aspect (l/d) ratio, interface properties between the fiber and polymer, fiber content, and fiber orientation affect the polymer composite's creep and stress relaxation.

10.2   CALCULATION OF INTERACTION PARAMETERS FROM THE DYNAMIC MECHANICAL ANALYSIS PROFILE

The reinforcing effect of the fibers/fillers with the polymer matrix can be calculated using Equations 10.1–10.4 from the dynamic mechanical analysis (DMA) profile. This gives an overview of the possible interactions between the fiber/filler and polymer matrix. From the E' profile, the extent of reinforcement of fibers with the polymer matrix can be calculated using Equation 10.1 [7].

$$\text{Co-efficient of effectiveness } (C) = \frac{(E'_G/E'_R)_{\text{Composite}}}{(E'_G/E'_R)_{\text{Resin}}} \tag{10.1}$$

where E'_G and E'_R are the E' of the composites in the glassy and rubbery regions, respectively. If the value of C is less than 1, it indicates the substantial effectiveness of reinforcement of the fibers with the polymer. Similarly, the mechanical fiber effectiveness (MFE) is given by Equation 10.2 [9].

$$\text{MFE} = 1 - \frac{(E'_G/E'_R)_{\text{Composite}}}{(E'_G/E'_R)_{\text{Resin}}} \tag{10.2}$$

The higher value of MFE, the higher the effectiveness of the filler.

The interfacial strength indicator (B) can be calculated from the tan δ profile using Equation 10.3 [10].

$$\tan \delta = \frac{\tan \delta_m}{(1+1.5B\phi)} \tag{10.3}$$

where tan δ and tan δ_m are the peak height of composite and neat polymer, respectively; ϕ is the volume fraction of the fiber. The positive value of B shows good interfacial strength between the fiber and polymer.

The parameter interfacial adhesion (β) can be calculated from the tan δ profile using Equation 10.4 [11].

$$\frac{\tan \delta_{\text{max } c}}{\tan \delta_{\text{max } m}} = 1 - \beta\phi \tag{10.4}$$

where tan $\delta_{\text{max } c}$ and tan $\delta_{\text{max } m}$ are the peak height of composite and neat polymer, respectively; ϕ is the volume fraction of the fiber.

10.3 VISCOELASTIC PROPERTIES OF BIODEGRADABLE THERMOPLASTICS-BASED BIOCOMPOSITES

Ali et al. [12] studied the creep behavior of sisal fiber reinforced with Mater-Bi® Y (consisting of starch and cellulose acetate blends) [13] and Mater-Bi® Z (comprising of poly(ε-caprolactone) (PCL) matrix) [13] by DMA. Sisal fibers were treated with NaOH for one hour at 80°C under continuous stirring, whereby the fiber loading was maintained as 20%. The composites were fabricated using a Haake model 9000 Rheocorder and the creep test was carried out at 30°C. The researchers observed that the fibers in the composites with the highest l/d ratio (126.2) exhibited the highest creep resistance. The results also showed that the presence of the fiber prevented the deformation of the polymer. Interestingly, the creep behavior of Mater-Bi® Z was different from Mater-Bi® Y due to the difference in the type of the polymer matrix. On the one hand, Mater-Bi® Y-based composites showed a rapid increase in instantaneous creep with applied load due to the high molecular mobility of the amorphous phase, which was followed by creep stabilization. On the other hand, Mater-Bi® Z (crystalline polymer) showed low creep, followed by a slow deformation due to structural rearrangement and stabilization.

Jacob et al. [14] studied the viscoelastic effects of sisal/oil palm fiber-reinforced hybrid natural rubber composites. Three different types of silanes were used to enhance the interface between the fiber and rubber matrix: (a) silane F8261, (b) silane A1100, and (c) silane A151. Irrespective of the type of silane used, the rubber composites showed improved E'. This could be ascribed to the silane treatment reducing the fibers' hydrophilicity and enhancing the fiber and rubber interface interactions. The good interfacial bond between the fiber and polymer may reduce the void formation, thus causing improved wettability of the rubber over the fiber. This, in turn, improved the stiffness and strength of the composites. The researchers also observed an increase in E'' peak height with the silane treatment of the fibers. It may be due to the heat released because of the friction between the filler and rubber during the DMA studies [15]. On the other hand, the damping was reduced with the addition of treated fibers. The drop in tan δ was owing to the reduction in the molecular mobility of the macromolecular chains at the interfacial zone. It could be due to the improved interface interaction between the treated fiber and polymer. The researchers also calculated the interfacial strength indicator (B) of the sisal/oil palm fiber-reinforced hybrid natural rubber composites. The value of B was the highest for the silane-treated fibers, thus, confirmed the excellent interface interaction between the treated fiber and polymer.

Li et al. [16] studied the viscoelastic properties of silk fiber fibroin-reinforced PCL composites using the DMA tester. The silk fiber was extracted from the cocoons of B-Mori. The concentrations of the silk fiber used to fabricate the composites were 0, 15, 25, 35, 45, 55, and 65 wt.%. From the studies, the E' was minimum for the virgin PCL (VPLA) and was increased with the incorporation of silk fibroin fiber. Before and after the glass transition temperature (T_g), the modulus was maximum for 65 wt.% silk fiber-modified composites. A sharp drop in E' was observed at the T_g for the neat PCL. However, a sharp drop in modulus was not observed for the composites with silk fibers. This showed that the fibers formed a network structure in PCL composites. Interestingly, the T_g of the composites was marginally reduced; it was ascribed to the lesser interaction between the PCL chains in the presence of silk fiber fibroin.

Huda et al. [17] studied the viscoelastic properties of kenaf fiber/polylactic acid (PLA) composites. The kenaf fiber was treated with alkali (FIBNA), silane (FIBSI), and both alkali and silane (FIBNASI). A fiber loading of 40 wt.% was used for the fabrication of the composites. The E' and E'' of the kenaf/PLA composites are shown in Figure 10.1. Incorporation of kenaf into the neat PLA matrix (FIB) increased the E' by 41%, while the composites with treated fibers showed an increasing trend with 67, 87, and 161% increase in value for FIBNA, FIBSI, and FIBNASI composites, respectively, compared to the neat PLA matrix (Figure 10.1A). A sudden drop in E' was observed at the T_g due to the increased segmental mobility of the composites. From the E'' curve (Figure 10.1B), the peak height of FIB and FIBNA composites is higher than neat PLA, whereas it was lower than neat PLA for the FIBSI and FIBNASI composites. The lower peak height of the FIBSI and FIBNASI composites is believed to be due to the reduced macromolecular chain mobility caused by the

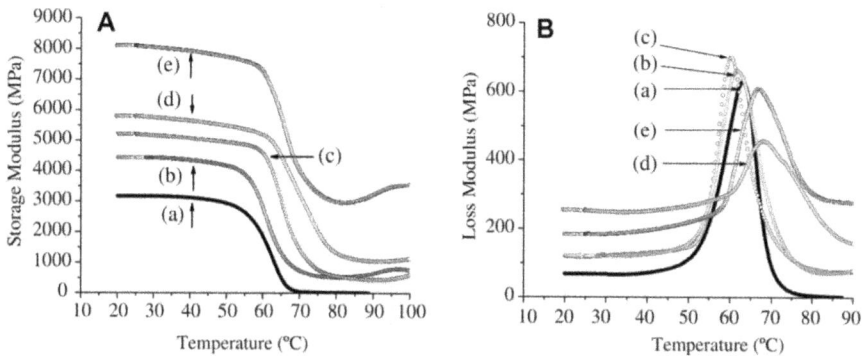

FIGURE 10.1 Kenaf/PLA composites (A) E', and (B) E'''. *(a) neat PLA (b) PLA/FIB (kenaf fiber) (c) PLA/FIBNA, (d) PLA/FIBSI, and (e) PLA/FIBNASI [17]. (Reproduced with thanks from Elsevier; License Number: 5137080166956.)

better interfacial adhesion between the polymer and fiber because of the coupling effect of the silane. Han et al. [18] investigated the synergistic effect of adding kenaf and exfoliated graphene nanoplatelets (xGnP) on the viscoelastic properties of the PLA matrix. Kenaf fiber loading was kept to 0, 20, 30, and 40 wt.%, while the xGnP was mixed at concentrations of 0, 1, 3, and 5 wt.%. From their observation, the E' of the composites showed an increasing trend with respect to the fiber loading, while there was only a moderate increase in the E' of the composites with the incorporation of xGnP. However, with the addition of kenaf and xGnP together, considerable improvement in E' was observed. The *tan δ* peak height was not affected by the addition of xGnP, while the kenaf fiber reduced the *tan δ* peak height of the PLA matrix. The presence of both xGnP and kenaf fiber in the PLA matrix increased the T_g of the composites by 3°C–5°C.

Gil-Castell et al. [9] studied the effect of hydrothermal aging on the viscoelastic performance of the PLA/sisal composites. The sisal fiber content used for the fabrication of the composites was 10, 20, and 30 wt.%. The coupling agent maleic anhydride (MAH) of 2.5 wt.% and dicumyl peroxide (DCP) of 0.3 wt.% were added into the matrix. The E' of VPLA, PLA/sisal composites (PLA10, PLA20, and PLA30) and MAH-DCP-PLA/sisal composites (PLA10C, PLA20C, and PLA30C) before and after hydrothermal aging are shown in Figure 10.2a and b. The E' of the PLA matrix increased with the incorporation of sisal fibers and the addition of compatibilizer (MAH-DCP). The better wettability between the fiber and matrix was primarily responsible for the enhancement in the E'. A sharp drop in E' was observed at the T_g followed by an increase in E' due to the cold crystallization of PLA. However, after the hydrothermal aging, the drop in E' at the T_g was not observed due to the complete crystallization of the PLA matrix. On the other hand, the E' of the aged composites showed lower E' value than the non-aged composites.

In recent work, Lila et al. [19] fabricated bagasse fiber/PLA composites with a fiber loading of 20 wt.%. The composites were subjected to accelerated weathering for 4, 8, and 12 weeks, respectively. The incorporation of bagasse fiber increased the E' of the composites in the glassy state. Composites subjected to accelerated weathering

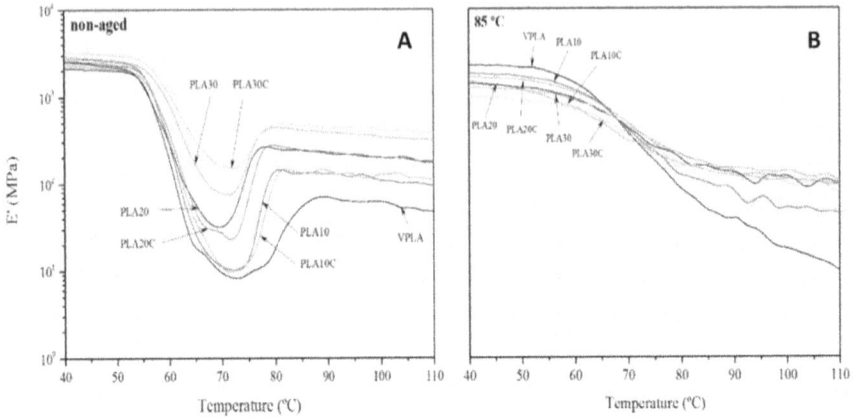

FIGURE 10.2 E' of the PLA/sisal composites and compatibilized PLA/sisal fiber composites (a) before aging and (b) after aging [9]. (Reproduced with thanks from Elsevier; License Number: 5137160986106.)

displayed lower E' in the glassy state and higher E' in the rubbery state. The increase in value of E' for the rubbery state could have been caused by the increased crystallinity of the polymeric composite exposed to accelerated weathering. Mannai et al. [11] studied the viscoelastic properties of polyvinyl alcohol/cellulose fibrous networks composites and styrene-butadiene rubber/cellulose fibrous networks composites. The composite specimen was cut with fibers oriented in vertical direction or fibers oriented in the longitudinal direction (VF) and fibers oriented in the horizontal or transverse direction (HF), as shown in Figure 10.3. VF composites possessed superior E' and higher interfacial adhesion represented by correction parameter (β) than the HF composites. The closer network in the vertical or longitudinal direction in the cellulose fibrous network-enabled better fiber–polymer interaction, which led to lower dissipation energy, thereby lower tan δ values for the VF composites. Their study highlighted the influence of fiber orientation on the thermomechanical properties of the composites.

10.4 VISCOELASTIC PROPERTIES OF BIOCOMPOSITES BASED ON TRADITIONAL THERMOPLASTICS

Zhang et al. [20] studied the DMA of rice husk (RH) char-modified high-density polyethylene (HDPE) composite. The concentrations of the biochar used were 30, 40, 50, 60, and 70 wt.%. The increase in biochar content in the composites increased E', E'', and T_g, and reduced tan δ. This was because the composite became rigid at high biochar content. The researchers stated that the rigidity of the composites was due to the filler–polymer interaction because the HDPE filled into the holes of biochar. The changes in E', E'', and tan δ of the composites at different frequencies (1, 2, and 10 Hz) were studied. With the increase in the frequency, the E' marginally increased, the peak height of E'' reduced, became narrow and shifted to higher temperatures,

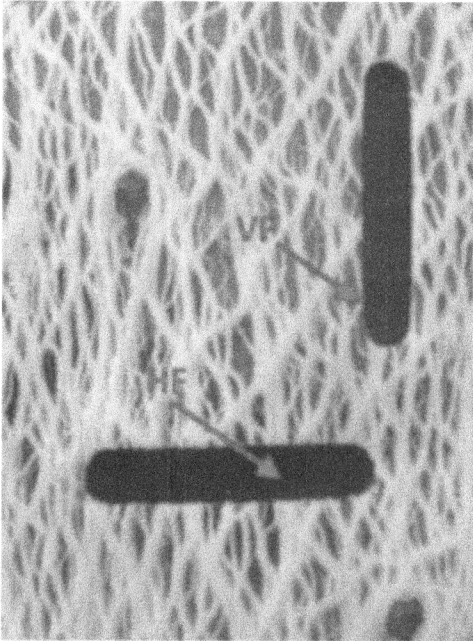

FIGURE 10.3 DMA specimens of the composite samples fabricated according to the vertical (longitudinal) and horizontal (transverse) fiber orientation [11]. (Reproduced with thanks from Springer nature; License Number: 5137180467802.)

while the tan δ also reduced and shifted to the higher temperature. The effect of char content on creep and stress relaxation at different temperatures was also studied, as shown in Figure 10.4a–e. Biochar reinforcement increased the creep resistance, enhanced the composites' structural stability, and increased the relaxation modulus; hence, anti-stress relaxation was improved. However, the creep resistance was lower at elevated temperatures because the free volume and thermal kinetic energy increased.

Yaghoobi et al. [5] studied the viscoelastic properties of polypropylene (PP)/ kenaf/polypropylene-graft-maleic anhydride (PP-g-MAH) and PP/kenaf/PP-g-MAH/MWCNTs composites. The concentration of kenaf and PP-g-MA used was 30, and 5 wt.%, respectively. The concentration of MWCNTs was varied between 0.5, 1, 1.5, and 2 wt.%. The incorporation of nanofiller improved the E', and the maximum E' was observed at 1.5 wt.% nanofiller. The drop at 2 wt.% of MWCNTs was due to the agglomeration of the nanofiller in the composite. The reinforcement of MWCNTs caused restricted polymer mobility resulting in stiff composites with a high E'. A typical decreasing trend was observed for the E' of the PP matrix with respect to the temperature, which is attributed to the increased molecular mobility of the polymer chain at higher temperatures [21]. The increase in E'' for the composites incorporated with MWCNTs represents the dissipation of more heat energy which is supposed to be due to the internal friction between the filler–filler and filler–polymer

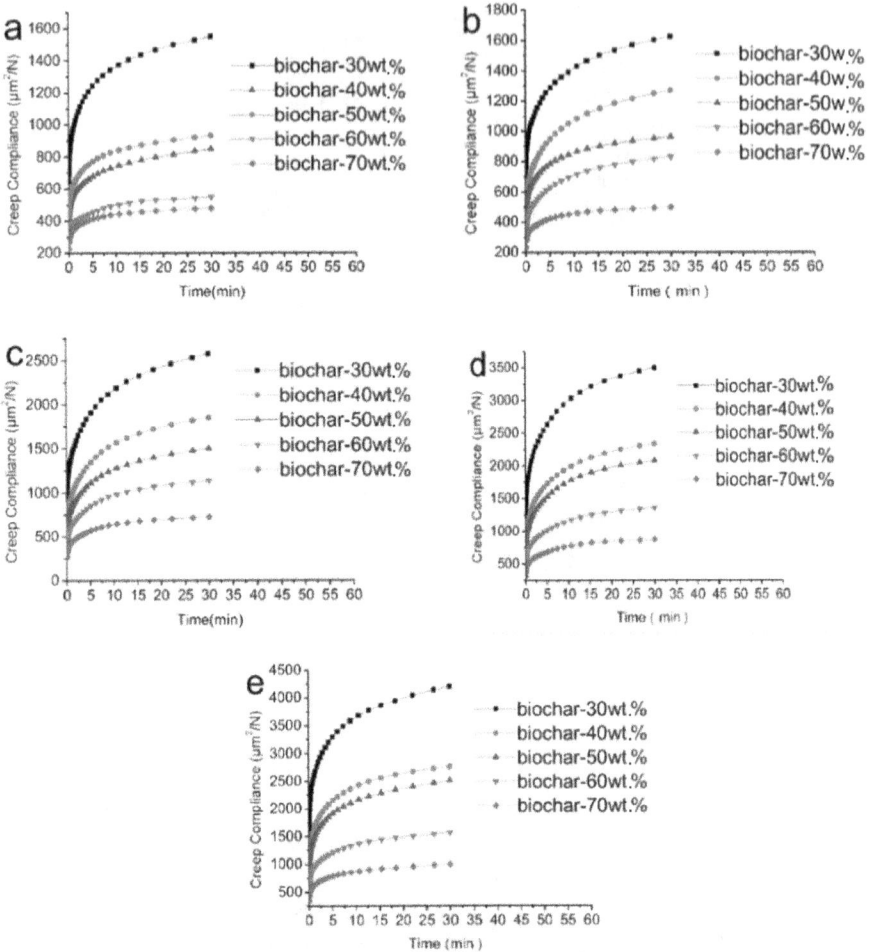

FIGURE 10.4 Creep behavior of rice husk char-modified HDPE composite at different temperatures (a) 25°C; (b) 35°C; (c) 45°C; (d) 55°C, and (e) 65°C [20]. (Open access.)

interactions during cyclic loading [22]. Only a marginal increase in T_g (1°C–2°C) and a decrease in damping factor were observed with the incorporation of MWCNTs. Hidalgo-Salazar et al. [23] studied and compared the performance of neat PP and RH incorporated PP composites. The weight % of RH used were 10, 20, and 30 wt.%. The reported E' of the PP/RH composites was higher compared to neat PP. Also, the E' of PP increased with the increasing content of RH. The high value of modulus of the composites indicated the reinforcing effect of RH in the PP matrix. All the composites showed a drop in E' after the T_g, due to the softening of the PP matrix. The E''' vs. temperature profiles presented an increase in T_g with the incorporation of RH filler. Also, the tan δ peak height is reduced with the incorporation of RH validating the better compatibility of RH with the PP matrix.

10.5   VISCOELASTIC PROPERTIES OF BIOCOMPOSITES BASED ON EPOXY THERMOSETS

Mittal et al. [24] fabricated coir fiber-reinforced epoxy and pineapple fiber-reinforced epoxy composites and investigated their viscoelastic characteristics. The composites were made with 17%, 23%, and 34% fibers by volume. The E', E'', and T_g increased with fiber content regardless of the type of fiber, whereas the tan δ peak height dropped. The reinforcing action of fiber with epoxy matrix caused these modifications in the composites' viscoelastic characteristics. The influence of frequency on the viscoelastic characteristics of composites containing 34 vol.% fiber was also investigated with a frequency of 1 and 5 Hz, respectively. In general, the E' was found to be higher at a higher frequency; however, it was shown to be lower at a high frequency for PALF/epoxy and coir/epoxy composites, possibly due to enhanced polymer–polymer interactions or polymer entanglements.

Kumar et al. [6] studied the viscoelastic properties of bioepoxy, bioepoxy + ramie, bioepoxy + flax, and hybrid bioepoxy + ramie + flax with fiber loading of 10, 20, and 30 wt.%. The hybrid composites were fabricated with an equal amount of ramie/flax fibers. When compared to clean bioepoxy matrix or single fiber composites, the E' of the composite was highest for the hybrid composites. This is due to the dense packing of fibers and polymer. In addition, the hybrid composites showed a considerable improvement in T_g. On the one hand, the E'' for hybrid composites was found to have the highest value, while the E'' for bioepoxy matrix was the lowest. According to the researchers, the high elasticity and energy absorption of fibers produced substantial energy dissipation during the DMA testing. On the other hand, the hybrid composites had the lowest damping due to the close packing of fibers and bioepoxy matrix.

Yorseng et al. [25] investigated the viscoelastic behavior of sisal fabric, kenaf fabric, and hybrid sisal/kenaf fabric-reinforced bioepoxy composites before and after the weathering test. The neat epoxy samples displayed a drastic reduction in E' at T_g. The composites, on the other hand, showed a slight decrease in modulus at T_g. This study demonstrated the fiber mats' reinforcing impact in bioepoxy composites. The E''' profiles revealed no change in T_g. However, observed a decrease in peak height and a widening of the peak, likely owing to the bioepoxy chains' limited segmental mobility in the presence of sisal and/or kenaf fiber fabrics. The samples were also tested for their viscoelasticity after being subjected to accelerated weathering. The E' profile of the weathered composites was identical to the E' profile of the composite samples before weathering. Surprisingly, after accelerated weathering, the E'' curves demonstrated that the T_g of the composites increased slightly, which confirms the composites' stability even after accelerated weathering. The reinforcing effect of the fabric was estimated, where the coefficient (c) was found to be the smallest for the composites and is less than 1 before and after the weathering test. This indicates the reinforcing effect of the fabric in the composites fabricated. Furthermore, Cole-Cole plot of the composite samples revealed that the epoxy fabric systems were relatively homogeneous.

Rangappa et al. [7] studied the DMA of poly (ethylene glycol)-block-poly (propylene glycol)-block-poly (ethylene glycol) triblock copolymer (TBCP)-modified sisal/epoxy composites. In the glassy state, the E' of the bioepoxy resin treated with TBCP revealed a minor decrease in the modules. This could be due to the fact that the TBCP has a lower modulus than the crosslinked bioepoxy thermoset. With the addition of TBCP, the T_g of bioepoxy was also decreased slightly. After the T_g, the E' reduced rapidly because the free volume was increased and hence the segmental mobility. The E''' profile of the TBCP-modified bioepoxy also showed a decrease in T_g, which agreed with the E' profile. The reduction in T_g suggested that TBCP and bioepoxy resin were miscible. Incorporating sisal fiber mats into a TBCP-modified bioepoxy system, on the other hand, resulted in a different outcome. In agreement with reference [25], the significant reduction in E' was not seen in sisal fiber-modified epoxy/TBCP blends following the T_g. It is also worth mentioning that the T_g of the bioepoxy/TBCP combination marginally increased in the presence of sisal fiber mats. This finding also supports the use of sisal fiber mats as reinforcement in epoxy/TBCP system. In another study, Yorseng et al. [26] investigated the reinforcing effect of neat epoxy thermoset and hybrid sisal/kenaf fabric-reinforced epoxy composites before and after accelerated weathering. It was observed that the addition of fabrics did not influence the E' of the epoxy matrix. However, there was a marginal increase in the T_g values with the incorporation of the fabrics. The composites showed higher modulus at the rubbery region due to high modulus natural fibers and the possible reinforcement with the epoxy matrix in the rubbery state. After the accelerated weathering, there was an increase in the modulus in the rubbery area, and using Equation 10.1, the reinforcing effect or the interaction between the fiber and the polymer was calculated. The composites possessed a C value of less than 1, indicating the fiber–polymer interaction. However, since the modulus in the glassy state does not rise with the addition of fiber fabric, it is possible that the presence of stiff natural fibers also contributed to the increase in modulus of the composites in the rubbery state.

Senthilkumar et al. [27] investigated the creep behavior of neat epoxy, and sisal/hemp-reinforced hybrid biocomposites before and after the weathering test. The test was conducted at a constant load of 5 N for 1,800 s, and the recovery was observed for 3,600 s. Due to its elastic nature, the bioepoxy displayed fast, immediate deformation, followed by viscous creep, followed by elastic recovery. Because a complete recovery is impossible, this resulted in residual stress and a permanent set. The creep strain was found to be highest for the neat epoxy samples, but it was lowered when hemp or sisal was added. This indicates that the inclusion of hemp or sisal enhanced the creep resistance. Regardless of the fiber sequences, creep deformation differed. All hemp fibers (HHHH(four hemp fiber mats in layers)) composites had the least instantaneous deformation of all the composites tested. This could be ascribed to the high stiffness of hemp and its interfacial interaction with the bioepoxy matrix. After the weathering test, the composites showed greater creep strain. This is because the fibers were rearranged during the accelerated weathering test. On the other hand, weathered composites outperformed neat bioepoxy (before and after weathering) in terms of creep resistance. The composites' creep behavior was in good agreement with the burgers model. The composites exhibited minimal creep, high elastic rebound, and low permanent set, and based on these results, the composites were recommended for low-load interior and exterior applications.

10.6 CONCLUSION

The viscoelastic characteristics of biocomposites are briefly explored in this chapter. Fully biodegradable composites, partly biodegradable composites, and epoxy thermoset have all been examined in detail. The following was concluded from the review:

- The inclusion of natural fibers improved the E' and T_g of composites in most published reports.
- In semi-structural outdoor applications, the results encourage using natural fiber in polymer materials rather than synthetic fibers.
- Even modified natural fibers, however, perform poorly when compared to synthetic fibers.
- Hence, more research is needed to improve the interfacial characteristics of the natural fiber and polymer matrix so that natural fibers can completely replace synthetic fibers in advanced applications.

REFERENCES

1. Senthilkumar, K., Ungtrakul, T., Chandrasekar, M., Kumar, T.S.M., Rajini, N., Siengchin, S., Pulikkalparambil, H., Parameswaranpillai, J. and Ayrilmis, N., 2021. Performance of sisal/hemp bio-based epoxy composites under accelerated weathering. *Journal of Polymers and the Environment*, 29(2), pp. 624–636.
2. Sanjay, M.R., Siengchin, S., Parameswaranpillai, J., Jawaid, M., Pruncu, C.I. and Khan, A., 2019. A comprehensive review of techniques for natural fibers as reinforcement in composites: Preparation, processing and characterization. *Carbohydrate Polymers*, 207, pp. 108–121.
3. Imoisili, P.E. and Jen, T.C., 2020. Mechanical and water absorption behaviour of potassium permanganate (KMnO₄) treated plantain (Musa Paradisiaca) fibre/epoxy bio-composites. *Journal of Materials Research and Technology*, 9(4), pp. 8705–8713.
4. Ferry, J.D., 1980. Viscoelastic properties of polymers. John Wiley & Sons.
5. Yaghoobi, H. and Fereidoon, A., 2018. Evaluation of viscoelastic, thermal, morphological, and biodegradation properties of polypropylene nano-biocomposites using natural fiber and multi-walled carbon nanotubes. *Polymer Composites*, 39, pp. E592–E600.
6. Kumar, S., Zindani, D. and Bhowmik, S., 2020. Investigation of mechanical and viscoelastic properties of flax-and ramie-reinforced green composites for orthopedic implants. *Journal of Materials Engineering and Performance*, 29, pp. 3161–3171.
7. Rangappa, S.M., Parameswaranpillai, J., Yorseng, K., Pulikkalparambil, H. and Siengchin, S., 2021. Toughened bioepoxy blends and composites based on poly (ethylene glycol)-block-poly (propylene glycol)-block-poly (ethylene glycol) triblock copolymer and sisal fiber fabrics: A new approach. *Construction and Building Materials*, 271, p. 121843.
8. Khan, T., Sultan, M.T.H., Jawaid, M., Safri, S.N.A., Shah, A.U.M., Majid, M.S.A., Zulkepli, N.N. and Jaya, H., 2021. The effects of stacking sequence on dynamic mechanical properties and thermal degradation of kenaf/jute hybrid composites. *Journal of Renewable Materials*, 9(1), pp. 73–84.
9. Gil-Castell, O., Badia, J.D., Kittikorn, T., Strömberg, E., Ek, M., Karlsson, S. and Ribes-Greus, A., 2016. Impact of hydrothermal ageing on the thermal stability, morphology and viscoelastic performance of PLA/sisal biocomposites. *Polymer Degradation and Stability*, 132, pp. 87–96.

10. Konnola, R., Joji, J., Parameswaranpillai, J. and Joseph, K., 2015. Structure and thermo-mechanical properties of CTBN-grafted-GO modified epoxy/DDS composites. *RSC Advances*, 5(76), pp. 61775–61786.

11. Mannai, F., Elhleli, H., Dufresne, A., Elaloui, E. and Moussaoui, Y., 2020. Opuntia (Cactaceae) fibrous network-reinforced composites: Thermal, viscoelastic, inter-facial adhesion and biodegradation behavior. *Fibers and Polymers*, 21(10), pp. 2353–2363.

12. Ali, R., Iannace, S. and Nicolais, L., 2003. Effect of processing conditions on mechanical and viscoelastic properties of biocomposites. *Journal of Applied Polymer Science*, 88(7), pp. 1637–1642.

13. Aldas, M., Rayón, E., López-Martínez, J. and Arrieta, M.P., 2020. A deeper micro-scopic study of the interaction between gum rosin derivatives and a Mater-Bi type bio-plastic. *Polymers*, 12(1), p. 226.

14. Jacob, M., Francis, B., Varughese, K.T. and Thomas, S., 2006. The effect of silane cou-pling agents on the viscoelastic properties of rubber biocomposites. *Macromolecular Materials and Engineering*, 291(9), pp. 1119–1126.

15. Parameswaranpillai, J., Joseph, G., Shinu, K.P., Sreejesh, P.R., Jose, S., Salim, N.V. and Hameed, N., 2015. The role of SEBS in tailoring the interface between the poly-mer matrix and exfoliated graphene nanoplatelets in hybrid composites. *Materials Chemistry and Physics*, 163, pp. 182–189.

16. Li, W., Qiao, X., Sun, K. and Chen, X., 2008. Mechanical and viscoelastic proper-ties of novel silk fibroin fiber/poly (ε-caprolactone) biocomposites. *Journal of Applied Polymer Science*, 110(1), pp. 134–139.

17. Huda, M.S., Drzal, L.T., Mohanty, A.K. and Misra, M., 2008. Effect of fiber surface-treatments on the properties of laminated biocomposites from poly (lactic acid) (PLA) and kenaf fibers. *Composites Science and Technology*, 68(2), pp. 424–432.

18. Han, S.O., Karevan, M., Bhuiyan, M.A., Park, J.H. and Kalaitzidou, K., 2012. Effect of exfoliated graphite nanoplatelets on the mechanical and viscoelastic properties of poly (lactic acid) biocomposites reinforced with kenaf fibers. *Journal of Materials Science*, 47(8), pp. 3535–3543.

19. Lila, M.K., Shukla, K., Komal, U.K. and Singh, I., 2019. Accelerated thermal age-ing behaviour of bagasse fibers reinforced poly (lactic acid) based biocomposites. *Composites Part B: Engineering*, 156, pp.121–127.

20. Zhang, Q., Cai, H., Ren, X., Kong, L., Liu, J. and Jiang, X., 2017. The dynamic mechani-cal analysis of highly filled rice husk biochar/high-density polyethylene composites. *Polymers*, 9(11), p. 628.

21. Parameswaranpillai, J., Joseph, G., Chellappan, R.V., Zahakariah, A.K. and Hameed, N., 2015. The effect of polypropylene-graft-maleic anhydride on the morphology and dynamic mechanical properties of polypropylene/polystyrene blends. *Journal of Polymer Research*, 22(2), pp. 1–11.

22. Parameswaranpillai, J., Joseph, G., Shinu, K.P., Jose, S., Salim, N.V. and Hameed, N., 2015. Development of hybrid composites for automotive applications: Effect of addi-tion of SEBS on the morphology, mechanical, viscoelastic, crystallization and ther-mal degradation properties of PP/PS–x GnP composites. *RSC Advances*, 5(33), pp. 25634–25641.

23. Hidalgo-Salazar, M.A. and Salinas, E., 2019. Mechanical, thermal, viscoelastic perfor-mance and product application of PP-rice husk Colombian biocomposites. *Composites Part B: Engineering*, 176, p. 107135.

24. Mittal, M. and Chaudhary, R., 2018. Effect of fiber content on thermal behavior and vis-coelastic properties of PALF/epoxy and COIR/epoxy composites. *Materials Research Express*, 5(12), p. 125305.

25. Yorseng, K., Rangappa, S.M., Pulikkalparambil, H., Siengchin, S. and Parameswaran-pillai, J., 2020. Accelerated weathering studies of kenaf/sisal fiber fabric reinforced fully biobased hybrid bioepoxy composites for semi-structural applications: Morphology, thermo-mechanical, water absorption behavior and surface hydrophobicity. *Construction and Building Materials*, 235, p. 117464.
26. Yorseng, K., Mavinkere Rangappa, S., Parameswaranpillai, J. and Siengchin, S., 2020. Influence of accelerated weathering on the mechanical, fracture morphology, thermal stability, contact angle, and water absorption properties of natural fiber fabric-based epoxy hybrid composites. *Polymers*, 12(10), p. 2254.
27. Senthilkumar, K., Subramaniam, S., Ungtrakul, T., Kumar, T.S.M., Chandrasekar, M., Rajini, N., Siengchin, S. and Parameswaranpillai, J., 2020. Dual cantilever creep and recovery behavior of sisal/hemp fibre reinforced hybrid biocomposites: Effects of layering sequence, accelerated weathering and temperature. *Journal of Industrial Textiles*, https://doi.org/10.1177/1528083720961416

Part 2

Viscoelastic Properties of
the Biocomposites

11 Viscoelastic Properties of Completely Biodegradable Polymer-Based Composites

Akarsh Verma
University of Petroleum and Energy Studies

Naman Jain
Meerut Institute of Engineering and Technology

M.R. Sanjay and Suchart Siengchin
King Mongkut's University of Technology North Bangkok

CONTENTS

11.1 INTRODUCTION

Today, the application of plastics occurs in many industries such as building commodities and materials, packaging, as well as in hygiene products due to their durability. Dumping of plastic waste obtained from petroleum products leads to environmental issues, because these materials are not generally renewable and are resistant to microbial deprivation. Due to these facts researchers are moving toward biodegradable polymers-based plastic, in particular biodegradable biopolymers. There are so

DOI: 10.1201/9781003173625-13

many ways to produce biodegradable polymers; one of the most famous is producing synthetic from biodegradable polymers. Biopolymers are found in large quantities from biodegradable resources in comparison to polymers (synthetic), which are produced from degradable petroleum resources. Biodegradation is done by the act of enzymes and/or chemical worsening linked with active organisms. Biodegradation is done in two different steps. The first one is the breakup of the biopolymers into smaller molecular mass components by means of biotic reactions, i.e., degradation with the help of microorganisms, or by abiotic reactions, i.e., oxidation, hydrolysis, or photodegradation. The second step is biodegradation of the polymer by mineralization. Biodegradability depends mainly on the environmental degrading conditions and its chemical structure. Many researchers have been able to review the main mechanisms and assessment methods of polymer biodegradation [1]. However, the chemical or structural composition of biopolymers play a vital role in the mechanical behavior of biodegradable materials [2,3], as well as the storage, processing, and production characteristics [4,5], the application and aging conditions [6].

11.2   BEHAVIOR OF VISCOELASTIC BIOCOMPOSITE

Biocomposites inherit the behavior of the matrix phase with which they are fabricated, making their properties (mechanical) sturdily reliant on the applied strain ratio; therefore, the viscoelastic and mechanical behavior of the fabricated products can be useful in automotive and construction applications. Material behavior is mostly affected by the dependence of temperature conditions and stress (applied) at that instance. Biocomposites are processed to their desired size, e.g., molded housings, extruded beams (or any merchandise) that has an application area of bending, tension, combinations, or simply the constant load. These combinations of continuous loading conditions on the bioplastic for a long period of time result in the failure of the product due to creep phenomena. The serviceability of biocomposite materials can restrict their applications, and the hard work of researchers in the direction of developing feasible materials for practical products are at risk. The natural fiber-based polymeric composites under steady load circumstances deform with increase in time (creep effect). It can be understood that in a structure of biocomposite load is being redistributed among the natural fibers and matrix during deformation when a constant load is being applied to the developed product for a long time. Moreover, the final deformation of the material can also be affected by various factors such as creep loading, fiber, the matrix, and the interfacial bonding between fiber–matrix. There are many applications that have achieved biocomposite polymers from extrusion process with an addition of 40%–60%; thermoplastics such as polyvinyl chloride (PVC), polypropylene (PP), high-density polyethylene (HDPE), and materials [7,8]. Products were developed by using biodegradable fibers obtained from dissimilar resources, acting as reinforcements or fillers. These biocomposites (thermoplastic) have many applications such as benches for railway sleepers, fences, decks, and so on. The viscoelastic or creep behavior/deformation becomes the crucial parameter to study when used under these requirements. Because the function of these materials is to work under such applied load over a long term (sometimes months or even years) that may induce creep. This type of phenomenon has been measured broadly in the

product of advanced biocomposites made up of thermosets. Presently, the researchers found that the visco-elastoplastic behavior of biocomposites can lead to failure
when these products are subject to undergo large extensions for a long term, under
the circumstances of static or dynamic loading with relatively higher working temperature environment. These products gradually accumulate plastic strain resulting
in internal failure due to fatigue/creep; together they become the source of collective
failure [9–11]. There are so many techniques or mathematical models which can be
used to associate the long-term effects at minor scales [12–14] due to nonlinearity,
which is a lasting distortion in nature. Several studies have been done on thermoplastic composites to understand the pattern of deformation, and many investigations
have also shown the strain-flounce compounds such as wood-plastic, having modification in both constituents and configurations [9,15,16]. It has also been found that
the effect of creep decreases with increased fiber content. Agricultural waste materials are also gaining interest in developing biocomposites for structural applications
but should have moderate mechanical properties, in particular the creep strength. It
has been found that variation in the properties of biocomposites is highly influenced
by the content of filler, coupling agents, type, and kinds of biopolymer used for fabrication [7,17]. However, molding techniques also play an important role in improving
the viscoelastic behavior of fabricated biocomposites [9,17–20]. At present, researchers show interest in the development of new thermoplastic biocomposites as sustainable products for future implementation and sustainability over time during their
application. The viscoelastic behavior of these biocomposites can be understood
with the help of a technique known as dynamic mechanical analysis (DMA). Much
research has been conducted on creep-recovery behavior of biocomposite materials made up of diverse fillers such as bamboo and wood flour and polymers such
as HDPE and polyvinyl virgin recycled vinyl. Researchers also succeeded in the
development of new mathematical models that can fit throughout all three common
stages of creep deformation, which helps to forecast long-term behavior. They also
found that models obtained good approximation in the secondary stage, i.e., linear
creep zone but difficult to predict the primary and tertiary creep stages. However,
only limited experiments have been performed in the tertiary creep, i.e., in the failure
region by using DMA in accelerated mode [7,17]. At consistent load under low environmental temperature, biocomposites show higher creep resistance than in normal
plastic structures. But, biocomposites normally display better temperature dependence. Numerous types of creep (Findley power regulation model, Burgers model,
and a version of easy energy law parameters) were applied to regulate the information waft. The principle of time-temperature superposition (TTSP) is generally used
for calculating long-term creep deformation, where it is essential to remember that
this technique is valid especially in the linear viscoelastic vicinity of the biocomposites, inclusive of errors in time, which has now complicated the process of developing sustainable biocomposites. Research has shown that the four-element model also
known as Burgers model and the energy law with parameters approximately regulate creep deformation waft curves of biocomposites [7,9]. Many authors have also
shown that PP-agglomerate composites display special behaviors and align with the
processing situations, i.e., with growing fiber content material [16,18]; research is not
derived from expressions that genuinely include the drift residences of the fiber and

matrix of their fashions, nor getting old, or different factors related to the nature of herbal fibers. Therefore, the consistent creep curve of those mathematical models is completely distinct for biocomposites and simple and effective for these composites mainly and the circumstances enforced in recent times. There may be no entire version/model that may absolutely predict the viscoelastic performance of biocomposites with high accuracy. However, DMA can be used for a short time period effect in the industry and try to find programs for the improvement of latest sustainable merchandise using biocomposites.

11.3 LINEAR VISCOELASTICITY

On the basis of the response of the biocomposites under the applied load it can be used to classify the material as elastic solid and viscous solid but almost all plastics show combined behavior of visco-elasticity. Studies show that both the mechanical and viscoelastic nature of material depend on the application/timing of loading. Some of the important factors that affect are loading condition, environmental temperature, fiber–matrix interfacial bonding, micro and nano mechanism that occur in the biocomposites, or the load transfer. Hook's law and Newtonian law of viscosity cannot be applied directly on biocomposites because it shows combined viscoelastic behavior and highly depends on the environmental temperature, particularly if the functioning temperature goes beyond the glass transition temperature. The biocomposites when loaded with constant stress should be considered as cold fluid as per the findings of Boltzmann. As the understanding of biocomposites with different fabrication techniques, the manufacturing rate of biocomposites increases exponentially. As the demand for sustainable developed products increases day by day, fabrication and application of biocomposites increased significantly. Therefore, the behavior of biocomposites should be analyzed from an engineering point of view when loaded with different combinations of stress coupled with the understanding of temperature and humidity environment for the working range of these materials. Thus, from a mechanical point, all bioplastics have viscoelastic behavior. The viscoelastic behavior of the bioplastic is complex in nature and highly depends upon the type of reinforcement material used such as fiber, particle, or flakes. Viscoelastic behavior was studied experimentally by using various techniques in which the use of DMA is gaining interest in researchers nowadays. In the experiment in which DMA is employed, a constant stress is applied for a short term on the specimen and the strain ε produced is observed as a function of time. The typical flow curve or creep curve or viso-elastic behavior of bioplastics obtained from the DMA test is shown in Figure 11.1. The curve has three stages: stage 1 is known as primary creep in which instantaneous elastic response after concave downward curve is obtained; stage 2 is known as secondary creep in which deformation is time-dependent; and stage 3 is known as tertiary creep in which the strain rate is very high and the critical condition is reached, which results in creep rupture of the material.

In the creep or yield curve, the cloth indicates linear viscoelastic behavior; therefore, it is possible to put in force the superposition principle of time TTSP temperature. However, with the increase in load depth nonlinearity will also increase, i.e., high strain rate. Creep curve as a function of pressure fee might also suggest a switch from linear to non-linear behavior in creep experiments as shown in Figure 11.2.

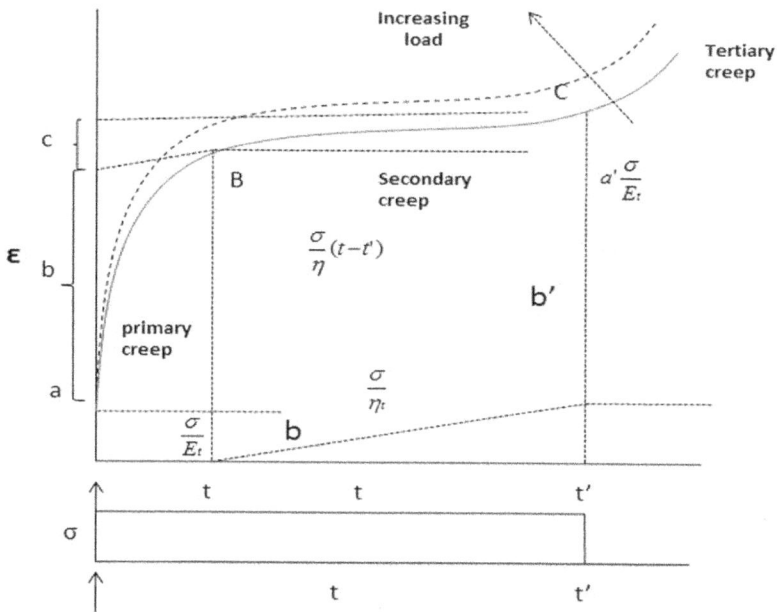

FIGURE 11.1 Snapshot of the creep behavior of biocomposites.

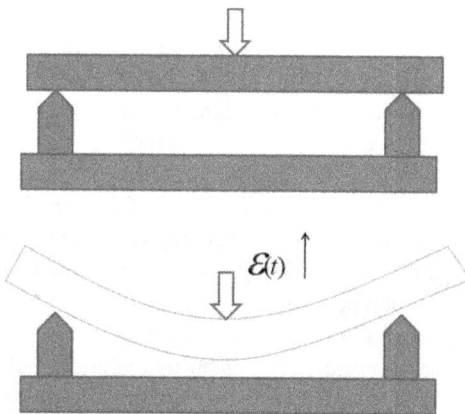

FIGURE 11.2 Biocomposite beam under bending loading to determine creep response.

While the biopolymer thermoplastic composite was subjected to continuous pressure for a long term, the glass transition temperature of the matrix is revealed. Irrespective of the direction of the applied load on a product, it gets compressed, bending, tension, or a combination of these stresses.

Strain rate or deformation measure over time can be articulated in terms of creep compliance D, as:

$$\varepsilon_t = \sigma_p D(t, T, \sigma_0) \tag{11.1}$$

As shown in Equation 11.1, the creep compliance $D(t)$ is the function of strain, stress, and time. When biocomposite response is being recorded for creep as the linear material then creep strain is independent of the level of load applied. The total strain at any particular instance of time $\varepsilon(t)$ in a short-term creep test of the biocomposite can be characterized by the sum of two deformations: (a) elastic deformation ε_E (i.e., the initial response in terms of instantaneous deformation of material under applied stress) and (b) ε_V viscoelastic deformation (i.e., time-dependent deformation). On the basis of the above deformations, creep compliance can be divided into viscous and elastic components, respectively. Therefore when biocomposites are subjected to constant loads, regardless of your working environment, there is always viscoelastic deformation of the material. Equation 11.2 shows a mathematical expression for calculating the compliance:

$$D_{(t)} = \frac{E_{(t)}}{\sigma_0} \tag{11.2}$$

Biocomposite products can be fabricated and designed on the primary of creep as a characteristic of time, while the specimen is subjected to steady stress in special working surroundings. This might also consist of cyclic loading, environmental, temperature, and other conditions. The biocomposites typically show creep behavior in all of the working surroundings at room temperature; particularly due to its micromechanical dating between the matrix and fiber. Therefore, design techniques are less difficult due to steady modules (except at excessive temperatures). However, the modulus of the polymer/composite fabric does not continue to be constant (as proven in Equation 11.2). Because the stiffness of the product that's without delay related to the deformation is a function of compliance and time by means of employing properly installed precise use of the biocomposites variation in creep deformation may be minimized. Or with the aid of editing the constituents of biocomposites and the addition of fillers as a reinforcement material the overall mechanical and viscoelastic performance is improved. For biocomposites, the goal is to evaluate the stress level that does not show everlasting deformation, fractures, or insupportable products. Intense deformation becomes a controlling component within the choice of labor attempt, concluding that it is crucial to succeed in specific quantity the deformation behavior of biocomposites, relying on temperature and time. A systematic representation of flow behavior (creep) is demonstrated in Figure 11.2; applied load indicates an outline of the four-factor bending biocomposite. The load at the side of gravity provides a steady attempt in biocomposites. Five days later, in this situation, no sizable adverse distortion occurs, but seven months later strain resulting from the attempt has elevated appreciably.

The biopolymer matrix composites product shows high sensitivity when applied under temperature and time, resulting in limited application such as the structure material. Products working under stress environment lose its functionality due to extreme deformation when this biocomposite is subjected to high stresses. Third stage (in which creep strain reaches to failure point) is known as the acceleration phase of creep and is also known as upper region. Material is designed in such a way that deformation does not lie in the tertiary region for a normal operation.

11.4 MATHEMATICAL MODELS

Data obtained from response (experimental) of a short-term creep test under tension, compression, or bending can be mathematically modeled with the help of spring and damper mechanical system. In these two mechanical systems, spring shows the elastic behavior of the biocomposites whereas damper represents viscous behavior-like viscous liquid. Spring follows the Hooke law under the application of applied stress, whereas the damper flow Newtonian law (i.e., stress is proportional to the strain rate). Biocomposites can be mathematically modeled with stress, strain, and time [21]. The mathematical model approximates the actual viscoelastic behavior of biocomposites, whereas these mechanical elements do not have any real similarities as compared to biocomposites. However, these mechanical elements in a certain combination show a mathematical understanding of the viscoelastic behavior of biocomposites. Presently, creep test is being accelerated in the laboratory, so that results are obtained in short duration. In this chapter, different mechanical models with their mathematical representation studied by many researchers are presented. When these models are applied to biocomposites to approximate their viscoelastic behavior, it allows researchers to analyze the creep deformation behavior at both short and long terms.

11.5 MATHEMATICAL MODELS FOR CREEP-RECOVERY BEHAVIOR

In Figure 11.3(a), the Maxwell model is shown on the left-hand side in which the spring and damper are connected in series, whereas in Figure 11.3(b), the Kelvin model (or Voigt) is shown on the right-hand side in which the spring is connected with the damper in parallel. In both, the system is characterized by time-dependent recovery where E_1 and E_2 represent the spring constant, i.e., elastic nature of the material, η_1 and η_2 represent visco-damper coefficient, i.e., time-dependent deformation of material, and η in the Maxwell model represents the relaxation time whereas response delay time for Kelvin model.

These two-parameter models can approximate the behavior of biocomposite but to analyze the flow behavior (creep) of biocomposites, it is likely to apply the four parameters model which itself is derived from the grouping of Kelvin model and Maxwell model in series, as shown in Figure 11.4.

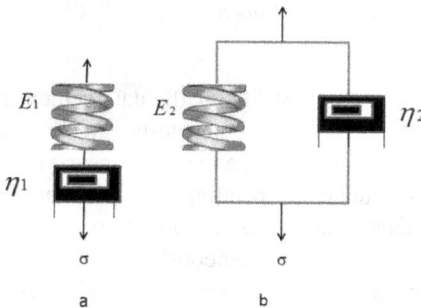

FIGURE 11.3 Two element (a) Maxwell model and (b) Kelvin model.

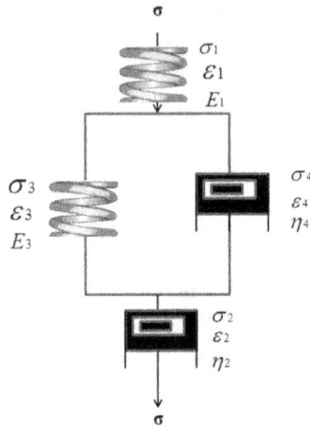

FIGURE 11.4   Burger model (four-element model).

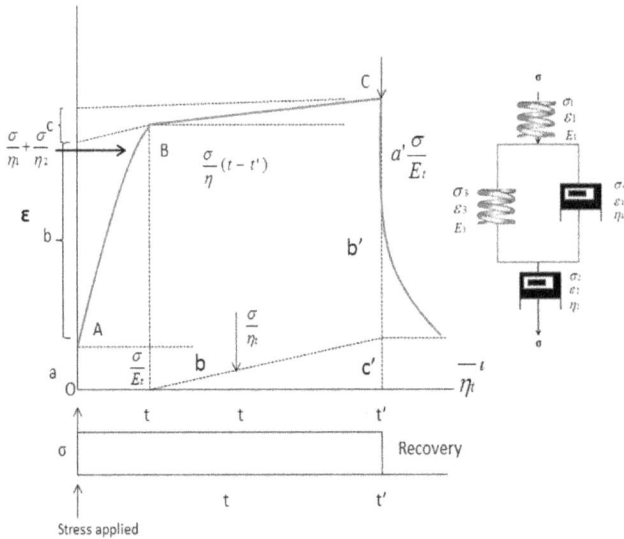

FIGURE 11.5   Creep curve fitting the Burger model.

The Burger model gives the best fits for the data achieved experimentally from the short-term creep test. Short-term creep tests can be obtained through DMA. Figure 11.5 represents a creep curve having the linear viscoelastic region of bio-compatible material subjected under bending at the constant load. The curve O to A represents the instantaneous reaction to the applied stress on creep curve, i.e., elastic response of material that occurs instantaneously. This behavior is followed by curve A to B representing time-dependent deformation. In this region, the creep strain rate decreases in a constant rate. As stressed is released, the instant elastic response (O to A) is completely recovered as represented from C to D, i.e., the distance a' = a.

After that, creep curve gets lower from point D to E showing steady-state recovery. However, recovery cannot occur up to the initial state, i.e., c' = c. This response is a measure of plastic flow and is completely unrecoverable.

Figure 11.5 shows the elastic reaction variations in A and A'. An appropriate law in which reaction elements are modeled is the Hooke law. The Voigt-Kelvin version resembles modifications in b' (trade image) within the state of creep (c' representing plastic deformation of damper). Figure 11.6 suggests four-element models (Burger model) for demonstrating manifestation of float curve.

The Burger model is an assemblage of Kelvin-Voigt model and Maxwell model, in which the latter element is time-dependent. Figure 11.6 represents the different zones of Burger model: zone 1 represents the idle system without load. In zone 2, constant three-point bending load/stress is applied, which results in an instantaneous extension in spring E_1 to the curve point 'a', i.e., σ/E_1 = a. After that, zone 3 (shown in Figure 11.6) shows that the fluence rate decreases until E_2 is extended fully with a steady rate increment in load-bearing capacity of spring (E_2), but still damping (η_2) does not carry any load. As soon as spring E_2 gets fully extended, the creep load reaches the solid phase denoted by the constant η_1 damper, representing the plastic flow of the biocomposite material, i.e., linear viscoelastic region. The damper deforms until the load is removed, leaving behind the permanent deformation in material as illustrated by zone 4 in Figure 11.6. Now, in zone 5, load is

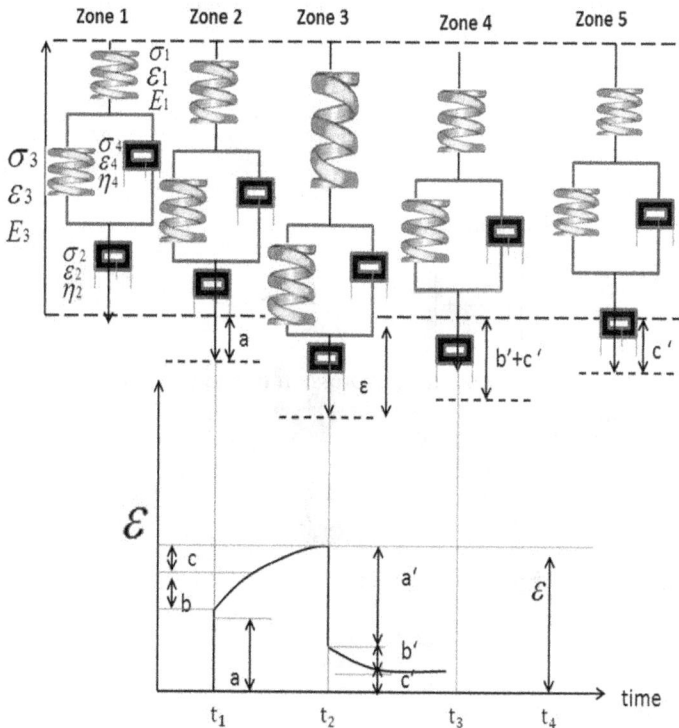

FIGURE 11.6   Creep response of biocomposite fitted by four-parameter model.

removed from the material, which results in quick retraction of spring E_1 to their original position, i.e., a = a', and then the time-dependent recovery period is b'. Through this period, initially the spring (E_2) restricts the motion of damper (η_2) and maintains its original position representing an anelasticity or delayed elastic response. But the final position of damper (η_3) remains in extended form as it is not affected by the spring; this is shown in zone 5 of Figure 11.6. Thus, the plastic flow which is non-recoverable in nature is equal to c' = σ_t/η_3. This Burger model is fully characterized by the following behaviors of biocomposites: elastic, then inelastic, and finally viscous deformation. The natural fibres are used as fillers to improve the strength of the biocomposites. However the strength may be affected by the interfacial bonding between the fiber–matrix interface, which in turn may vary the damping coefficient η_1.

In the four-parameter model (given the balance of forces), we can mark down the following expressions (Equations 11.3 and 11.4), with respect to the deformation and effort:

$$\sigma = \sigma_1 = \sigma_2 = \sigma_3 + \sigma_4 \tag{11.3}$$

$$\varepsilon = \varepsilon_1 + \varepsilon_2 + \varepsilon_{(3,4)} = \varepsilon_1 + \varepsilon_2 + \varepsilon_k \ldots \varepsilon_3 = \varepsilon_4 \tag{11.4}$$

where the Kelvin-Voigt model strain is represented by ε_k. Different stress-strain relations such as Hook's law (for spring) and Newton's law (damper) are presented in Equation 11.5:

$$\sigma_1 = E_1, \ \varepsilon_1\sigma_2 = \eta_2\left(\frac{d\varepsilon_2}{dt}\right), \ \sigma_3 = E_3\varepsilon_3, \ \sigma_4 = \eta_4\left(\frac{d\varepsilon_4}{dt}\right) \tag{11.5}$$

And the equation that relates and models the behavior of viscoelastic-plastic biocomposite holistically can be approved as follows (Equation 11.6):

$$\frac{\eta_1\eta_2}{E_1E_2}\left[\frac{d^2}{dt^2}\right] + \frac{\eta_1}{E_1} + \left[\frac{\eta_1+\eta_2}{E_2}\right]\frac{d\sigma}{dt} + \sigma = \frac{\eta_1\eta_2}{E_1}\frac{d^2}{dt^2} + \eta_1\frac{d\varepsilon}{dt} \tag{11.6}$$

Equation 11.7 shows the mathematic expression that approximates the creep deformation of biocomposites:

$$\epsilon(t) = \frac{\sigma_0}{E_1} + \frac{\sigma_0}{\eta_1}t + \frac{\sigma_0}{E_2}\left(1 - e^{\frac{E_2 t}{\eta_2}}\right) \tag{11.7}$$

And Equation 11.8 shown below approximates the creep-recovery behavior:

$$\epsilon(t) = \frac{\sigma_0}{\eta_1}t + \frac{\sigma_0}{E_2}\left(e^{\frac{E_2 t}{\eta_2}} - 1\right)e^{\frac{E_2 t}{\eta_2}} \tag{11.8}$$

where σ_0 is the initial applied stress, $\varepsilon(t)$ is the creep strain rate, and t is the time. E_1 and E_2 represent the elastic modulus of the Kelvin and Maxwell spring elements, respectively. η_1 and η_2 represent the viscosity coefficients of the Kelvin and Maxwell damper elements, respectively. $\tau = \eta_2/E_2$ usually denotes delayed time, i.e., the time required to generate 63.2% strain on the Kelvin unit [22].

In Equation 11.7, the first expression, i.e., $\dfrac{\sigma_0}{E_1}$, is the instantaneous elastic response to the external applied load. The second expression $\dfrac{\sigma_0}{\eta_1}t$ shows the initial zone of creep strain, and it mainly occurs due to relaxation and extension mechanisms of the biopolymer chain. Biocomposite's creep deformation is closely related to the matrix-fiber interfacial bonding. Long-term creep strain is represented by the last term expression in Equation 11.7. The data obtained from the creep test are curve fitted with respect to Equation 11.7 by adjusting parameters E_1, E_2, η_1, and η_2 so that overall creep behavior of biocomposites can be predicted. From Equation 11.7, creep strain rate of the biocomposites can be derived, and the strain rate can be obtained from Equation 11.9:

$$\frac{d\varepsilon}{dt} = \frac{\sigma_0}{\eta_1}t + \frac{\sigma_0}{\eta_2}\left(e^{\frac{-E_2 t}{\eta_2}}\right) \tag{11.9}$$

The creep compliance $D(t)$ is defined by Equation 11.10 below:

$$D_{(t)} = \frac{1}{E_1} + \frac{1}{\eta_1} + \frac{1}{E_2}\left[1 - e^{\frac{-E_2 t}{\eta_2}}\right] \tag{11.10}$$

Biocomposite shows different time delay viscoelastic behaviors; therefore, to increase the accuracy of the Burger model, the number of Kelvin units should be increased and attached with one another in a series as shown in Figure 11.7.

By solving the differential equation with appropriate load important viscoelastic properties like relaxation or creep compliance can be determined. For example, compliance can be determined from the creep test performed on biocomposites and would be represented by Equation 11.11:

$$D_{(t)} = \frac{1}{E_1} + \frac{1}{\eta_1} + \frac{1}{E_2} \tag{11.11}$$

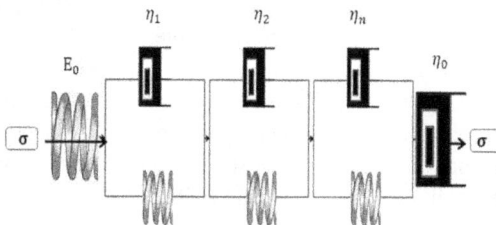

FIGURE 11.7   Multiple models for adjusting creep curves.

11.6　TIME-DEPENDENT DEFORMATION BEHAVIOR OF BIOCOMPOSITE

Filler materials, that also act as reinforcement, play a vital role in the development of biocomposites. Study of creep behavior as a function of environment temperature help in determining the nature of the biocomposites. DMA technique can be employed to perform the creep and creep-recovery tests in a short duration with a high level of accuracy. To evaluate the creep-recovery behavior of bioproducts using dynamic mechanical analyzer, a constant load is applied on the standard specimen for a particular period of time (short or long term). Correspondingly for the same interval of time, recovery behavior is studied with the same specimen in a continuous process. Salazar [23] fabricated bioproducts using polyethylene as matrix and aluminum as reinforcement. Creep and recovery behavior of materials was studied. During the experiment, a constant stress of 1.2 MPa was applied in a three-point bending condition at 25°C.

With a proper evaluation of the value of elements of Burger model, we can approximate the creep and recovery behavior of biocomposites. Moreover, by adjusting the values of Burger model element, a researcher can study the effect of reinforcement and also the effect of nano-reinforcements used in the development of biocomposites. A similar type of study has been done by Salazar [23] on the behavior of an LDPE-fabricated biocomposite Al-Fique.

11.7　STRAIN RATE OF BIOCOMPOSITES

To evaluate the strain rate of biocomposites exposed to a constant load, the Burger model can be employed to approximate the experimental data obtained using the mathematical model presented in Equation 11.6. Nowadays, the focus of researchers is shifted to study the variation in the value and relationship between elements of Burger model when biocomposites are incorporated with fibers; effect of surface treatment such as application of coupling agent to increase interfacial bonding; nano-reinforcements; etc. One can observe from the data obtained from the creep test done by Salazar [23] that it is approximated by both Burger model (four-element model) and n parameters model. When compared to the Burger model, the biocomposites with filler and surface treatment n-parameter model provides the best fit of the creep curve. But due to increase in calculation complexity, Burger model is highly preferred. We can observe the reaction of viscous parameter n_1 as shown in the work of Salazar [23], which permits the evaluation of strain rate for the creep tests. This experiments were performed by altering the wt% of natural fibers from fique biocomposites. Thus, deepening the study of effect of fillers or reinforcements to biocomposites. Salazar [23] also evaluated the values of the parameters of Burger model on two biocomposites, with dissimilar volumes of biodegradable sisal fiber reinforcement. Salazar [23] showcased the effect of error inclusion on adjustment of the various parameters for a six parameters model, Salazar [23] demonstrated the effect of error inclusion on parameter adjustment for a six-parameter model, which is the approximation of least squares adjustment parameters and can be calculated using the weighted sum of squares (WSS), using Equation 11.12:

$$\text{WSS} = \sum_{i1}^{n} w_i \left[\varepsilon(t_i) - \varepsilon^{\wedge}(t_i) \right]^2 \tag{11.12}$$

where $\varepsilon(t_i)$ symbolizes the experimental deformation observed at time (t_i), and ε is predicted by the respective model. A larger value indicates an inferior fit WSS model to the experimental data. In addition, w_i is the difference between two samples of time deformation.

Salazar [23] observed micrographs of an LDPE-Al-Fique biocomposite attained by the scanning electron microscopy, where we were able to observe that fiber possesses a hydrophobic property/nature, not adhering properly to the polymer domain, while aluminum has a little adhesion. On the other hand, the compression molding manufacturing technique helps the adhesion between respective faces that can be reflected in a decrease of viscoelastic and mechanical performances of resulting biocomposites.

Presently, it is acknowledged that those defects may be improved by surface treatments in fibers. The use of coupling agents and changes in the polymers can lead to better micromechanical bonding and lesser strain rate.

11.8   APPLICATIONS

Most renewable polymers are developed through the traditional plastic-processing strategies with a few changes of processing situations and alterations of equipment. Blow molding, injection molding, thermoforming, and film extrusion are a few of the fabrication strategies utilized. The three important sectors where renewable polymers had been brought in are agriculture, medicine, and packaging. Renewable polymer applications include not only matrices for controlled-release devices, enzyme immobilisation, and pharmacological devices. [24]; however, also as brief prostheses, therapeutic gadgets, porous structure for tissue engineering. Because biopolymers have a low water solubility and a significant water uptake, they can be used as absorbent substances in healthcare, agricultural, and horticulture packages [25]. Packaging waste has brought growing environmental concerns. Expansion of packaging materials (biodegradable) has acquired growing interest [26]. Table 11.1 regroups a few materials with their uses.

Various applications are as follows:

1. Medicine and pharmacy
2. Natural or bacterial polymers
3. Synthetic polymers
4. Packaging
5. Agriculture
6. Other fields including automotive, electronics, construction, sports, and leisure.
7. Biotechnological applications: Disposability and short-term life character applications.
8. Uncommon packages: There are numerous different applications which are no longer healthy in any of the previous classes. Hence, pens (Green Pen® from Yokozuna and Begreen® from Pilot Pen), combs, and mouse pads fabricated from renewable polymers. Renewable polymers may be utilized to amend food textures. Because of its non-toxicity nature, alginate is used as

TABLE 11.1
Various Materials with Their Applications

Product	Applications	Applications/Polymers	Society
Polynat®	Containers	Flower	Roverc'h (France)
Mater-Bi®	Collection bags for green waste, agricultural films, disposable items	Starch and polyester	Novamont (Italy)
Biopol®	PHB/PHV razors, bottles		Goodfellow (Great Britain)
Ecofoam®	Starch wrapping plastics		American Excelsior Company (USA)
Bio-D®	Agricultural films	Proteins extracted from cotton seed	Cirad (France)
Eco-PLA®	Sport clothes, sanitary products, packaging and conditioning	PLA	Cargill Dow (USA)
Eastar Bio®	Agricultural films	Co-polyester	Eastman (Great Britain)
BAK 1095®	Disposable items, flower containers	Polyester amide	Bayer (Germany)
Ecoflex®	Agricultural films	Co-polyester	BASF (Germany)

a thickener in ice creams, salad dressings, and also as additive in meals. Chitosan and chitin are used as feed and meal additives [27]. PLA (semi-synthetic polymers) is used in compostable meals.

11.9 CONCLUSIONS

Scrutiny on the viscoelastic properties of the completely biodegradable polymer-reinforced composite materials has been reviewed in this chapter. Various research articles and books have reported the trends in the configuration modifications and mechanical properties of composites. The authors have also described in detail the mathematical models that describe the time-dependent deformation behavior of bio-composite and the strain rate. Furthermore, many mechanisms have been covered that occur inside the biodegradable polymer-based composites domain when subjected to creep phenomenon.

ACKNOWLEDGMENT

Monetary and academic support from the University of Petroleum and Energy Studies SEED-Grant, India is highly appreciated.

CONFLICTS OF INTEREST

There are no conflicts of interest to declare by the authors.

REFERENCES

1. Lucas, N., Bienaime, C., Belloy, C., Queneudec, M., Silvestre, F. and Nava-Saucedo, J.E., 2008. Polymer biodegradation: Mechanisms and estimation techniques–A review. *Chemosphere*, *73*(4), pp. 429–442.

2. Willett, J.L., 1994. Mechanical properties of LDPE/granular starch composites. *Journal of Applied Polymer Science*, *54*(11), pp. 1685–1695.

3. Cho, J.W., Woo, K.S., Chun, B.C. and Park, J.S., 2001. Ultraviolet reflective and mechanical properties of polyethylene mulching films. *European Polymer Journal*, *37*(6), pp. 1227–1232.

4. Jaserg, B., Swanson, C., Nelsen, T., Doane, W., 1992. Mixing polyethylene-poly (ethylene-co-acrylic acid) copolymer starch formulations for blown films. *Journal of Polymer Materials 9*, 153–162.

5. Lawton, J.W., 1996. Effect of starch type on the properties of starch containing films. *Carbohydrate Polymers*, *29*(3), pp. 203–208.

6. Briassoulis, D., 2006. Mechanical behaviour of biodegradable agricultural films under real field conditions. *Polymer Degradation and Stability*, *91*(6), pp. 1256–1272.

7. Xu, Y., Wu, Q., Lei, Y. and Yao, F., 2010. Creep behaviour of bagasse fiber reinforced polymer composites. *Bioresource Technology*, *101*(9), pp. 3280–3286.

8. Cheung, H.Y., Ho, M.P., Lau, K.T., Cardona, F. and Hui, D., 2009. Natural fibre-reinforced composites for bioengineering and environmental engineering applications. *Composites Part B: Engineering*, *40*(7), pp. 655–663.

9. Suardana, N.P.G., Piao, Y. and Lim, J.K., 2011. Mechanical properties of hemp fibers and hemp/pp composites: Effects of chemical surface treatment. *Materials Physics and Mechanics*, *11*(1), pp. 1–8.

10. Martins, C., Pinto, V., Guedes, R.M. and Marques, A.T., 2015. Creep and stress relaxation behaviour of PLA-PCL fibres: A linear modelling approach. *Procedia Engineering*, *114*, pp. 768–775.

11. Dobah, Y., Bourchak, M., Bezazi, A., Belaadi, A., 2013. Static and fatigue strength characterization of sisal fiber reinforced polyester composite material. In: *9th International Conference on Composite Science and Technology: 2020-Scientific and Industrial Challenges (ICCST/9)*, 24–26 April 2013, Sorrento, Naples, Italy.

12. Jabbar, A., Militky, J., Madhukar Kale, B., Rwawiire, S., Nawabb, Y. and Baheti, V. 2016. Modeling and analysis of the creep behaviour of jute/green epoxy composites incorporated with chemically treated pulverized nano/micro jute fibers. *Industrial Crops and Products*, *84*, pp. 230–240.

13. Kiguchi, M., 2007. Latest market status of wood and wood plastic composites in North America and Europe. In: *The Second Wood and Wood Plastic Composites Seminar in the 23rd Wood Composite Symposium*, Kyoto, Japan, pp. 61–73.

14. Belaadi, A., Bezazi, A., Maache, M. and Scarpa, F. 2014. Fatigue in sisal fiber reinforced polyester composites: Hysteresis and energy dissipation. *Procedia Engineering*, *74*, pp. 325–328.

15. Haq, S. and Srivastava, R., 2016. Measuring the influence of materials composition on nano scale roughness for wood plastic composites by AFM. *Measurement*, *91*, pp. 541–547.

16. Alkbir, M.F.M., Sapuan, S.M., Nuraini, A.A. and Ishak, M.R., 2016. Fibre properties and crashworthiness parameters of natural fibre-reinforced composite structure: A literature review. *Composite Structures*, *148*, pp. 59–73.

17. Acha, B.A., Reboredo, M.M. and Marcovich, N.E., 2007. Creep and dynamic mechanical behaviour of PP–jute composites: Effect of the interfacial adhesion. *Composites Part A: Applied Science and Manufacturing*, *38*(6), pp. 1507–1516.

18. Premalal, H.G., Ismail, H. and Baharin, A., 2002. Comparison of the mechanical properties of rice husk powder filled polypropylene composites with talc filled polypropylene composites. *Polymer Testing, 21*(7), pp. 833–839.

19. Nuñez, A.J., Marcovich, N.E. and Aranguren, M.I., 2004. Short-term and long-term creep of polypropylene-woodflour composites. *Polymer Engineering & Science, 44*(8), pp. 1594–1603.

20. Samuel, J.B., Jaisingh, S.J., Sivakumar, K., Mayakannan, A.V. and Arunprakash, V.R., 2020. Visco-elastic, thermal, antimicrobial and dielectric behaviour of areca fibre-reinforced nano-silica and neem oil-toughened epoxy resin bio composite. *Silicon,* volume 13, pages 1703–1712 (2021).

21. Ascione, L., Berardi, V., D'Aponte, A., 2012. Creep phenomena in FRP materials. *Mechanics Research Communications, 43*, pp. 15–21.

22. Pothan, L.A., Oommen, Z. and Thomas, S., 2003. Dynamic mechanical analysis of banana fiber reinforced polyester composites. *Composites Science and Technology, 63*(2), pp. 283–293.

23. Salazar, M.A.H., 2016. Viscoelastic performance of biocomposites in book composites from renewable and sustainable materials. *IntechOpen*, 303–331, doi:10.5772/66148.

24. Castro, G.R., Panilaitis, B. and Kaplan, D.L., 2008. Emulsan, a tailorable biopolymer for controlled release. *Bioresource Technology, 99*(11), pp. 4566–4571.

25. Kiatkamjornwong, S., Chomsaksakul, W. and Sonsuk, M., 2000. Radiation modification of water absorption of cassava starch by acrylic acid/acrylamide. *Radiation Physics and Chemistry, 59*(4), pp. 413–427.

26. Petersen, K., Nielsen, P.V., Bertelsen, G., Lawther, M., Olsen, M.B., Nilsson, N.H. and Mortensen, G., 1999. Potential of biobased materials for food packaging. *Trends in Food Science & Technology, 10*(2), pp. 52–68.

27. Agulló, E., Rodríguez, M.S., Ramos, V. and Albertengo, L., 2003. Present and future role of chitin and chitosan in food. *Macromolecular Bioscience, 3*(10), pp. 521–530.

12 Influencing Effects of Hybridization and Addition of Fillers on the Viscoelastic Properties of Reinforced Polymer-Based Composites

Sameena Shaik, Dharma Raghu Raj Reddy,
Anwar Kornipalli, and Ashish Kasam
Hindustan Institute of Technology & Science

D. Aravind
Madurai Kamaraj University
Kalasalingam Academy of Research and Education

Senthilkumar Krishnasamy
King Mongkut's University of Technology North Bangkok

Chandrasekar Muthukumar
Hindustan Institute of Technology & Science

Senthil Muthu Kumar Thiagamani
Kalasalingam Academy of Research and Education

CONTENTS

DOI: 10.1201/9781003173625-14

12.1 INTRODUCTION

Mankind, which depends on nature for everything, will face restrictions for their energy needs procured through different phases. Fossil fuels are the primary energy sources. This is because of their demand for sophistication in the current modern lifestyle. The need diminishes the availability of fossil fuels at a very rapid phase than its creation, which has taken millions of years. Besides, fossil fuels highly pollute the environment, and their utilization leaves out wasteful remains also. Thus, all these problems are becoming a global issue [1,2].

Polymers are the real alternate solution for traditional materials such as steel, aluminum, etc., in numerous applications because of their better performance and competitive costs compared to their counterparts [3–8].

Polymer production always includes energy as an essential element of cost, but the expected shortfall in fossil fuels denotes severe conservation issues [9]. As most of the industries depend on it, the fuel crisis has focused fuel on the conservation agenda. The polymer industry especially the plastic sector, depends not only on energy but also on raw material. This sector has been aware of its requirement of fossil hydrocarbons for a long. It is known that although the engineers have deployed energy-efficient techniques in individual plants, the true energy demand assessment could be carried out with proper consideration of all the production steps from the extraction of raw materials to the final polymer [10]. This strengthens a prevailing argument that puts the actual cost of plastics produced to be calculated using natural resources.

Also, the production of polymers, mainly synthetic, enhances pollution starting from the synthesis. The end products are also believed to cause plastic pollution [11]. Earth, which possesses a complex ecology, is blessed to have naturally recycling processes that use solar system radiation. However, any deviation between resources and their utilization can affect its availability even by a small ratio. Hence, it is essential to configure new methods and technologies to achieve competitiveness in cost and enhance performance [12]. This is influenced by corporate sectors as well.

12.2 VISCOELASTIC PROPERTIES

From the term, viscoelastic (viscous + elastic) refers to exhibiting both viscous characteristics and elastic characteristics. This phenomenon occurs while undergoing deformation. An example of viscous material is water, which resists shear flow and strains linearly with respect to time on the application of stress. An example of elastic material is rubber, which strains on stretching and gets back to its original form immediately after removing stress. The viscoelastic materials will have both viscous and elastic behavior. They manifest time-dependent strains. They exhibit elasticity due to orderly stretching of bonds along planes in solid that is crystallographic in nature an ordered solid, whereas viscosity occurs due to the diffusion of atoms/molecules within an amorphous structure [5,13].

Polymer-based composites are highly viscoelastic since they contain longer molecules that get entangled with neighboring ones. In simple terms, viscoelastic behavior can be given as follows:

- A characteristic of materials having both fluid and elastic properties simultaneously.
- It is caused due to a link between fiber-like particles that are temporary.

Any material having viscoelastic property will have the following:

- Hysteresis in the stress–strain curve.
- Relaxation in stress happens as step constant strain decreases the stress.
- Creep happens as step constant stress increase the strain.
- Stiffness is dependent on
  (i) the strain rate, or
  (ii) the stress rate.

12.3 NEED FOR POLYMER-BASED HYBRID COMPOSITES

Kesle reported that resources for the future are directly proportional to the ratio between recorded resources and their annual consumption [14]. Also, recycling minimizes energy by restricting new synthesis; thereby, energy utilized becomes minimal to process an already synthesized material. This contributes to the conservation of energy [15,16].

Despite all environmental concerns over polymer-related industries, particularly synthetic ones, they use only 4% of the total global petroleum production. This is at a minimal best when compared to various other sectors. All these rising concerns, including environmental concerns and the disadvantages of the polymer as mentioned, have resulted in finding polymer-based composites as an alternate solution [17].

The polymer-based composites are fabricated using different combinations, which include (a) resins, (b) fibers, (c) fillers, and (d) additives. The incorporating materials could be synthetic or natural or a combination of both. The reinforcements are also combined with bio-additives and biofillers to get a more stable structure. Fillers, as the name suggests, are used to occupying the resin partially to enhance cost-effectiveness. Some of the widely used fillers are calcium carbonate, talc, and mica. In general, additives act as stabilizers for heat, impact modifiers, retardants for flame, coloring agents, lubricating aids, anti-blocking and anti-static materials, antioxidants, etc. The nature of incorporation depends on required material properties like mechanical, thermal, viscoelastic, wear, friction, and lubrication properties. Natural-based polymer composites involve natural compositions to answer environmental-related problems.

The entire structure of pure natural composites, including additives and fillers, should be of bio-origin or from renewable sources [18–21]. However, since reinforcements between synthetic and natural such as glass/carbon fiber, jute fiber, kenaf fiber, hemp fiber, bamboo fiber, wheat straw, sisal fiber, agave fiber, abaca fiber, cottonseed fiber, coir fiber, wood fiber, etc., or incorporation between synthetic/natural fiber with synthetic/natural polymers and/or synthetic/natural filler and/or synthetic/natural additives have given performance-enhancing results [22–24]. Sometimes, nanoclay compositions as per requirement contribute to the fabrication of nanocomposites [25].

Frequently used matrix resins include epoxy, polyimides, and bismaleimides. They command advanced polymer due to their increasing demand in automobile, marine, aerospace, and construction [26,27]. These ever-rising demands have added new dimensions and forms, which have brought evolutionary changes in polymer-based composites.

12.4 STRUCTURE OF HYBRID POLYMER-BASED COMPOSITES

Most commonly, the structure of polymer-based composites is hybrids with different combinations to integrate individual advantages as a single structure [28]. The structure has a resin matrix reinforced with fibers, which contributes to the strength of the matrix. The resin matrix is responsible for holding the fibers together [29]. Thus, the final composite will have advantages of all-inclusive individual materials. Moreover, in advanced composites, continuous fibers with resin matrix incorporation provide high strength and stiffness properties. The selection of composite materials depends on choosing suitable reinforcing fiber and the matrix in the right proportion according to the requirements [30]. Practically, polymer matrix composites are widely used. These fiber-reinforced polymers are also commonly called plastics.

12.4.1 FIBER REINFORCEMENTS OF POLYMER-BASED HYBRID COMPOSITES

Fibers occupy around 30%–70% by volume in any hybrid polymer-based composites. They are generally chopped, woven, stitched, and/or braided and are then subjected to treatment with starch, gelatin, oil, or wax. These treatment agents act as sizing elements as well as binders. The sizing elements are responsible for bonding, and binders are accountable for improving handling. Some widely used fibers include fiberglass, aramid (human-made by spinning fiber from the chemical blend), and carbon [31]. Table 12.1 gives the significant differences between the fiberglass, carbon, aramid fiber, and other fibers.

12.4.2 RESIN SYSTEMS OF HYBRID POLYMER-BASED COMPOSITES

Resin is another essential element to form a matrix in the composite. They are of two types, namely thermoplastics and thermosets. The former is solid at standard conditions and could be melted on heating, whereas the latter does not [40]. The

TABLE 12.1
Widely Used Fibers

Fiberglass	Carbon	Aramid Fibers	Other Fibers
It is the cheapest [32,33]	The most expensive one [34]	The cost of these is similar to low grades of carbon fiber [35]	Being economically competitive, these are in high demand currently [36]. Example: Boron, natural fibers jute kenaf, etc.
They have a low thermal expansion coefficient compared to steel, but their strength and modulus decrease with increased temperature [37]	They have a low thermal expansion coefficient when compared with the glass and the aramid fibers. Their transverse modulus magnitude is less than their longitudinal modulus with very high fatigue and creep resistance. Their tensile strength decreases with an increase in modulus. Hence, their strain at rupture will be significantly less. Their brittleness at high modulus makes it critical for joint and connection due to high-stress concentrations. Hence, these are suitable for adhesive bonding, which eliminates mechanical fastening [38]	They have high specific strengths due to their low density, excellent impact resistance, good fatigue resistance, and better creep resistance. However, there will be restrictions in fiber–resin bonding [39]	They have high strength and high modulus [36]

thermoplastics cannot be cured due to long-chain, which cannot form cross-link chemically. However, thermosets can be cured permanently [41]. Thus, there always exists a preference for its structural stability. The most preferred resins are unsaturated polyesters, epoxies, and vinyl esters, while the least preferred resins are polyurethanes and phenolics.

12.5 INFLUENCING FACTORS OF VISCOELASTIC PROPERTIES OF POLYMER-BASED COMPOSITES

The polymer-based hybrid composites inherit the characteristic of the matrix during manufacturing, which makes the performance of composites dependent on the strain that is applied over them. Hence, the viscoelastic properties are strongly affected by the time-dependent applied stress and the temperature. When these polymer-based hybrid composites take their required dimension as per application, for instance, a beam or a mold, load on them could cause specific effects due to constant load. These effects, like a creep, impact on the matrix, the fiber, and the interface, could cause product failures. This creep or viscoelastic deformation could become an issue since the loads are applied for a prolonged period.

 Although various factors influence viscoelastic properties such as (a) frequency at excitation, (b) temperature, (c) dynamic strain rate, (d) pre-load at static condition, (e) creep and relaxation, which is time-dependent, (f) aging, and (g) other irreversible effects, our chapter is restricted to discuss regarding two influential parameters, namely

 (i) hybridization and
 (ii) addition of wood fillers.

12.5.1 Influencing Effects of Hybridization of Various Natural Fibers

Hybridization of polymer-based composites is an effective tool to enhance the performance of composites, especially in solving practical difficulties. Thus, hybridization facilitates the adoption of this class of composites for many applications. Table 12.2 reports on some hybridization carried out with polymer-based composites and their influence on their viscoelastic properties.

TABLE 12.2
Influencing Effects of Hybridization on Polymer-Based Composites

Hybridization	Effect of Hybridization
Palmyra Palm Leaf Stalk Fiber/jute fiber reinforced polyester hybrid composites [42]	• The storage modulus enhanced for composite with high jute content. • A positive shift of tan δ peaks at high temperatures. • Higher static and dynamic mechanical performance was achieved.
Biodegradable green epoxy hybrid composites with hemp and sisal [24]	• The hybrid combination was less effective on storage modulus. • The values of tan δ obtained showed only a small shift in temperature for the hybrid combination compared to pure forms.
Epoxy hybrid composites reinforced with 2-hydroxy ethyl acrylate-treated oil palm empty fruit bunch and jute fibers [43]	• Treated fiber-based hybrid composites showed higher storage modulus as they have (a) increased surface area and (b) better cross-connections inside matrix–fiber bonding. • The treated forms showed higher E value than untreated forms above tan δ, which shifted toward lower temperature.
Kevlar/*Cocos nucifera* sheath (CNS) reinforced epoxy hybrid composites [44]	• The storage and loss modulus were increased for hybrid combination 75 Kevlar/25 CNS because of improved interfacial interactions and efficient stress transfer rates. • Also, viscoelastic damping was better for the same combination.
Hybrid bio-composite with polylactic acid (PLA) reinforced with flax and jute [45]	• The storage modulus obtained was noteworthy with the value of 2,524 MPa. The reinforcement in hybrid form served to improve the solidness. Also, a higher loss modulus curve was achieved by hybrids when compared to neat form. • The tan δ value obtained was lower than perfect PLA. • The hybrid Jute/PLA also achieved a lower damping factor (0.65). Additionally, a higher T_g value of 68.68°C was achieved by Flax/PLA as demonstrated by damping bends compared to neat form.

12.5.2 Influencing Effects of Addition of Wood Fillers

At present, the need for composite materials that could be recycled, that are friendly to the environment, and more importantly, that are cost-effective have dramatically increased. With these as objectives, the researchers have started searching for wood sources for using polymers. These composites that also use waste wood particles are eco-friendly, low in cost, and enhance mechanical and physical properties. Most of the time, in these composites, waste wood particles were filled into the polymers. Fabrication techniques such as extrusion, hand layup, compression molding, injection molding, and additive manufacturing (currently 3D printing) were deployed for manufacturing [46]. The addition of wood fillers into polymer-based composites is, in itself, a kind of hybridization.

The viscoelastic performance of these composites depends on the interactions between the polymer and the filler [47]. Table 12.3 reports on the addition of some wood fillers into polymer-based composites and their influence on their viscoelastic properties.

TABLE 12.3
Influencing Effects of Addition of Wood Fillers into Polymer-Based Composites

Addition of Wood Fillers	Effect of Addition of Wood Fillers
Cork-wood filler system for polypropylene and polylactic acid hybrid composite [48]	• Measurements were done at i. the range of −50°C to 150°C, ii. the constant strain of 0.01%, iii. frequency of 1 Hz, and iv. Heating rate was set to 3°C/min. • Changes in storage modulus were evident in PP composites with the use of cork-wood filler. • Changes in storage modulus were evident in PLA composites with the use of cork-wood filler. • Higher stiffness of the pure PLA resins with cork filler decreases the storage modulus values in a small way, but storage modulus increases slightly for wood flour-based materials. • An increase in storage modulus after the cold crystallization region projects improvement in mechanical behavior and thermal resistance.
Biochar filler system for glass fiber-reinforced hybrid composite [49]	• The results of three-point bend tests showed i. less damping (300 MPa) and ii. higher storage moduli (4000 MPa). • High stiffness for 10% char composite in comparison with 5% char and without char composite. • The stiffness of activated charcoal was higher than biochar.
Reinforcing biofiller system for wood–plastic hybrid composite [50]	• Maximum storage modulus value for biofiller system for wood–plastic hybrid composite because of strong interaction of carbon with the matrix. • An increase in tan δ as the temperature increase with a decrease in storage modulus and an increase in loss modulus.

(Continued)

TABLE 12.3 (*Continued*)
Influencing Effects of Addition of Wood Fillers into Polymer-Based Composites

Addition of Wood Fillers	Effect of Addition of Wood Fillers
Wood-based flour filler system for epoxy-based thermosetting polymer hybrid composite [51]	• Investigation was carried out in nitrogen atmosphere at i. temperature range from RT to 200°C, ii. heating rate of 5°C iii. fixed frequency of 1 Hz iv. with a static strain of 0.2% v. dynamic strain of 0.1% • A significant increase in storage modulus when compared with neat form confirms good reinforcement of fillers. • Reduction in storage modulus values as the temperature increases because of molecular mobility, breaking cross-links between the molecular chain.
Wood filler system for high-density polyethylene (HDPE)–wood hybrid composite [52]	• With the addition of wood filler percentage increases, the composite becomes more elastic than viscous.

All these studies revealed that the effects inherited by polymer-based composites attribute to transitions underwent by their matrices, and always fiber matrix-filled showed that the effect creep is always positive for these composites. Although, when exposing the composite to constant loads with respect to time and at the above glass transition temperature, the composites behave like a super-cooled fluid rather than an elastic solid.

12.6 HYBRIDIZATION OF POLYMER-BASED COMPOSITES WITH BIO-ORIGIN SUBSTANCES FOR SUSTAINABILITY

Sustainability is termed as an interaction that we, the human beings, make with nature for an effective and need only based utilization of resources available in nature. Apart from giving us the required properties such as mechanical, thermal, and viscoelastic properties, to name a few, hybridization of polymer-based composites with bio-origin materials helps in two ways:

(i) minimizes the volume of polymers used because some of the polymers are of environmental concern and
(ii) more importantly, it reduces the risk of hazardous debris from entering the ecosystem, thereby reducing the risk of disturbing the balance of the ecosystem.

Also, hybridization of polymer-based composites with bio-origin composition would help to improve waste management as bio-origin substances are involved. In addition, using suitable plastics of bio-origin or renewable plastics facilitates degradability that will not disturb the ecosystem. Apart from these, there could be commendable energy conservation if the composites could be subjected to recycling as reuse minimizes new production.

The prospects of achieving sustainability in the long term are dangerous and should be addressed shortly as we are in a situation to face the rapid depletion of natural resources. When synthesizing or recycling is costly, it is mandatory to develop and deploy new technologies that are economical and take us toward. But often find that only the quantitative data support the concerns while more substantial magnitudes have always been missing.

Hybridization of polymer-based composites with bio-origin substances gives a material formed by a polymer matrix and a filler or reinforcement, with the characteristic and advantages of both the matrix and the filler or reinforcement. However, more importantly, at least one becomes a biological origin. Thus, the concept of sustainability has now initiated industries to look for alternate materials that are sustainable also like using natural fillers for reinforcement or using natural filler for even industrial requirements. Hence, sincere efforts are required to use wood-based substances with synthetic polymers to support sustainability. Bio-origin hybrid composites with fibers or fillers from natural sources are highly accepted, are viable alternatives to polymers, and are also sustainable. We have started using numerous biological and plastic matrices to strengthen them because of the low cost involved compared with synthetic fibers [53].

12.7 CONCLUSION

With the necessity of knowing the viscoelastic characteristics of polymer-based composites, the real challenge lies in choosing natural fibers or fillers from vast choices on any scale and the matrices according to different applications. The structural rigor of automotive, aerospace, construction, marine, and packaging sectors requires a prominent relationship between the shape of the products and the stresses, which will act on the products. In addition, other factors such as change in temperature, environmental parameters, and sometimes physico-chemical factors should also be considered as they could affect the viscoelastic performance of polymer-based composites.

REFERENCES

1. Andrady AL, Neal MA. Applications and societal benefits of plastics. *Philos Trans R Soc B Biol Sci* 2009;364:1977–84. doi: 10.1098/rstb.2008.0304.
2. Chandrasekar M, Ishak MR, Sapuan SM, Leman Z, Jawaid M. A review on the characterisation of natural fibres and their composites after alkali treatment and water absorption. *Plast Rubber Compos* 2017. doi: 10.1080/14658011.2017.1298550.
3. Subramanian K, Krishnasamy S, Thiagamani SMK, Pradeepkumar C, Dhandapani A, Muthukumar C, et al. Tribo performance analysis on polymer-based composites. In *Polymer-Based Composites*. CRC Press: Boca Raton, FL, 2021, pp. 115–30.
4. Senthilkumar K, Saba N, Chandrasekar M, Jawaid M, Rajini N, Siengchin S, et al. Compressive, dynamic and thermo-mechanical properties of cellulosic pineapple leaf fibre/polyester composites: Influence of alkali treatment on adhesion. *Int J Adhes Adhes* 2021;106:102823.

5. Senthil Muthu Kumar T, Senthilkumar K, Chandrasekar M, Rajini N, Siengchin S, Varada Rajulu A. Characterization, thermal and dynamic mechanical properties of poly(propylene carbonate) lignocellulosic Cocos nucifera shell particulate biocomposites. *Mater Res Express* 2019;6. doi: 10.1088/2053-1591/ab2f08.

6. Senthilkumar K, Saba N, Chandrasekar M, Jawaid M, Rajini N, Alothman OY, et al. Evaluation of mechanical and free vibration properties of the pineapple leaf fibre reinforced polyester composites. *Constr Build Mater* 2019. doi: 10.1016/j.conbuildmat.2018.11.081.

7. Senthilkumar K, Kumar TSM, Chandrasekar M, Rajini N, Shahroze RM, Siengchin S, et al. Recent advances in thermal properties of hybrid cellulosic fiber reinforced polymer composites. *Int J Biol Macromol* 2019;141:1–13.

8. Chandrasekar M, Siva I, Kumar TSM, Senthilkumar K, Siengchin S, Rajini N. Influence of fibre inter-ply orientation on the mechanical and free vibration properties of banana fibre reinforced polyester composite laminates. *J Polym Environ* 2020; 28(11):2789–2800.

9. Lambert S. Environmental risk of polymer and their degradation products. 2013:1–198.

10. Ojeda T. Polymers and the environment. *Polym Sci* 2013:1–34. doi: 10.5772/51057.

11. Stanton T, Kay P, Johnson M, Chan FKS, Gomes RL, Hughes J, et al. It's the product not the polymer: Rethinking plastic pollution. *Wiley Interdiscip Rev Water* 2021;8:1–12. doi: 10.1002/wat2.1490.

12. Schmaltz E, Melvin EC, Diana Z, Gunady EF, Rittschof D, Somarelli JA, et al. Plastic pollution solutions: Emerging technologies to prevent and collect marine plastic pollution. *Environ Int* 2020;144. doi: 10.1016/j.envint.2020.106067.

13. Alothman OY, Jawaid M, Senthilkumar K, Chandrasekar M, Alshammari BA, Fouad H, et al. Thermal characterization of date palm/epoxy composites with fillers from different parts of the tree. *J Mater Res Technol* 2020;9:15537–46.

14. Kesle SE. Mineral supply and demand into the 21st century. *Proceedings of a workshop that followed the 31st International Geological Congress*, Rio de Janeiro, Brazil, August 18–19, 2000, US Geological Survey Circular 1294, 55–62.

15. Stein RS. Environmental aspects of polymer science and engineering. *J Plast Film Sheeting* 2015;31:355–62. doi: 10.1177/8756087915596304.

16. Shahroze RM, Chandrasekar M, Senthilkumar K, Kumar TSM, Ishak MR, Rajini N, et al. Mechanical, interfacial and thermal properties of silica aerogel-infused flax/epoxy composites. *Int Polym Process* 2021;36:53–9.

17. Scott G. Plastics packaging and coastal pollution. *Int J Environ Stud* 1972;3:35–6. doi: 10.1080/00207237208709489.

18. Atiqah A, Chandrasekar M, Kumar TSM, Senthilkumar K, Ansari MNM. Characterization and interface of natural and synthetic hybrid composites, 2019.

19. Thiagamani SMK, Krishnasamy S, Muthukumar C, Tengsuthiwat J, Nagarajan R, Siengchin S, et al. Investigation into mechanical, absorption and swelling behaviour of hemp/sisal fibre reinforced bioepoxy hybrid composites: Effects of stacking sequences. *Int J Biol Macromol* 2019;140:637–46. doi: 10.1016/j.ijbiomac.2019.08.166.

20. Krishnasamy S, Muthukumar C, Nagarajan R, Thiagamani SMK, Saba N, Jawaid M, et al. Effect of fibre loading and Ca(OH)$_2$ treatment on thermal, mechanical, and physical properties of pineapple leaf fibre/polyester reinforced composites. *Mater Res Express* 2019;6:085545. doi: 10.1088/2053-1591/ab2702.

21. Chandrasekar M, Shahroze RM, Ishak MR, Saba N, Jawaid M, Senthilkumar K, et al. Flax and sugar palm reinforced epoxy composites: Effect of hybridization on physical, mechanical, morphological and dynamic mechanical properties. *Mater Res Express* 2019;6(10):1–28.

22. Kesava M, Dinakaran K. Natural-fiber-reinforced epoxy and USP resin composites. In Mittal V. (Ed.) *Spherical and Fibrous Filler Composites*. John Wiley & Sons: Boca Raton, FL, 2016, pp. 127–56. doi: 10.1002/9783527670222.ch5.

23. Senthilkumar K, Ungtrakul T, Chandrasekar M, Kumar TSM, Rajini N, Siengchin S, et al. Performance of sisal/hemp bio-based epoxy composites under accelerated weathering. *J Polym Environ* 2021;29:624–36.

24. Krishnasamy S, Thiagamani SMK, Muthukumar C, Tengsuthiwat J, Nagarajan R, Siengchin S, et al. Effects of stacking sequences on static, dynamic mechanical and thermal properties of completely biodegradable green epoxy hybrid composites. *Mater Res Express* 2019;6:105351. doi: 10.1088/2053-1591/ab3ec7.

25. Tolinski M. Foams seek sustainability. *Plast Eng* 2009;65:6–8. doi: 10.1002/j.1941-9635.2009.tb00389.x.

26. Hatti-Kaul R, Nilsson LJ, Zhang B, Rehnberg N, Lundmark S. Designing biobased recyclable polymers for plastics. *Trends Biotechnol* 2020;38:50–67. doi: 10.1016/j.tibtech.2019.04.011.

27. Neşer G. Polymer based composites in marine use: History and future trends. *Procedia Eng* 2017;194:19–24. doi: 10.1016/j.proeng.2017.08.111.

28. Civgin F. Analysis of composite bars in Torsion. 2005:141.

29. Mallick PK. *Fiber-Reinforced Composites: Materials, Manufacturing, and Design.* CRC Press: Boca Raton, FL, 2007.

30. Ratwani MM. Composite materials and sandwich structures: A primer. *Rto-En-Avt* 2010;156:1–16.

31. Ekşi S, Genel K. Comparison of mechanical properties of unidirectional and woven carbon, glass and aramid fiber reinforced epoxy composites. *Acta Phys Pol A* 2017;132:879–82. doi: 10.12693/APhysPolA.132.879.

32. Oladele IO, Ayanleye OT, Adediran AA, Makinde-Isola BA, Taiwo AS, Akinlabi ET. Characterization of wear and physical properties of pawpaw-Glass fiber hybrid reinforced epoxy composites for structural application. *Fibers* 2020;8. doi: 10.3390/FIB8070044.

33. Jagannatha T, Harish G. Mechanical properties of carbon/glass fiber reinforced epoxy hybrid polymer composites. *Int J Mech Eng Robot Res* 2015;4:131–7.

34. Huang X. Fabrication and properties of carbon fibers. *Materials (Basel)* 2009;2:2369–403. doi: 10.3390/ma2042369.

35. Denchev Z, Dencheva N. Manufacturing and properties of aramid reinforced composites. In Bhattacharyya D, Fakirov S (Eds.) *Synthetic Polymer-Polymer Composites.* Carl Hanser Verlag GmbH & Company KG: Munich, Germany, 2012. doi: 10.3139/9781569905258.008.

36. Ahmad Nadzri SNZ, Hameed Sultan MT, Shah AUM, Safri SNA, Basri AA. A review on the kenaf/glass hybrid composites with limitations on mechanical and low velocity impact properties. *Polymers (Basel)* 2020;12:1–13. doi: 10.3390/POLYM12061285.

37. Kanitkar YM, Kulkarni AP, Wangikar KS. Characterization of Glass Hybrid composite: A Review. *Mater Today Proc* 2017;4:9627–30. doi: 10.1016/j.matpr.2017.06.237.

38. Morgan P. Properties of carbon fibers. In Handerson, L. (Ed.) *Carbon Fibers and Their Composite Materials.* MDPI: Basel, Switzerland, 2020, pp. 831–900. doi: 10.1201/9781420028744-24.

39. Prasad N, Dev S, Khande K, Chandra G, Prakash P, Sen K, et al. Study on aramid fibre and comparison with other composite materials. *IJIRST-Int J Innov Res Sci Technol* 2014;1:303–6.

40. Zaki A. *Thermoplastic: Composite Materials.* IntechOpen: London, 2012. doi: 10.5772/2637.

41. Xie F, Huang L, Leng J, Liu Y. Thermoset shape memory polymers and their composites. *J Intell Mater Syst Struct* 2016;27:2433–55. doi: 10.1177/1045389X16634211.

42. Shanmugam D, Thiruchitrambalam M. Static and dynamic mechanical properties of alkali treated unidirectional continuous Palmyra Palm Leaf Stalk Fiber/jute fiber reinforced hybrid polyester composites. *Mater Des* 2013;50:533–42. doi: 10.1016/j.matdes.2013.03.048.

43. Jawaid M, Alothman OY, Saba N, Tahir PM, Khalil HPSA. Effect of fibers treatment on dynamic mechanical and thermal properties of epoxy hybrid composites. *Polym Compos* 2015;36:1669–74.

44. Naveen J, Jawaid M, Zainudin ES, Sultan MTH, Yahaya R, Abdul Majid MS. Thermal degradation and viscoelastic properties of Kevlar/Cocos nucifera sheath reinforced epoxy hybrid composites. *Compos Struct* 2019;219:194–202. doi: 10.1016/j.compstruct.2019.03.079.

45. Manral A, Ahmad F, Chaudhary V. Static and dynamic mechanical properties of PLA bio-composite with hybrid reinforcement of flax and jute. *Mater Today Proc* 2019;25:577–80. doi: 10.1016/j.matpr.2019.07.240.

46. Khan MZR, Srivastava SK, Gupta MK. A state-of-the-art review on particulate wood polymer composites: Processing, properties and applications. *Polym Test* 2020;89:106721. doi: 10.1016/j.polymertesting.2020.106721.

47. Askanian H, Verney V, Commereuc S, Guyonnet R, Massardier V. Wood polypropylene composites prepared by thermally modified fibers at two extrusion speeds: Mechanical and viscoelastic properties. *Holzforschung* 2015;69:313–9. doi: 10.1515/hf-2014-0031.

48. Andrzejewski J, Szostak M, Barczewski M, Patrycja Ł. Cork-wood hybrid filler system for polypropylene and poly (lactic acid) based injection molded composites. *Struct Eval Mech Perform* 2019;163:655–68. doi: 10.1016/j.compositesb.2018.12.109.

49. Dahal RK, Acharya B, Saha G, Bissessur R, Dutta A, Farooque A. Biochar as a filler in glassfiber reinforced composites: Experimental study of thermal and mechanical properties. *Compos Part B Eng* 2019;175:107169. doi: 10.1016/j.compositesb.2019.107169.

50. Wang X, Yu Z, McDonald AG. Effect of different reinforcing fillers on properties, interfacial compatibility and weatherability of wood-plastic composites. *J Bionic Eng* 2019;16:337–53. doi: 10.1007/s42235-019-0029-0.

51. Kumar R, Kumar K, Bhowmik S. Mechanical characterization and quantification of tensile, fracture and viscoelastic characteristics of wood filler reinforced epoxy composite. *Wood Sci Technol* 2018;52:677–99.

52. Tazi M, Sukiman MS, Erchiqui F, Imad A, Kanit T. Effect of wood fillers on the viscoelastic and thermophysical properties of HDPE-wood composite. *Int J Polym Sci* 2016;2016. doi: 10.1155/2016/9032525.

53. Salazar MÁH. Viscoelastic performance of biocomposites. *Compos Renew Sustain Mater* 2016. doi: 10.5772/66148.

13 Viscoelastic Properties of Polymer-Based Bionanocomposites Reinforced with Inorganic Fillers

W.S. Chow and Z.E. Ooi
Universiti Sains Malaysia

CONTENTS

13.1 INTRODUCTION

Bionanocomposite (BNC) is a composite material comprising of biopolymer matrix and nanofillers with at least one dimension in the nanoscale (Sheng et al., 2019). The combination and/or hybridization of materials is a viable approach to improve the mechanical, thermal, gas barrier, optical, biodegradability, and biocompatibility properties of BNC. The increasing demand for biopolymers is attributed to the plastic waste problem and environmental issues. Petroleum-based plastics bring some convenience and advantages over other materials due to their lightweight and durability. However, if the petroleum-based plastics are used for single-used plastics application, or "simply" disposed to the landfill (without considering the potential of recycling), then it would bring a negative impact to the environment (Flaris and Singh, 2009; Zaverl et al., 2012). The biopolymer synthesized from renewable resources (especially the agricultural waste for non-food application) can be used to minimize the environmental impact (George et al., 2020). Thus, besides recycling, the BNC could be one of the feasible strategies to achieve sustainable development in the polymer industry, reduce the carbon footprint, and lessen the burden of Mother Earth.

Recently, there are various biopolymers and biodegradable polymers have been developed, including polycaprolactone (PCL), polylactic acid (PLA), polyhydroxy butyrate, and polybutylene succinate. In this chapter, the focus is on PLA. PLA is a promising biopolymer due to its high strength, high transparency, good eco-friendliness, good biodegradability, and biocompatibility. PLA can be synthesized from renewable resources, for example, corns and starches (Schreck and Hillmyer, 2007). Further, PLA can be processed using film extrusion, blow molding, injection molding, and thermoforming. The above-mentioned properties make PLA an ideal polymer, which can be used in various applications, e.g., packaging and automotive (Chen et al., 2018; Hassan and Koyama, 2010). The recycling of PLA can further increase its life cycle and thus make it a pro-environmental and more sustainable material (Żenkiewicz et al., 2009). Everything has its pros and cons; PLA has some fantastic features; however, some of its limitations should be mentioned, include low impact strength, low heat distortion temperature, and slow crystallization rates. The disadvantages of PLA as mentioned would not always bring problems to the manufacturing and applications. For example, in some processes it may allow slow crystallization, whereas for some applications it may require low heat distortion temperature. Nonetheless, if for a process/application that needs a high crystallization rate, high heat distortion temperature, or high impact strength, then modification of PLA can be tailored using different approaches, e.g., blending with rubbery materials, adding flexible polymers, copolymerization, and using nanotechnology (Bitinis et al., 2014; Pande and Sanklecha, 2017; Teamsinsungvon et al., 2017; Hareesh et al., 2018; Li et al., 2020). The incorporation of nanofillers (e.g., nanotubes, nanoclay, and nanoplatelets) often enhances the mechanical properties, thermal stability, and crystallization of PLA (Chow et al., 2018). Ternary nanocomposite concepts and adding of coupling agent/compatibilizer/toughening agent can be used to adjust the final properties of PLA (Chow et al., 2013). The viscoelastic properties of the PLA nanocomposites can be changed with the variations in their molecular architecture. In this chapter, the viscoelastic behavior of the PLA BNCs is discussed. The information on the viscoelastic properties of the PLA BNCs can be used for product design and widen the application of PLA BNCs.

13.2 VISCOELASTIC PROPERTIES OF POLYMERIC MATERIALS AND POLYMER BIONANOCOMPOSITES

Polymer is a viscoelastic material, and its viscoelasticity is related to time, frequency, and temperature (Lakes, 2004). Viscoelasticity refers to a combination of "visco" behavior (associates with the spring-like behavior) and "elastic" behavior (associates with the flow-like behavior) of a material (Kulik and Boiko, 2018). Viscoelasticity is an essential feature that can be used to understand the phase transformation and molecular mobility of polymeric materials, and further, it can be used to develop and design functional devices (e.g., instrument mounts, vibration isolation, and shock absorber). The viscoelastic properties of polymeric materials can be illustrated using the Maxwell model and Kelvin-Voigt model (Delgado-Reyes et al., 2013; Chen et al., 2015). According to Yuya and Patel (2014), in a dynamic nano-indentation analysis, tan $\delta < 1$ indicates a predominantly solid-like behavior, while tan $\delta > 1$ indicates a predominantly viscous or fluid-like response.

In general, the viscoelasticity of a polymeric material is time, temperature, and frequency dependent. The responses at a different time or frequency scales are often associated with the relaxation of the molecular structure with different size scales. The percolation networks due to nanofillers, as well as the interaction between nano-fillers and polymers, could influence the viscoelastic behavior of polymer nano-composites (Ding et al., 2016). The use of nanofillers could refine and stabilize the polymer blend structure. The preferential localization of nanofillers in a polymer blend-based nanocomposite involves the changes in the interfacial properties, affect-ing the viscosity ratio, and influencing the morphological evolution. Consequently, the viscoelastic behavior and mechanical properties of the BNC could be controlled by a proper selection of nanofiller, binary or ternary blend system. Nofar et al. (2021) reported that when nanofillers are localized at the interface in a polymer nanocom-posite, the relaxation of droplets is prolonged. The long relaxation in the polymer blend nanocomposites, as well as the nanofiller self-interaction affecting the rheo-logical and viscoelastic properties, especially for the elastic modulus measured at low frequencies.

According to Wang et al. (2017), a compatibilizer can be used to strengthen the interface of polymer blend-based material. A compatibilizer is able to reduce inter-facial tensions, enhance interfacial adhesion, suppress particle coalescence, and achieve a finer phase morphology. Subsequently, it would influence the final local-ization of the nanofiller in the polymer nanocomposites, rheological properties, pro-cessability, and viscoelastic of the material. Recently, it was found that nanofiller can modify the rheological and viscoelastic of polymer nanocomposites effectively, due to their nano-size structure, high aspect ratio, and high surface area per unit volume. Some chemically treated or functionalized nanofiller (e.g., organo-modified nanofiller) can modify the morphological properties of a polymer blend when there are preferentially or selectively located at one of the phases, and furthermore, they can change the viscosity ratios and migrate to the interface of the blends. This phe-nomenon can strengthen the interfacial adhesion and reduce the interfacial tensions. Thus, the modified nanofillers can be used to control the viscoelasticity of the poly-mer BNCs (Salehiyan et al., 2014).

Dynamic mechanical analyzer (DMA) can be used to characterize the elastic and viscous behavior of polymeric materials under an oscillating load as a function of temperature, frequency, and time. Dynamic mechanical analysis (DMA) is an excellent tool to obtain information on their viscoelasticity because it measures the stiffness and damping properties of polymers (Karger-Kocsis, 2017). Viscoelastic materials often experience relaxation, transition (e.g., glass transition temperature and melting temperature), and properties changes when subjected to the temperature. In general, a sample is subjected to an applied oscillating force, and deformation is carried out in a sinusoidal mode. The resultant strain due to the sinusoidal load is always governed by the elastic and viscous properties of the polymeric material. According to Hosur et al. (2017), the storage modulus (E') represents the energy stored in a polymer composite material after the deformation under a specified load. The E' reflects the elastic modulus of the polymer composite, which determines the recoverable strain energy in the deformed sample. The loss modulus (E'') is associated with the energy lost attributed to the energy dissipation as heat (Shakuntala et al., 2014). According to Jose and Thomas (2014), the effectiveness of nanofillers on the modulus of a polymer composite can be determined from the coefficient (c), as shown in Equation 13.1. The lower the value of coefficient (c) suggests the higher the effectiveness of the nanofiller.

$$c = \left(E'_G \ / \ E'_R \right)_{\text{composite}} \Big/ \left(E'_G \ / \ E'_R \right)_{\text{resin}} \qquad (13.1)$$

where E'_G is the storage modulus value below the glass transition temperature and E'_R is the storage modulus above the glass transition temperature.

The storage modulus (E' – as an in-phase component), the loss modulus (E'' – as an out-phase component), and the tan δ (ratio of E''/E' – as a measurement of the capacity of a material to dissipate energy) determined by DMA can be used to understand the viscoelastic behavior of a polymeric material (Costa et al., 2016). In general, tan δ is dependent on the viscoelastic behavior in the T_g region, where polymer molecular chains tend to vibrate in phase with external deformation. The tan δ is associated with molecular relaxation, intra-molecular forces, matrix–filler internal friction, and filler–filler internal friction. A polymer nanocomposite that is having a high tan δ value often exhibits good energy dissipation performance. The enhanced energy dissipation can be attributed to the frictional sliding damping mechanism that occurs in the polymer/nanofiller system (Gibson, 2010). However, bear in mind that for some of the polymer nanocomposites, using nanofillers would constrain the physical mobility of the polymer macromolecular chain, causing lesser internal friction and hence reducing the tan δ. Consequently, the nanofillers decreased the damping behavior of the polymer nanocomposites (Wei et al., 2019). The shift (or widening) in the glass transition temperature and the decrease in the height of the tan δ peak, which can be used as an indication of the load-bearing of an elastic solid material, could be associated with a stronger interfacial adhesion in the microstructure of a polymer nanocomposite.

There are several rheological techniques for characterizing the viscoelastic behavior of polymer materials, including oscillatory testing, stress relaxation, and creep testing. A small amplitude oscillatory shear test is used to determine the viscoelastic properties, and this normally can be measured using a rotational rheometer. Viscoelastic materials have a phase angle higher than 0° but smaller than 90°. The phase angle provides us the important information about the viscoelasticity of a material. The tan δ is the ratio $G'':G'$. It describes the viscoelastic balance of the material and is called the damping factor. When the $G'' = G'$, then tan $\delta = 1$; it indicates the gel point. The viscoelastic liquid can be observed when $G'' > G'$, and tan $\delta > 1$. The viscoelastic solid is detectable when $G' > G''$ and tan $\delta < 1$. In the oscillatory dynamic test, using an amplitude sweep test (the amplitude is ramped while the frequency and the temperature are held constant) can determine the linear viscoelastic region, dynamic yield point, and flow point for a polymeric material. For a frequency sweep test, the frequency of oscillation is ramped, while the amplitude and the temperature are held constant. The frequency sweep test can be used to quantify zero shear viscosity for viscoelastic liquid, as well as provides information on molecular weight distribution, degree of crosslinking, and stability of the material. The complex modulus (G^*) is related to the resistance to the material stiffness and deformation. The loss modulus (G'') represents the energy loss, which is related to the viscous component contribution. Understanding the relaxation time of polymeric materials can be used to predict the material's viscoelastic response stressed for a given time.

The rheological properties characterization can be used to assess the state of dispersion of nanofiller in the polymer matrix, the interaction between polymer and nanofiller, and the viscoelastic behavior (Singh et al., 2012). From the rheology measurement, the elastic component is represented by the storage modulus (G'), while the viscous component is represented by the loss modulus (G'') (Delgado-Reyes et al., 2013). There is a wide range of rheological tests that can be done on polymer and polymer nanocomposites. For example, the dynamic oscillatory shear test that performed at small amplitudes at various frequencies in the linear elastic range can be used to characterize the rheological and viscoelastic properties of polymer nanocomposites. In some studies, the low-frequency range is quite interesting, because it is most sensitive to melt elasticity and network formation.

Rheology is a feasible tool to characterize the microstructure transitions of the multi-component polymer systems, the formation of network structures, and the change of droplet shape (Zheng et al., 2007). The morphology and structure of polymer subjected to the dynamic rheological test are not disrupted and often cannot be destroyed under small-strain amplitude. The second plateau, which appears at the low-frequency region (termed as terminal region) is a unique phenomenon involving the deviation of dynamic viscoelastic functions. This phenomenon is related to the heterogeneous characteristics of polymer materials, which include phase separation, the polydispersity of molecular weight, and network structure formation.

The loss factor tan δ can be plotted in addition to the G' and G'' curves if there is a phase transition in a polymer sample. The sol-gel transition point, gel point, and viscosity changes of the polymer nanocomposite can be determined. The changes in the viscosity and modulus could be attributed to the filler content, as well as the filler network formation in a polymer nanocomposite. Often, as the nanofiller loading increases, the filler–filler interaction could influence the viscoelastic behavior of the polymer, attributed to the rheological percolation structure-induced transient network. In general, a power-law relationship can be used to determine the threshold of rheological percolation. Thus, rheological measurement is a good tool to understand the viscoelastic behavior of polymer, as well as polymer BNCs.

13.3 PLA BIONANOCOMPOSITES: EFFECTS OF INORGANIC FILLER ON THE VISCOELASTIC PROPERTIES

PLA BNCs are gaining attention attributed to the properties tunability of the PLA with various nanofiller (with and without chemical functionalization). The adding of nanofiller could modify the mechanical, thermal, and viscoelastic properties of the PLA (Zhang et al., 2014; Lai and Hsieh, 2016; Avolio et al., 2018). Some of the commercially available inorganic nanofillers are proven their usefulness in enhancing the properties of polymeric materials (e.g., PLA), for example, nanoclay, halloysite nanotube, nano-calcium carbonate, and nanosilica. In this chapter, the focus is on PLA/nanoclay and PLA/halloysite nanotube nanocomposites.

13.3.1 PLA/NANOCLAY

Nanoclay is a naturally occurring mineral (layered mineral silicates from the smectite family) and can be categorized according to its chemical composition and structure, such as montmorillonite, kaolinite, bentonite, and hectorite. Montmorillonite is one of the most commonly used nanoclays in the development of polymer nanocomposites (thermoplastic-based, thermoset-based, and elastomer-based), attributed to their plate-like nanostructure, high aspect ratio, intercalation/exfoliation-ability, and high reinforcing efficiency. Adding a small amount of nanoclay (especially the surface-treated or organically modified nanoclay) often enhanced selected properties (modulus, strength, heat distortion temperature, and barrier) of polymer nanocomposites. Organically modified nanoclay (also called organoclay) enable expansion of the d-spacing of clay and allows for exfoliation into a high aspect ratio single layer (Chow and Lok, 2018; Bartel et al., 2017). Various strategies are used to develop polymer nanocomposites using nanoclay in order to widen their applications. Thus, it is quite reasonable that a similar approach has been attempted for the research and development of PLA-based BNCs.

Hong and Kim (2013) investigated the thermo-mechanical properties of PLA/cellulose nano-whisker/maleic anhydride-grafted PLA with and without organically modified montmorillonite nanoclay (Cloisite® 20A). The storage modulus (E') and tan δ of the PLA nanocomposites were measured using a dynamic mechanical analyzer (testing parameters: sinusoidal displacement = 20 µm, heating rate = 5°C/minutes, and frequency = 1 Hz). Adding nanoclay increased the storage modulus of the PLA nanocomposites. This indicates the nanoclay can act as a good reinforcing agent. Pirani et al. (2014) prepared PLA/Cloisite 30B nanoclay nanocomposites using a twin-screw micro-compounder followed by injection molding. The viscoelastic properties of the PLA nanocomposites were studied using a dynamic mechanical analyzer (testing parameters: double cantilever mode and temperature range = 30°C–90°C). The storage modulus (at 30°C) of the PLA was increased significantly by the addition of 3 wt.% nanoclay. Recall that the storage modulus represents the stiffness of viscoelastic material and is proportional to the energy stored during a loading cycle. The storage modulus of the polymer nanocomposite often increased with increasing temperature, which can be attributed to the increase in segmental polymer chain motion as the temperature rises (Ashori et al., 2019). Moreover, the incorporation of nanoclay shifted the tan δ of the PLA nanocomposites to a higher value. This is attributed to the fact that the nanoclay hinders the PLA molecular chain movement. The magnitude of the tan δ peak is directly associated with the interaction between the polymer matrix and the nanofiller.

As'habi et al. (2013) produced PLA/linear low-density polyethylene (LLDPE)/nanoclay (Cloisite® 30B) nanocomposites using a counter-rotating twin-screw extruder. The rheological behavior of the PLA nanocomposites was measured using

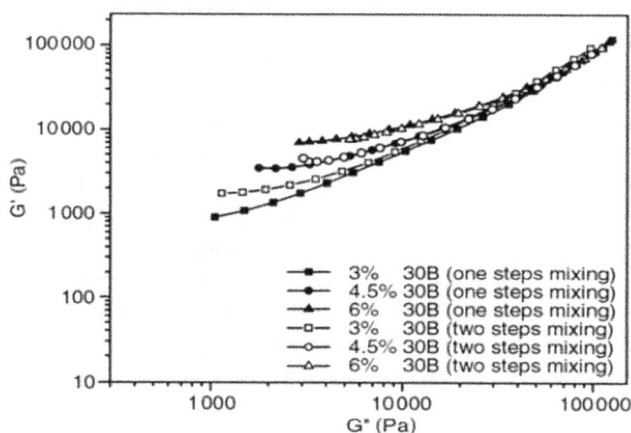

FIGURE 13.1 Cole-Cole plots for PLA/LLDPE/Cloisite 30B nanocomposites prepared via different mixing methods (As'habi et al., 2013). (Permission obtained from Express Polymer Letters, BME-PT.)

small amplitude oscillatory frequency sweeps and temperature sweeps. Cole-Cole equation is often used to describe relaxation in the polymer. The author used the Cole-Cole plot to characterize the rheological properties of the PLA/LLDPE/nanoclay (30B) nanocomposites prepared using different mixing methods (c.f. Figure 13.1). The development of a plateau in G' at lower G'' is detected at lower nanoclay loading (i.e., 3 wt.% 30B) for the PLA nanocomposites prepared using two-step mixing approach, which indicates the formation of network structure and solid-like behavior.

Darie et al. (2014) investigated the dynamic rheological properties of PLA/nanoclay using a parallel-plate rheometer (oscillatory frequency sweeps ranging from 0.05 to 500 rad/seconds). The storage modulus (G') of the PLA measured at low frequencies increased with the increasing of nanoclay loading. The solid-like viscoelastic response is attributed to the percolation network formation in the polymer–clay nanocomposites. At lower angular frequency, the G' is smaller than the loss modulus (G''), suggesting the liquid-like response of the molten PLA polymer. The viscous behavior ($G'' > G'$) at low angular frequencies is noticeable for both PLA and PLA/nanoclay nanocomposites. The PLA–nanoclay interfacial adhesion causes an increase in the complex viscosity and storage modulus at low frequencies. A predominant shear-thinning behavior was observed at a higher frequency. This indicates the good dispersibility of nanoclay in the PLA matrix.

13.3.2 PLA/HALLOYSITE NANOTUBE

Halloysite nanotube is a naturally occurring clay mineral with a predominantly hollow tubular nanostructure (Ferrante et al., 2015; Murariu et al., 2012). The length of Halloysite Nanotube (HNT) is ~50 to 5,000 nm, while its external diameter and internal diameter are in the range of 20–200 nm and 10–70 nm, respectively (Makaremi et al., 2017; Pasbakhsh et al., 2013). HNT exhibits special surface chemical property attributed to the multi-layered structure with hydroxyl groups on its surface (Yeniova et al., 2012). HNT is widely used in various applications, such as drug delivery systems, biomimetic nanoreactors, and reinforcing fillers (Tham et al., 2015). HNT is widely used to reinforce polymer due to its rod-like morphology, tubular geometry, high aspect ratio, high stiffness, unique surface properties, and good dispersibility. The incorporation of HNT can enhance the modulus, strength, and ductility of polymer simultaneously. Surface treatment and functionalization of HNT are essential to enhance the mechanical and thermal properties (Wu et al., 2015; Risyon et al., 2016; Suppiah et al., 2019). Adding HNT into PLA often improves mechanical and thermal properties due to the reinforcing efficiency, good dispersibility, and interfacial interaction between PLA and HNT (Lim et al., 2019).

Kaynak and Kaygusuz (2016) investigated the effects of HNT on the storage modulus using dynamic mechanical analyzer (testing parameters: three-points bending mode, heating rate = 2°C/min, and frequency = 1 Hz). The storage modulus (E') of PLA/HNT measured at 25°C and 50°C is higher than that of unfilled PLA. For example, the E' of PLA ($T = 25°C$) is 2.78 GPa, while the E' of PLA/HNT-10 wt.% (at $T = 25°C$) is 3.89 GPa. The improvement of E' of the PLA is ~111% by adding 10 wt.% of HNT. This is due to the high reinforcing efficiency, nano-tubular structure, and high aspect ratio of the HNT.

Prashantha et al. (2013) used dynamic mechanical analyzer (testing parameters: tensile mode, displacement amplitude = 10 μm, static force = 1.0 N, and heating rate = 3°C/minutes; frequency = 1 Hz) to determine the storage modulus and tan δ of PLA/quaternary ammonium salt-modified HNT nanocomposites. The tan δ peak of the PLA/surface-treated HNT nanocomposites broadened and shifted to a higher temperature compared to unfilled PLA. This is attributed to the polymer chain mobility restriction, which is associated with the better interfacial interaction between the surface-treated HNT and PLA matrix.

Rashmi et al. (2015) prepared PLA/polyamide 11/HNT using a twin-screw extruder followed by injection molding. The viscoelastic properties of the PLA/HNT nanocomposites were determined using a dynamic mechanical analyzer in a tension mode (testing parameters: strain amplitude = 2%, heating rate = 3°C/min, and frequency = 1 Hz). The storage modulus of the PLA was increased by the incorporation of HNT. The higher the loading of HNT (from 2 to 6 wt.%), the higher the storage modulus. This is due to two factors: (a) chain mobility restriction of the PLA matrix and (b) reinforcing effects of the HNT. The tan δ peak of the PLA nanocomposites was lowered by the addition of HNT (c.f. Figure 13.2). The reduction of sharpness and the height of the tan δ peak are often associated with the reduction in damping behaviors. In their study, the lowering of the damping behavior of the PLA nanocomposites could be attributed to the HNT loading as well as its localization in polyamide 11 that hinders the segmental motions of the PLA chains during the transition. In the presence of a third component or compatibilizer, the nanofiller can be preferentially located at the interface and influence the morphology and viscoelastic properties of polymer nanocomposites (Chow and Mohd Ishak, 2015).

Montava-Jorda et al. (2019) investigated the viscoelastic properties of PLA/HNT (with polyvinyl acetate compatibilizer) using an oscillatory rheometer (testing parameters: torsion–shear conditions, heating rate = 2°C/minutes, maximum shear/torsion deformation = 0.1%, and oscillation frequency = 1 Hz). The storage modulus

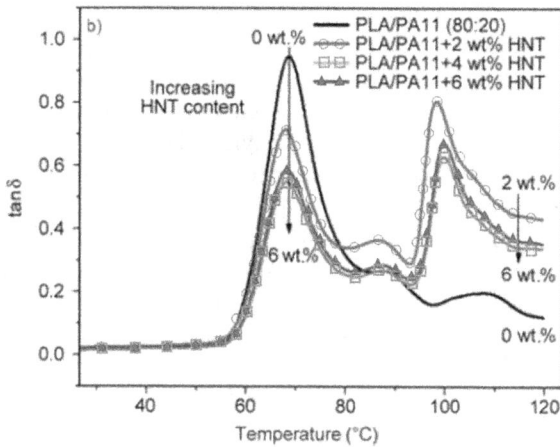

FIGURE 13.2 tan δ versus temperature for PLA/polyamide 11/HNT system (Rashmi et al., 2015). (Permission obtained from Express Polymer Letters, BME-PT.)

of the PLA (G' at 30°C = 1,599 MPa) increased as the HNT loading increases (from 3 to 9 wt.%). The storage modulus (G' at 30°C) of PLA/HNT 9 wt.% is 1,843 MPa (ca. 15% improvement from the unfilled PLA). The G' of the PLA and its PLA/HNT nanocomposites remained constant below 50°C. At the temperature range from 55°C to 70°C, the G' decreased due to the glass transition. Interestingly, the G' increased again in the temperature range of 80°C–100°C. This is due to the cold crystallization process and thus leads to the packed-ordered structure formation, which is responsible for the increase in the storage modulus (G').

13.4 STRATEGIES TO MODIFY VISCOELASTIC PROPERTIES OF PLA BIONANOCOMPOSITES

Various strategies are used to tune the viscoelasticity of the polymer matrix, which include polymer blending, using hollow fiber, adding nano-reinforcement, and utilizing semi-interpenetrating networks (Karger-Kocsis and Mohd Ishak, 2017). Similarly, there are several approaches to modify the viscoelastic properties of PLA BNCs, for example, by adding rubber, polymer blending, compatibilizer, toughening agent, core-shell structure, and thermoplastic elastomer technology (Li et al., 2006; Tham et al., 2016).

Balakrishnan et al. (2012) prepared PLA/octadecylamine-modified montmorillonite nanoclay nanocomposite toughened with an ethylene copolymer (Biomax Strong 100). The DMA test (parameters: three-point bending mode, heating rate = 3°C/minutes, and frequency = 1 Hz) results showed that the tan δ peak of the PLA was shifted to a lower temperature by the incorporation of the ethylene copolymer. On the other hand, the tan δ peak of the PLA was shifted to a higher temperature by the nanoclay. Eng et al. (2013) prepared PLA/PCL/nanoclay using melt blending. The dynamic mechanical thermal properties of the PLA/PCL nanocomposites were determined using a dynamic mechanical analyzer (testing parameters: bending mode, heating rate = 2°C/minutes, and frequency = 1 Hz). The area underneath the tan δ peak often associated with the damping ability of the polymeric materials. In their study, it was found that the area underneath the tan δ peak demonstrated that the PLA/PCL (with and without nanoclay) exhibited similar damping abilities.

Yeniova et al. (2016) investigated the viscoelastic properties of PLA/HNT nanocomposites with two types of plasticizer/toughening agent (i.e., polyethylene glycol and thermoplastic polyurethane) prepared using micro-injection molding. The E', E'', and tan δ of the PLA/HNT nanocomposites was characterized using dynamic mechanical analyzer (testing parameters: tensile mode, heating rate: 1°C/ minutes, and frequency = 1 Hz). From their study, it was found that the E' was increased with the increasing loading of nanofiller. This is associated with the higher energy that is stored per cycle of oscillation, as well as the reinforcing effects of the HNT. The stiffening effects of the HNT are more prominent at temperatures below T_g (glass transition temperature). On the contrary, adding polyethylene glycol and thermoplastic polyurethane reduces the storage modulus of PLA/ HNT nanocomposites.

Bijarimi et al. (2013) investigated the viscoelastic properties of PLA/liquid natural rubber/Cloisite C30B nanoclay nanocomposites using dynamic mechanical analyzer (testing parameters: three-point bending mode, constant strain = 0.02%, and heating rate = 5°C/minutes; frequency = 1 Hz). The storage modulus of the PLA nanocomposites is higher than the unfilled PLA, manifesting the significant effects of nanoclay on the elastic properties. Adding of liquid natural rubber increased the tan δ peak height and lowered the glass transition temperature of the PLA nanocomposites, suggests that the liquid natural rubber is able to increase the damping behavior. This gives us a hint that the viscoelastic properties of the PLA nanocomposites can be controlled by the suitable combination of an impact modifier/toughening agent and nanofiller (e.g., nanoclay and nanotube).

Jalalifar et al. (2020) investigated the effects of HNT and polyolefin elastomer-grafted maleic anhydride (POE-g-MA) compatibilizer on the dynamic mechanical and rheological properties of PLA/POE blends. The DMA test was performed at a heating rate of 3°C/minutes under a bending mode deformation (frequency = 1 Hz). From the DMA results, the increase in the storage modulus indicates the reinforcing effect of HNT as well as the improved interfacial adhesion due to the presence of POE-g-MA compatibilizer. The rheological properties were characterized by a frequency sweep test (strain = 1%) using a rotational rheometer with a 25 mm parallel-plate geometry. The linear viscoelastic range was determined using an amplitude sweep test (strain from 0.01% to 100% at a constant frequency of 10 rad/seconds). The notable changes in the rheological properties of the PLA blends-nanocomposites (with 4 wt.% HNT) suggest the transition from the liquid-like to the viscoelastic solid-state, attributed to the formation of an interconnected polymer network in the presence of POE-g-MA and HNT. By analyzing the current research work done on the PLA/nanofiller, it is reasonable to mention that the melt and solid viscoelastic properties of the PLA nanocomposites can be controlled by modifying the interfacial adhesion, compatibility, and toughening of the system. This can be achieved by using compatibilizers, coupling agents, surface-treated nanoreinforcements, and toughening agents.

13.5 FUTURE PERSPECTIVE AND CONCLUSION

The viscoelastic properties (storage modulus, loss modulus, tan δ, and damping behavior) of PLA nanocomposites are governed by several factors, i.e., the types and loading of nanofiller, the dispersibility and intercalation/exfoliation-ability of the nanofiller, the effects of surface functionalization of the nanofiller, the effects of surface modifier/toughening agent, etc. Understanding the controlling factors allows us to design the PLA nanocomposites with desired viscoelastic properties (e.g., high storage modulus or high damping properties). The viscoelastic behaviors should be considered when designing a polymer BNC (e.g., PLA/inorganic nanofiller BNCs) in order to understand how their microstructure influencing the flow behavior (processability) and performance during service life. Various strategies can be used to tune the viscoelastic properties of the PLA BNCs, include polymer blending (thermoplastic-thermoplastic, thermoplastic-elastomer, or thermoplastic-thermoset), nanofiller hybridization, coupling methods, core-shell structure, interpenetrating network structure, and toughening approach. Overall, the viscoelasticity balance is the key to achieve the desired properties and performance of the PLA bionanocomposites.

REFERENCES

As'habi L., Jafari S.H., Khonakdar H.A., Boldt R., Wagenknecht U., Heinrich G. "Tuning the processability, morphology and biodegradability of clay incorporated PLA/LLDPE blends via selective localization of nanoclay induced by melt mixing sequence." *Express Polymer Letters* 7 (2013): 21–39.

Ashori A., Jonoobi M., Ayrilmis N., Shahreki A., Fashapoyeh M.A. "Preparation and characterization of polyhydroxybutyrate-co-valerate (PHBV) as green composites using nano reinforcements." *International Journal of Biological Macromolecules* 136 (2019): 1119–1124.

Avolio R., Castaldo R., Avella M., Cocca M., Gentile G., Fiori S., Errico M.E. "PLA-based plasticized nanocomposites: Effect of polymer/plasticizer/filler interactions on the time evolution of properties." *Composites Part B: Engineering* 152(2018): 267–274.

Balakrishnan H., Masoumi I., Yussuf A.A., Imran M., Hassan A., Wahit M.U. "Ethylene copolymer toughened polylactic acid nanocomposites." *Polymer-Plastics Technology and Engineering* 51 (2012): 19–27.

Bartel M., Remde H., Bohn A., Ganster J. "Barrier properties of poly (lactic acid)/cloisite 30B composites and their relation between oxygen permeability and relative humidity." *Journal of Applied Polymer Science* 134 (2017): 44424/1–44424/10.

Bijarimi M., Ahmad S., Rasid R. "Mechanical, thermal and morphological properties of poly(lactic acid)/natural rubber nanocomposites." *Journal of Reinforced Plastics and Composites* 32 (2013): 1656–1667.

Bitinis N., Fortunati E., Verdejo R., Armentano I., Torre L., Kenny J.M., López-Manchado M.Á. "Thermal and bio-disintegration properties of poly(lactic acid)/natural rubber/organoclay nanocomposites." *Applied Clay Science* 93–94 (2014): 78–84.

Chen J.Z., Zhang X.Y., Huang Q.Y., Zhu S.R. "Prediction of viscoelastic behaviour of polymer matrix composites based on polymer matrix." *Materials Research Innovations* 19 (2015): S5/123–S5/126.

Chen L., Hu K., Sun S.T., Jiang H., Huang D., Zhang K.Y., Pan L., Li Y.S. "Toughening poly (lactic acid) with imidazolium-based elastomeric ionomers." *Chinese Journal of Polymer Science* 36 (2018): 1342–1352.

Chow W.S., Lok S.K. "Flexural, morphological and thermal properties of poly(lactic acid)/organo-montmorillonite nanocomposites." *Polymers & Polymer Composites* 16 (2018): 263–270.

Chow W.S., Mohd Ishak Z.A. "Polyamide blend-based nanocomposites: A review." *Express Polymer Letters* 9 (2015): 211–232.

Chow W.S., Tham W.L., Seow P.C. "Effects of maleated-PLA compatibilizer on the properties of poly(lactic acid)/halloysite clay composites." *Journal of Thermoplastic Composite Materials* 26 (2013): 1349–1363.

Chow W.S., Tham W.L., Poh B.T., Mohd Ishak Z.A. "Mechanical and thermal oxidation behavior of poly(lactic acid)/halloysite nanotube nanocomposites containing N, N'-ethylenebis (stearamide) and SEBS-g-MA." *Journal of Polymers and the Environment* 26 (2018): 2973–2982.

Costa C.S.M.F., Fonseca A. C., Serra A.C., Coelho, J.F.J. "Dynamic mechanical thermal analysis of polymer composites reinforced with natural fibers." *Polymer Reviews* 56 (2016): 362–383.

Darie R.N., Pâslaru E., Sdrobis A., Pricope G.M., Hitruc G.E., Poiată A., Baklavaridis A., Vasile C. "Effect of nanoclay hydrophilicity on the poly(lactic acid)/clay nanocomposites properties." *Industrial and Engineering Chemistry Research* 53 (2014): 7877–7890.

Delgado-Reyes V.A., Ramos-Ramírez E.G., Cruz-Orea A., Salazar-Montoya J.A. "Flow and dynamic viscoelastic characterization of non-purified and purified mucin dispersions." *International Journal of Polymer Analysis and Characterization* 18 (2013): 232–245.

Ding K.S., Wei N.X., Zhou Y.N., Wang Y., Wu D.F., Liu H.Y., Yu H., Zhou C., Chen J.X., Chen C. "Viscoelastic behavior and model simulations of poly(butylene adipate-co-terephthalate) biocomposites with carbon nanotubes: Hierarchical structures and relaxation." *Journal of Composite Materials* 50 (2016): 1805–1816.

Eng C.C., Ibrahim N.A., Zainuddin N., Ariffin H., Wan Yunus W.M.Z., Then Y.Y., Teh C.C. "Enhancement of mechanical and thermal properties of polylactic acid/polycaprolactone blends by hydrophilic nanoclay." *Indian Journal of Materials Science* 2013 (2013): 816503/1–816503/1.

Ferrante F., Armata N., Lazzara G. "Modeling of the halloysite spiral nanotube." *The Journal of Physical Chemistry C* 119 (2015): 16700–16707.

Flaris V., Singh G. "Recent developments in biopolymers." *Journal of Vinyl and Additive Technology*, 15 (2009): 1–11.

George A., Sanjay M.R., Srisuk R., Parameswaranpillai J., Siengchin S. "A comprehensive review on chemical properties and applications of biopolymers and their composites." *International Journal of Biological Macromolecules* 154 (2020): 329–338.

Gibson R.F. "A review of recent research on mechanics of multifunctional composite materials and structures." *Composite Structures* 92 (2010): 2793–2810.

Hareesh A., Muruli M.S., Ramesha A., Ranganath N., Panchakshari H.V. "Characterization of mechanical and thermal properties of biopolymer nanocomposites." *International Journal of Engineering Research & Technology* 7 (2018): 89–96.

Hassan M.M., Koyama K. "Thermomechanical and viscoelastic properties of green composites of PLA using chitin micro-particles as fillers." *Journal of Polymer Research* 27 (2010): 1–11.

Hong J., Kim D.S. "Preparation and physical properties of polylactide/cellulose nanowhisker/nanoclay composites." *Polymer Composite* 34 (2013): 293–298.

Hosur M., Mahdi T.H., Islam M.E., Jeelani S. "Mechanical and viscoelastic properties of epoxy nanocomposites reinforced with carbon nanotubes, nanoclay, and binary nanoparticles." *Journal of Reinforced Plastics and Composites* 36 (2017): 667–684.

Jalalifar N., Kaffashi B., Ahmadi S. "The synergistic reinforcing effects of halloysite nanotube particles and polyolefin elastomer-grafted-maleic anhydride compatibilizer on melt and solid viscoelastic properties of polylactic acid/ polyolefin elastomer blends." *Polymer Testing* 91 (2020): 106757/1–106757/11.

Jose J.P., Thomas S. "Alumina–clay nanoscale hybrid filler assembling in cross-linked polyethylene based nanocomposites: Mechanics and thermal properties." *Physical Chemistry Chemical Physics* 16 (2014): 14730–14740.

Karger-Kocsis J. "Loss factor from dynamic mechanical analysis (DMA): The right prediction tool?" *Express Polymer Letters* 11 (2017): 243.

Karger-Kocsis J., Mohd Ishak Z.A. "Composites' damping: An actual challenge." *Express Polymer Letters* 11 (2017): 935.

Kaynak C., Kaygusuz I. "Consequences of accelerated weathering in polylactide nanocomposites reinforced with halloysite nanotubes." *Journal of Composite Materials* 50 (2016): 365–375.

Kulik V.M., Boiko A.V. "Physical principles of methods for measuring viscoelastic properties." *Journal of Applied Mechanics and Technical Physics* 59 (2018): 874–885.

Lai S.M., Hsieh Y.T. "Preparation and properties of polylactic acid (PLA)/silica nanocomposites." *Journal of Macromolecular Science Part B* 55(2016): 211–228.

Lakes R.S. "Viscoelastic measurement techniques." *Review of Scientific Instruments* 75 (2004): 797–810.

Li T.N., Turng L.S., Gong S.Q., Erlacher K. "Polylactide, nanoclay, and core–shell rubber composites." *Polymer Engineering and Science* 46 (2006): 1419–1427.

Li Z.X., Shi S.W., Yang F., Cao D.F., Zhang K.Y., Wang B., Ma Z., Pan L., Li Y.S. "Supertough and transparent poly(lactic acid) nanostructure blends with minimal stiffness loss." *ACS Omega* 5 (2020): 13148–13157.

Lim K.M., Chow W.S., Pung S.Y. "Accelerated weathering and UV protection-ability of poly(lactic acid) nanocomposites containing zinc oxide treated halloysite hanotube." *Journal of Polymers and the Environment* 27 (2019): 1746–1759.

Makaremi M., Pasbakhsh P., Cavallaro G., Lazzara G., Aw Y.K., Lee S.M., Milioto S. "Effect of morphology and size of halloysite nanotubes on functional pectin bionanocomposites for food packaging applications." *ACS Applied Materials & Interfaces* 9 (2017): 17476–17488.

Montava-Jorda S., Chacon V., Lascano D., Sanchez-Nacher L., Montanes N. "Manufacturing and characterization of functionalized aliphatic polyester from poly(lactic acid) with halloysite nanotubes." *Polymers (Basel)* 11 (2019): 1314/1–1314/21.

Murariu M., Dechief A.L., Peeterbroeck S., Bonnaud L., Dubois P. "Polylactide (PLA)-halloysite nanocomposites: Production, morphology and key-properties." *Journal of Polymers and the Environment* 20 (2012): 932–943.

Nofar M., Salehiyan R., Ray S.S. "Influence of nanoparticles and their selective localization on the structure and properties of polylactide-based blend nanocomposites." *Composites Part B: Engineering* 215 (2021): 108845.

Pande V.V., Sanklecha V.M. "Bionanocomposite: A review." *Austin Journal of Nanomedicine and Nanotechnology* 5 (2017): 1045/1–1045/3.

Pasbakhsh P., Churchman G.J., Keeling, J.L. "Characterisation of properties of various halloysites relevant to their use as nanotubes and microfibre fillers." *Applied Clay Science* 74 (2013): 47–57.

Pirani S.I., Krishnamachari P., Hashaikeh R. "Optimum loading level of nanoclay in PLA nanocomposites: Impact on the mechanical properties and glass transition temperature." *Journal of Thermoplastic Composite Materials* 27 (2014): 1461–1478.

Prashantha K., Lecouvet B., Sclavons M., Lacrampe M.F., Krawczak P. "Poly(lactic acid)/halloysite nanotubes nanocomposites: Structure, thermal, and mechanical properties as a function of halloysite treatment." *Journal of Applied Polymer Science* 128 (2013): 1895–1903.

Rashmi B.J., Prashantha K., Lacrampe M.F., Krawczak P. "Toughening of poly(lactic acid) without sacrificing stiffness and strength by melt-blending with polyamide 11 and selective localization of halloysite nanotubes." *Express Polymer Letters* 9 (2015): 721–735.

Risyon N.P., Othman S.H., Basha R.K., Talib R.A. "Effect of halloysite nanoclay concentration and addition of glycerol on mechanical properties of bionanocomposite films." *Polymers and Polymer Composites* 24 (2016): 795–802.

Salehiyan R., Yoo Y., Choi W.J., Hyun K. "Characterization of morphologies of compatibilized polypropylene/polystyrene blends with nanoparticles via nonlinear rheological properties from FT-rheology." *Macromolecules* 47 (2014): 4066–4076.

Schreck K.M., Hillmyer M.A. "Block copolymers and melt blends of polylactide with nodax microbial polyesters: preparation and mechanical properties." *Journal of Biotechnology* 132 (2007): 287–295.

Shakuntala O., Raghavendra G., Samir Kumar A. "Effect of filler loading on mechanical and tribological properties of wood apple shell reinforced epoxy composite." *Advances in Materials Science and Engineering* 2014 (2014): 538651/1–538651/9.

Sheng K., Zhang S., Qian S. Fontanillo Lopez C.A. "High-toughness PLA/bamboo cellulose nanowhiskers bionanocomposite strengthened with silylated ultrafine bamboo-char." *Composites Part B: Engineering* 165 (2019): 174–182.

Singh S., Ghosh A.K., Maiti S.N., Raha S., Gupta R.K., Bhattacharya S. "Morphology and rheological behavior of polylactic acid/clay nanocomposites." *Polymer Engineering & Science* 52(2012): 225–232.

Suppiah K., Teh P.L., Husseinsyah S., Rahman R. "Properties and characterization of carboxymethyl cellulose/halloysite nanotube bio-nanocomposite films: Effect of sodium dodecyl sulfate." *Polymer Bulletin* 76 (2019): 365–386.

Teamsinsungvon A., Jarapanyacheep R., Ruksakulpiwat Y., Jarukumjorn K. "Melt processing of maleic anhydride grafted poly(lactic acid) and its compatibilizing effect on poly(lactic acid)/poly(butylene adipate-co-terephthalate) blend and their composite." *Polymer Science Series A* 59 (2017): 384–396.

Tham W.L., Poh B.T., Mohd Ishak Z.A., Chow W.S. "Water absorption kinetics and hygrothermal aging of poly(lactic acid) containing halloysite nanoclay and maleated rubber." *Journal of Polymers and the Environment* 23 (2015): 242–250.

Tham W.L., Poh B.T., Mohd Ishak Z.A., Chow W.S. "Epoxidized natural rubber toughened poly(lactic acid)/halloysite nanocomposites with high activation energy of water diffusion." *Journal of Applied Polymer Science* 133 (2016): 428501–42850/9.

Wang H.T., Fu Z., Zhao X.W., Li Y.J., Li J.Y. "Reactive nanoparticles compatibilized immiscible polymer blends: Synthesis of reactive SiO_2 with long poly(methyl methacrylate) chains and the in situ formation of Janus SiO_2 nanoparticles anchored exclusively at the interface." *ACS Applied Materials & Interfaces* 9 (2017): 14358–14370.

Wei W., Zhang Y.J., Liu M.H., Zhang Y.F., Yin Y., Gutowski W.S., Deng P.Y., Zheng C.B. "Improving the damping properties of nanocomposites by monodispersed hybrid POSS nanoparticles: Preparation and mechanisms." *Polymers* 11 (2019): 647/1–647/5.

Wu W., Cao X., Lin H., He G., Wang M. "Preparation of biodegradable poly(butylene succinate)/halloysite nanotube nanocomposite foams using supercritical CO_2 as blowing agent." *Journal of Polymer Research* 22 (2015): 1–11.

Yeniova C.E., Ozkoc G., Yilmazer U. "Use of halloysite nanotubes for the production of poly(lactic acid) nanocomposites." *MRS Online Proceedings Library* 1504 (2012): 1–9.

Yeniova C.E., Ozkoc G., Yilmazer U. "Effects of halloysite nanotubes on the performance of plasticized poly(lactic acid)-based composites polymer composites." Polymer Composites 37 (2016): 3134–3148.

Yuya P.A., Patel N.G. "Analytical model for nanoscale viscoelastic properties characterization using dynamic nanoindentation." *Philosophical Magazine* 94 (2014): 2505–2519.

Zaverl M., Seydibeyoğlu M.Ö., Misra M., Mohanty A. "Studies on recyclability of polyhydroxybutyrate-co-valerate bioplastic: Multiple melt processing and performance evaluations." *Journal of Applied Polymer Science*, 125 (2012): E324–E331.

Żenkiewicz M., Richert J., Rytlewski P., Moraczewski K., Stepczyńska M., Karasiewicz T. "Characterisation of multi-extruded poly (lactic acid)." *Polymer Testing* 28 (2009): 412–418.

Zhang Y., Deng B.Y., Liu Q.S. "Rheology and crystallisation of PLA containing PLA-grafted nanosilica." *Plastics, Rubber and Composites* 43 (2014): 309–314.

Zheng Q., Zuo M., Peng M., Shen L., Fan Y. "Rheological study of microstructures and properties for polymeric materials." *Frontiers of Materials Science in China* 1 (2007): 1–6.

14 Influence of Fiber Treatment on the Viscoelastic Properties of Biocomposites

Sabarish Radoor
King Mongkut's University of Technology North Bangkok

Jasila Karayil
Government Women's Polytechnic College

Aswathy Jayakumar
King Mongkut's University of Technology North Bangkok

Jyotishkumar Parameswaranpillai
Mar Athanasios College for Advanced
Studies Tiruvalla (MACFAST)

Suchart Siengchin
King Mongkut's University of Technology North Bangkok
and
Technische Universität Dresden

CONTENTS

14.1 INTRODUCTION

Plants are one of the common sources of natural fibers and are generally extracted from roots, stems, leaves, barks, and fruits (Pappu et al. 2015, Shahid et al. 2013, Senthilkumar et al. 2020). Environment-friendly and renewable nature of natural

DOI: 10.1201/9781003173625-16

fibers makes it a promising candidate to fabricate several materials. The other important features of natural fibers, which attracted their search community are high strength, easy availability, lightweight, recyclability, and enhanced energy recovery (Elanchezhian et al. 2018, Senthilkumar et al. 2018). Natural fiber-based composites have been commercialized in various sectors such as building, automotive, food packaging, marine, and sports. (Cruz and Fangueiro 2016, Ali et al. 2016, Kumar and Sekaran 2014, Thiagamani et al. 2019). Natural fibers are mainly composed of cellulose, lignin, hemicellulose, and water-soluble substance. The main drawback of natural fibers is their hydrophilic nature that results in poor interfacial interaction with a polymer matrix (Mukhopadhyay and Fangueiro 2009, Sanjay et al. 2019, Li et al. 2007). Physical or chemical treatment (plasma treatment, thermal treatment, alkali treatment, benzoylation, and silane treatment) will enhance the surface roughness and improve the interfacial adhesion between the fiber and matrix (Senthilkumar et al. 2019, Krishnasamy et al. 2019). As a result, the overall properties such as mechanical, thermal, and viscoelasticity of fibers will be enhanced (Saheb and Jog 1999, Srinivas 2017, Mohammed et al. 2015, Ahmad et al. 2015, Sathishkumar et al. 2013, Kumar et al. 2021).

An idea of storage and loss modulus, stress relaxation modulus, dynamic viscosity, and damping characteristic of natural fiber is required for identifying its suitable application. Dynamic mechanical analysis is generally employed to understand the viscoelastic properties of polymer (Alothman et al. 2020). The storage modulus (E') measures the elastic response of the material. Meanwhile, loss modulus (E'') is related to the viscous response or the tendency of material to dissipate energy. The dampening or tan δ (ratio of the loss modulus to the storage modulus) is related to the internal friction. The viscoelastic property of material is also influenced by the nature of constituents, presence of fiber content, fiber orientation, and filler/additives (Pothan et al. 2003, Ashok et al. 2019, Bledzki and Zhang 2016, Chandrasekar et al. 2019).

14.2 EFFECT OF NATURAL FIBER ON THE DYNAMIC MECHANICAL PROPERTIES OF POLYMER COMPOSITE

Rana et al. (1999) studied the influence of fiber and compatibilizer on the viscoelastic properties of polypropylene (PP) composite. It can be observed that on increasing the percentage of fiber, the storage and flexural modulus also increases. The addition of a compatibilizer further enhances the value of storage modulus. This was attributed to the strong reinforcement between natural fiber and polymer matrix. Meanwhile, both fiber and compatibilizer content has a positive impact on the retention value. However, with enhancement in the fiber content tan δ (loss factor or damping efficiency) value decreases. The introduction of fiber restricts the movement of the polymer chain and thereby decreases its damping efficiency. These observations were in accordance with the reports of Doan et al. (2007) for jute fiber/PP composite. Here, they employed maleic anhydride (MAH) as a coupling agent. Chemical and physical treatment of fiber is found to influence the viscoelastic property of fiber-reinforced composite. For instance, Ray et al. (2002) reported that alkali-treated jute fiber/vinyl ester composite exhibits high storage modulus than untreated composite. This is due to good mechanical bonding between fiber and matrix. They further observed that upon increasing the jute fiber the E' also increases which was in agreement with reports of previous studies. High storage modulus of the composites can be taken as a piece of strong evidence for the better compatibility between fiber and the polymer. Meanwhile, with increase in fiber loading, the peak of loss modulus gets broadened and is related to the high energy of absorption. The tan δ was found to be high for neat resin. This is ascribed to low storage modulus of resin. However, the composite exhibits low value for tan δ. Karaduman and Onal (2012) employed enzyme and alkali treatment to improve the properties of jute fiber. Dynamic mechanical analysis indicates an enhancement in the storage modulus of the composite after enzyme and sodium hydroxide (NaOH) treatment. It can be noted that NaOH treatment is superior to enzyme treatment and is reflected in the storage modulus value. The storage modulus for NaOH-treated composite is 2,699.404 MPa, whereas for enzyme-treated composite it is 2,433.760 MPa. This was attributed to the improvement in the fiber–matrix adhesion after treatments. The fiber treatment removes the amorphous and disordered polymers (lignin, hemicellulose, and pectin) from the surface of the fibers and thereby creates several sites for fiber–matrix adhesion. Furthermore, the fiber treatments reduce the formation of local stress concentration regions in the fiber structure and thereby enhance the load distribution capability of the composite. SEM analysis was complementary to dynamic mechanical analysis and confirmed high fiber–matrix interaction. The fiber treatment influences the loss modulus of composite, however, it has no effect on the damping parameter (tan δ). The same group (Karaduman et al. 2014) studied the properties of alkali-treated jute/PP composite. They found that on increasing the jute content, the storage and loss modulus of the composites also increase. However, the jute content has a negative effect on the damping parameter (tan δ) probably due to high fiber–matrix interaction. These observations were further supported by theoretical analyses such as adhesion efficiency factor (A) and reinforcement effectiveness coefficient (C). The effect of stacking sequence on dynamic mechanical properties of composites was studied. The MXMX stacking sequence (jute fiber were aligned alternatively in cross-machine and machine direction) yields high storage modulus than other sequences. This result was in compliance with the previous studies and was explained based on effective stress transfer from polymer to jute fibers. The effect of temperature on the viscoelastic properties was also studied. With an increase in temperature, the storage modulus decreases which is due to temperature-induced polymer softening. Meanwhile, the tan δ value of neat PP is shifted to high temperature, which could be due to the high thermal stability of composites (Figure 14.1).

Li et al. (2008) reported that the addition of silk fibroin (SF) fiber increases the storage and loss modulus of SF/poly (ε-caprolactone) (PCL) composites. Introduction of SF fiber into PCL matrix reduces its molecular interactions, which results in the enhancement of its mobility. Consequently, the glass transition temperature was shifted to a lower temperature range. The fiber content also affects the rheological behavior of composite. Upon addition of fiber, the composite change from Newtonian (liquid-like) to non-Newtonian behavior (solid-like). This was due to the formation of network structure in the composite, which acts as reinforcement phase. The rheological study further shows a weak interaction between PCL and SF fiber, which is attributed to the low compatibility between hydrophobic polymer and hydrophilic fiber. Morphological analysis supports this observation. Hidalgo-Salazar and Salinas

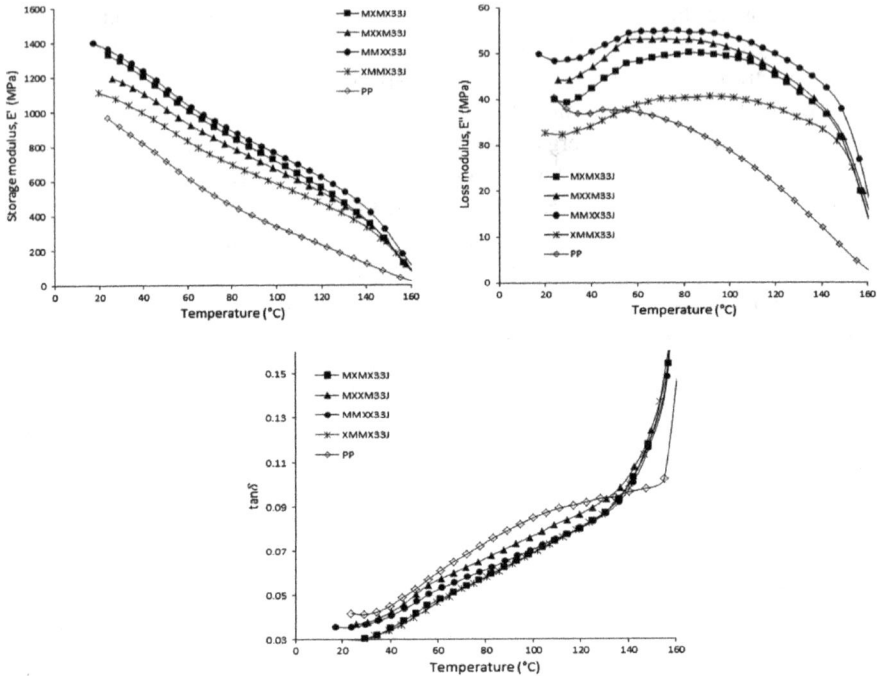

FIGURE 14.1 Variation of E′, E″, and tan δ of different stacking sequences of jute/polypropylene composites with respect to temperature. (Reproduced with permission from Elsevier, License Number: 5105780065939.)

(2019) studied the effect of Colombian rice husk (RH) on the viscoelastic properties of PP biocomposite. The PP-RH composite displayed higher storage modulus than the neat PP. It was also noted that on increasing the fiber content from 10% to 30%, the storage modulus value increases and a maximum value of 3,431, 2,598, 3,805, and 1,116 MPa were obtained at 0°C, 25°C, 50°C, and 80°C, respectively (Table 14.1). Meanwhile, the T_g value of composite was shifted to higher temperature due to the restricted mobility of the PP chains in PP-RH composite.

TABLE 14.1

T_g and Storage Modulus of Neat PP and PP-RH Biocomposites

Sample	Storage Modulus (MPa) at 50°C	Storage Modulus (MPa) at 0°C	Storage Modulus (MPa) at 25°C	Storage Modulus (MPa) at 80°C	T_g (°C)
PP	3,192	2,330	1,487	521	4.3
PP-RH 10	3,166	2,819	1,958	689	6.7
PP-RH 20	3,346	2,695	1,883	784	6.6
PP-RH 30	3,805	3,431	2,598	1,116	7.5

Reproduced with permission from Elsevier, License Number: 5105780532746.

Joseph et al. (2003) reported that fiber length, fiber loading, and surface modification improve the viscoelastic property of sisal fiber/PP composite. The viscoelastic properties (storage and loss modulus) were highest for chemically modified composites. Also, the composite with high fiber content displayed better viscoelastic properties. The effect of surface treatment (MAH-PP) on jute fiber/PP composite was carried out by Bledzki and Zhang (2016). The result showed that MAH-treated sisal fiber possesses better viscoelastic properties than untreated composite. This is due to improved adhesion between jute fiber and PP. Hydrothermal aging is found to influence the viscoelastic property of composites such as storage, loss modulus, and damping. Gil-Castell et al. (2016) in their study observed that aged composite has high T_g than non-aged samples. This is due to the fact that during aging the mobility of amorphous polymer chains gets restricted. Hydrothermal aging also adversely affects the storage modulus of the composite. However, the introduction of coupling agents (maleic acid anhydride, MAH, and DCP) improve the fiber–matrix interface and the stored value reaches a maximum of 3,300 MPa for such composites (Figure 14.2). Thus, it can be concluded that the coupling agent enhances the viscoelastic property of even-aged samples and thereby minimizes the effect of aging.

Righetti et al. (2019) analyzed the dynamic mechanical properties of polylactic acid (PLA)/potato pulp biocomposite. The poor interaction between hydrophobic polymer matrix and hydrophilic potato results in low viscosity. On the other hand, an improvement in viscosity value was noted for wax-treated composites. Pothen et al. (2003) reported that the viscoelastic property of banana fiber-reinforced polyester composites is highly influenced by fiber content. Upon increasing the fiber

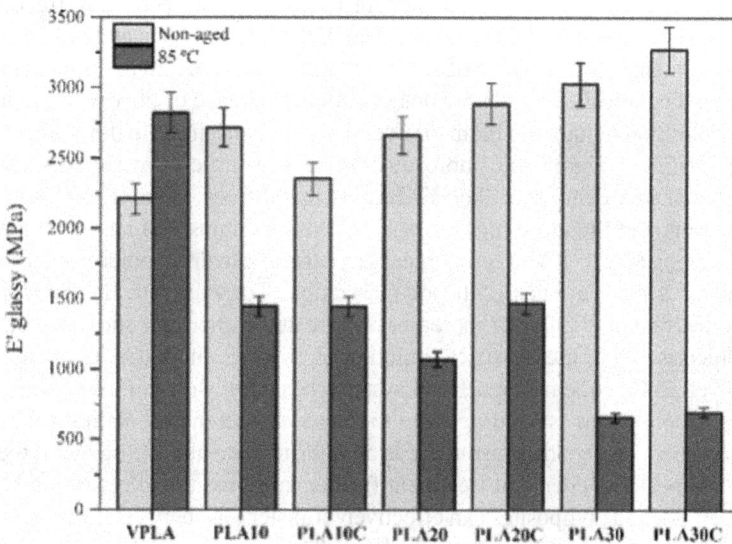

FIGURE 14.2 Storage modulus (E′) with respect to fiber content with non-aged and hydrothermal aging at 85°C. (Reproduced with permission from Elsevier, License Number: 5105780859974)

content, the crystallinity of the composite increases whereas the polymer mobility gets reduced. Thus, composite loaded with 40% fiber display high E′, low loss storage, and damping value. Furthermore, Cole-Cole plots and T_g value indicate a good interfacial adhesion at high fiber content. The same group reported (Pothan et al. 2006) that they noted an improvement in the damping value and storage modulus of the composite after chemical treatment (silane A174 and NaOH). Manikandan Nair et al. (2001) studied the effect of temperature, surface modification, and fiber characteristics such as length, orientation, and content on the storage modulus and damping behavior of short sisal fiber-reinforced polystyrene composites. They noticed that with addition of fiber, the storage modulus of composite increases. However, at high fiber loading, the storage modulus of composite becomes constant. Thermal energy triggers the segmental mobility of the polymer and therefore an inverse relation was observed between temperature and storage modulus. Three surface treatment methods namely benzoylation, acetylation, and M-polystyrene maleic anhydride (PSMA) were employed to improve the fiber–matrix interaction. Benzylated, as well as acetylated sisal fiber, has a rough surface with low hydrophilic character. This will lead to an enhancement in its interaction with the hydrophobic polymer. Another factor that improves the bonding between fiber and polymer is the formation of hydrogen between MAH group of PSMA and hydroxyl group of the sisal fiber. Thus, surface-treated fiber composites have higher viscoelastic property than that of untreated fiber composites. Martínez-Hernández et al. (2007) developed keratin fibers from chicken feathers and employed them to reinforce polymethyl methacrylate (PMMA). The effect of fiber content on the PMMA matrix was investigated using dynamic mechanical properties. At low fiber content (1% and 2%), the fiber has reinforcing effect and therefore an increasing trend in storage modulus with fiber content was observed. However, when the fiber content reaches 3%, a decreasing trend in the storage modulus was noted. On the other hand, the loss modulus and tan δ decrease with fiber content. The low value of damping or tan δ (loss modulus/storage modulus) in reinforced composite is probably due to better adherence of fiber with the matrix. The incorporation of jute into PP matrix results an enhancement in the storage modulus (55.89 wt.%). A significant improvement in viscoelastic property was noted for chemically treated composite. The $KMnO_4$ is found to be superior to other chemical treatments and therefore high storage and loss modulus was noted for $KMnO_4$-treated composite. The $KMnO_4$ treatment increases the hydrophobicity and roughens of jute fiber. It also initiates graft polymerization. Consequently, the compatibility between jute fiber and polymer increases. All treated composites show low damping effect which is due to the constricted motion of polymer chain (George et al. 2012). Varghese et al. (1994) employed HRH system (hexamethylenetetramine, resorcinol, and fine particle silica) as bonding agent to improve adherence of natural rubber with short sisal fiber. The bonding agent act as an intermediate and thereby bind the polymer and rubber. The chemical treatment further improves the effect of the bonding agent. Well bonded composites can effectively transfer load between fiber and matrix and therefore possess high storage modulus. The orientation of sisal fiber is also found to influence the viscoelastic property of composite. In the case of longitudinal fibers, the fibers are orientated in the same direction of load. However, the direction of load and fiber will be different in the case of transverse fiber. Hence, the longitudinally oriented fiber composite displayed high storage modulus than transverse fiber orientation. The viscoelastic property, however, decreases with increase in temperature. Han et al. (2011) studied the combined effect of exfoliated graphite nanoplatelets (xGnP) and kenaf fiber on the viscoelastic properties of PLA composite. The introduction of XGnP increases the dispersion of kenaf fiber and also reduces the fiber–polymer interfacial contact area. This is responsible for the low viscosity value of the composite. It is also reported that the cooperative effect of xGnP and fiber is responsible for enhancing the storage modulus. About 97% improvement in the storage modulus was observed when the composite was loaded with 5 wt.% XGnP and 40% kenaf fiber. However, XGnP has only minimal effect on the damping effect of composite. Shanmugam and Thiruchitrambalam (2013) evaluated the viscoelastic nature of alkali-treated palmyra palm leaf stalk fiber (PPLSF)/jute fiber-reinforced hybrid polyester composites. Introduction of jute fiber into pure resin increases its storage modulus from 2.15 to 3.39 MPa. This is attributed to the high Young modulus of jute fiber. The addition of jute fiber also increases the loss modulus of the composite. It is a known fact that composites with good adhesion have low damping value. Therefore, low damping was observed for treated composites. It was also noticed that the damping value further decreases with an increase in fiber content (Figure 14.3).

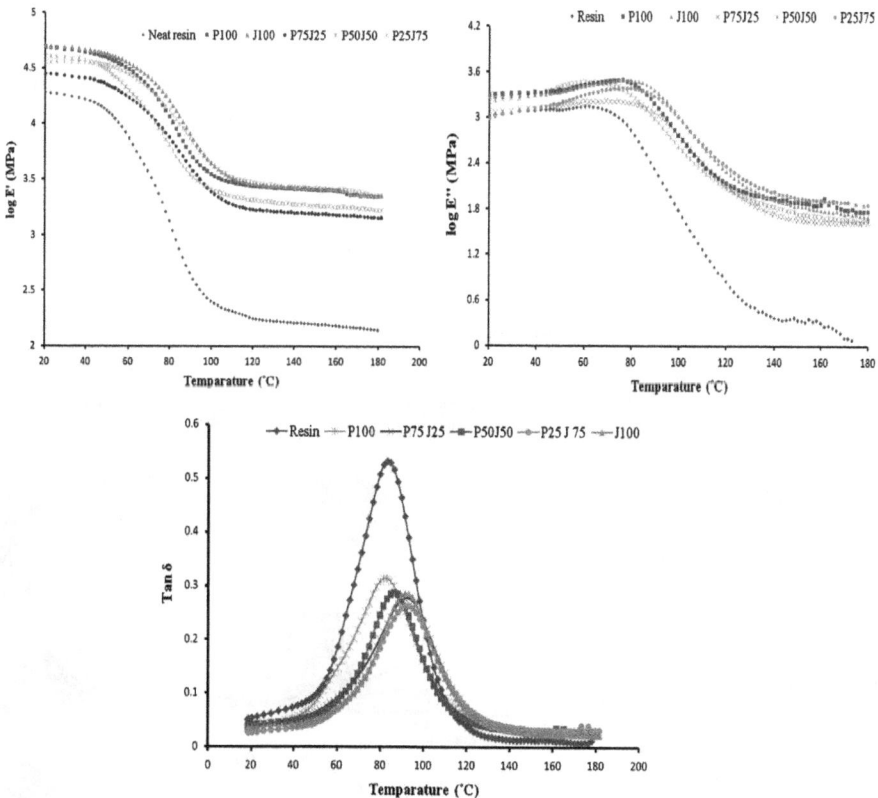

FIGURE 14.3 The variation of E′, E″, and tan δ as function of temperature for palmyra palm leaf stalk fiber (PPLSF)/jute fiber hybrid composites. (Reproduced with permission from Elsevier, License Number: 5105781249418.)

Pineapple leaf fiber (PALF) is commonly used to reinforce polymer composites. The effect of fiber (PALF and coir) on the viscoelastic property of epoxy resin was studied by Mittal and Chaudhary (2018).

They observed that upon addition of coir and pineapple leaf fiber, the E′ and E″) of composite is shifted to the higher temperature. Meanwhile, a reduction in tan δ value was observed for fiber-treated composites. Huda et al. (2012) observed that chemical-treated (silane and alkali) PALF is superior reinforcing agent than untreated PALF for PLA composite. Chemical treatment removes hemicelluloses, lignin, and waxy substance from the fiber surface and consequently enhances its roughness. Therefore, a better mechanical interlocking between fiber and matrix is possible. This is responsible for the high mechanical and viscoelastic properties of the composite. The comparison of storage modulus of neat PLA and chemically treated composites revealed that the storage modulus was almost double for chemically treated composite. The same group (Huda et al. 2008) reported the effect of treated kenaf fibers on the viscoelastic behavior of PLA composite. In this work, three treatment methods were adapted: (a) NaOH solution (FIBNA), (b) silane coupling agent (FIBSIL), and (c) NaOH treatment followed by silane coupling agent (FIBNASIL). The results showed that the storage modulus or stiffness of the FIB, FIBNA, and FIBSI composites increased to 41%, 67%, and 87%, respectively, with respect to pure PLA (Figure 14.4).

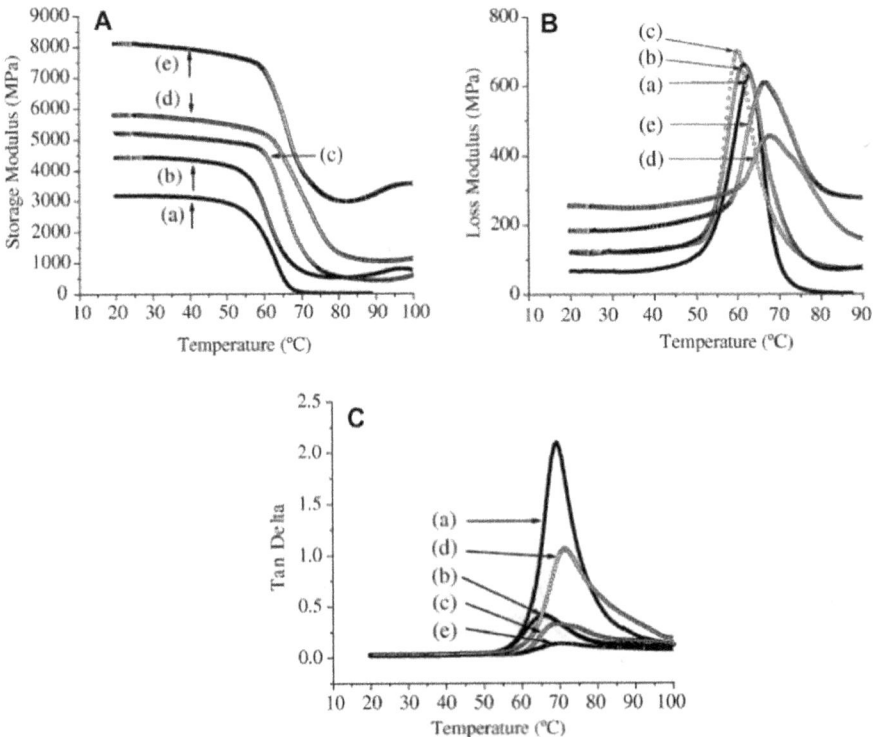

FIGURE 14.4   Variation of E′, E″, and tan δ of neat PLA and its composites. (Reproduced with permission from Elsevier, License Number: 5105840115878.)

This result was further strengthened by SEM analysis. Strong fiber–matrix interaction was observed for surface-modified composites.

Two different fibers can be hybridized and could be used to reinforce polymer composites. In the case of hybrid composite, better dispersion of fiber is attained. Also, stress could be transferred from one fiber to another without destructing the polymer matrix. Banana and sisal fibers have high cellulose content, smaller microfibrillar angle, and comparable elongation at break. This led Idicula et al. (2005) to select banana and sisal fiber to reinforce polyester matrix. Better fiber–matrix compatibility was observed when the volume fraction of fibers reaches 0.4, which was later confirmed from SEM micrographs. Furthermore, a positive shift in T_g value was also noted at 0.4 volume fraction of fiber. A minimum effectiveness coefficient C was obtained for 40% fiber volume fraction. At high fiber loading, the fiber-to-fiber contact becomes dominant and thereby decreases the effective stress transfer between the fiber and matrix. The composite with 3:1 (banana:sisal) fiber ratio exhibits high storage modulus and damping behavior. Meanwhile, Romanzini et al. (2013) developed a polyester hybrid composite from glass and ramie fiber. The results indicate that the incorporation of fiber will enhancement the E′ and E″. The loss modulus of composite increases with addition of fiber for all glass fiber:ramie fiber ratio. The peak height of loss modulus is related to the weight percentage of glass fiber. With increase in glass fiber content, the free volume and number of chain segments in the composite increase. This inhibits the relaxation process and consequently, the peak height increases. The presence of fiber also affects the tan δ value of the composites. This is due to fiber-induced stiffness which restricts the free movement of the polymer chains (Figure 14.5). Saw et al. (2011) investigated the effect of layering patterns on the viscoelastic behavior of short bagasse/coir fibers/epoxy novolac composites. In their work four different layering patterns were used: bilayer (bagasse/coir), trilayer ((bagasse/coir/bagasse) and (coir/bagasse/coir)), and intimate mix. The reinforcement efficiency of the composite is represented by coefficient C, which is related to the storage modulus. Larger the C, lower will be its reinforcement efficiency. The C value for bilayer, trilayer (bagasse/coir/bagasse) and (coir/bagasse/coir), and intimate mix are 0.234, 0.170, 0.178, and 0.168, respectively. The result thus indicates that trilayer and intimate mix arrangement provide greater reinforcement than bilayer. The storage modulus is found to be low for C/B/C pattern; this can be explained based on the low tensile strength of coir fiber. The C/B/C therefore has low stiffness and is less effective to transfer stress between fiber and matrix.

14.3 CONCLUSION

Natural fibers are ecofriendly material that is commonly used to reinforce polymer composites. PALF, sisal, jute, kenaf fibers, etc. are some of the widely explored reinforcing materials in composites. The introduction of fibers into composites improves their physical and chemical properties. A significant improvement in the thermomechanical property is also noted. The viscoelastic property of the composites namely storage, loss modulus, and damping behavior are highly influenced by the fiber reinforcement. The dynamic mechanical properties of the composites could be tuned by fiber characteristics such as fiber nature, distribution, orientation,

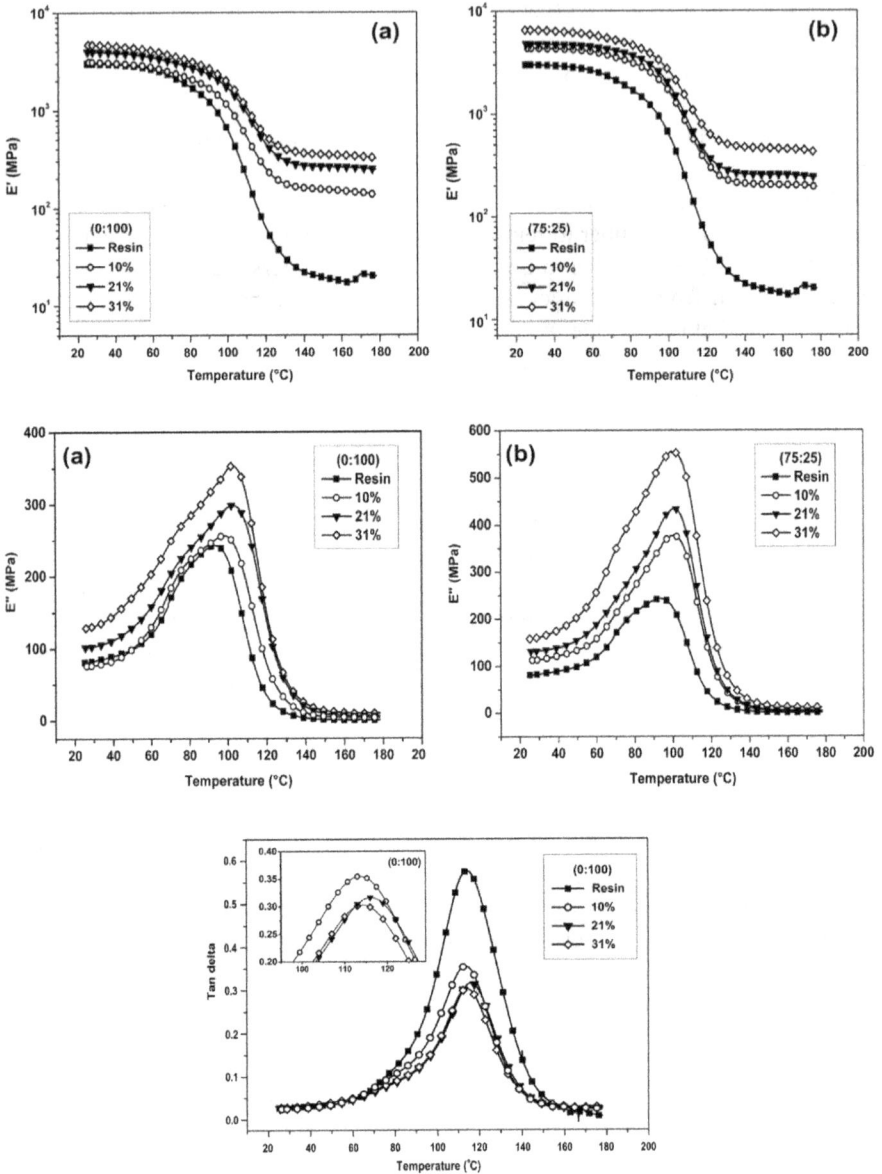

FIGURE 14.5 The variation of E′, E″, and tan δ with respect to temperature for glass/ramie fiber hybrid composites. (Reproduced with permission from Elsevier, License Number: 5105781249418.)

and layering pattern. Fiber loading is another factor that affects fiber–matric adherence and thereby affects the reinforcing action of fiber. The chemical treatment of fibers improves the fiber–matrix interaction and hence a better viscoelastic property was attained for treated composites. The temperature is another factor that can influence the viscoelastic parameter of natural fiber-reinforced polymer composites. Lastly, we also shed light on hybrid composites. The incorporation of two fibers will generally lead to strong reinforcement and a high storage modulus was attained in such cases.

ACKNOWLEDGMENT

Authors gratefully thank King Mongkut's University of Technology North Bangkok (KMUTNB), Thailand for the financial support through the Post-Doctoral Program (Grant No. KMUTNB-63-Post-03 and KMUTNB-64-Post-03 to SR) and (Grant No. KMUTNB-BasicR-64-16).

REFERENCES

Ahmad, F., H.S. Choi, and M.K. Park. 2015. "A review: Natural fiber composites selection in view of mechanical, light weight, and economic properties." *Macromolecular Materials and Engineering* 300(1):10–24. doi: 10.1002/mame.201400089.

Ali, A., K. Shaker, Y. Nawab, M. Jabbar, T. Hussain, J. Militky, and V. Baheti. 2016. "Hydrophobic treatment of natural fibers and their composites—A review." *Journal of Industrial Textiles* 47(8):2153–2183. doi:10.1177/1528083716654468.

Alothman, O.Y., M. Jawaid, K. Senthilkumar, M. Chandrasekar, B.A. Alshammari, et al. 2020. "Thermal characterization of date palm/epoxy composites with fillers from different parts of the tree." *Journal of Materials Research and Technology* 9(6):15537–15546.

Ashok, R.B., C.V. Srinivasa, and B. Basavaraju. 2019. "Dynamic mechanical properties of natural fiber composites—A review." *Advanced Composites and Hybrid Materials* 2(4):586–607. doi: 10.1007/s42114-019-00121-8.

Bledzki, A.K., and W. Zhang. 2016. "Dynamic mechanical properties of natural fiber-reinforced epoxy foams." *Journal of Reinforced Plastics and Composites* 20(14–15):1263–1274. doi: 10.1106/5pe2–8mnk-mtb5-k2d3.

Chandrasekar, M., R.M. Shahroze, M.R. Ishak, N. Saba, M. Jawaid, et al. 2019. "Flax and sugar palm reinforced epoxy composites: Effect of hybridization on physical, mechanical, morphological and dynamic mechanical properties." *Materials Research Express* 6(10):105331.

Cruz, J., and R. Fangueiro. 2016. "Surface modification of natural fibers: A review." *Procedia Engineering* 155:285–288. doi: 10.1016/j.proeng.2016.08.030.

Doan, T.-T.-L., H. Brodowsky, and E. Mäder. 2007. "Jute fibre/polypropylene composites II. Thermal, hydrothermal and dynamic mechanical behaviour." *Composites Science and Technology* 67(13):2707–2714. doi: 10.1016/j.compscitech.2007.02.011.

Elanchezhian, C., B. Vijaya Ramnath, G. Ramakrishnan, M. Rajendrakumar, V. Naveenkumar, and M.K. Saravanakumar. 2018. "Review on mechanical properties of natural fiber composites." *Materials Today: Proceedings* 5(1):1785–1790. doi:10.1016/j.matpr.2017.11.276.

George, G., E.T. Jose, D. Åkesson, M. Skrifvars, E.R. Nagarajan, and K. Joseph. 2012. "Viscoelastic behaviour of novel commingled biocomposites based on polypropylene/jute yarns." *Composites Part A: Applied Science and Manufacturing* 43(6):893–902. doi: 10.1016/j.compositesa.2012.01.019.

Gil-Castell, O., J.D. Badia, T. Kittikorn, E. Strömberg, M. Ek, S. Karlsson, and A. Ribes-Greus. 2016. "Impact of hydrothermal ageing on the thermal stability, morphology and viscoelastic performance of PLA/sisal biocomposites." *Polymer Degradation and Stability* 132:87–96. doi: 10.1016/j.polymdegradstab.2016.03.038.

Han, S.O., M. Karevan, Md.A. Bhuiyan, J.H. Park, and K. Kalaitzidou. 2011. "Effect of exfoliated graphite nanoplatelets on the mechanical and viscoelastic properties of poly (lactic acid) biocomposites reinforced with kenaf fibers." *Journal of Materials Science* 47(8):3535–3543. doi: 10.1007/s10853-011-6199-8.

Hidalgo-Salazar, M.A., and E. Salinas. 2019. "Mechanical, thermal, viscoelastic performance and product application of PP- rice husk Colombian biocomposites." *Composites Part B: Engineering* 176. doi: 10.1016/j.compositesb.2019.107135.

Huda, M.S., L.T. Drzal, A.K. Mohanty, and M. Misra. 2008. "Effect of fiber surface-treatments on the properties of laminated biocomposites from poly (lactic acid) (PLA) and kenaf fibers." *Composites Science and Technology* 68(2):424–432. doi:10.1016/j.compscitech.2007.06.022.

Huda, M.S., L.T. Drzal, A.K. Mohanty, and M. Misra. 2012. "Effect of chemical modifications of the pineapple leaf fiber surfaces on the interfacial and mechanical properties of laminated biocomposites." *Composite Interfaces* 15(2–3):169–191. doi: 10.1163/156855408783810920.

Idicula, M., S.K. Malhotra, K. Joseph, and S. Thomas. 2005. "Dynamic mechanical analysis of randomly oriented intimately mixed short banana/sisal hybrid fibre reinforced polyester composites." *Composites Science and Technology* 65(7–8):1077–1087. doi:10.1016/j.compscitech.2004.10.023.

Joseph, P.V., G. Mathew, K. Joseph, G. Groeninckx, and S. Thomas. 2003. "Dynamic mechanical properties of short sisal fibre reinforced polypropylene composites." *Composites Part A: Applied Science and Manufacturing* 34(3):275–290. doi: 10.1016/s1359-835x(02)00020-9.

Karaduman, Y., and L. Onal. 2012. "Dynamic mechanical and thermal properties of enzyme-treated jute/polyester composites." *Journal of Composite Materials* 47(19):2361–2370. doi: 10.1177/0021998312457885.

Karaduman, Y., M.M.A. Sayeed, L. Onal, and A. Rawal. 2014. "Viscoelastic properties of surface modified jute fiber/polypropylene nonwoven composites." *Composites Part B: Engineering* 67:111–118. doi:10.1016/j.compositesb.2014.06.019.

Krishnasamy, S., C. Muthukumar, R. Nagarajan, S.M.K. Thiagamani, N. Saba, et al. 2019. "Effect of fibre loading and Ca(OH)^2 treatment on thermal, mechanical, and physical properties of pineapple leaf fibre/polyester reinforced composites." *Materials Research Express* 6(8). https://doi.org/10.1088/2053-1591/ab2702.

Kumar, K.P., and A. Shadrach Jeya Sekaran. 2014. "Some natural fibers used in polymer composites and their extraction processes: A review." *Journal of Reinforced Plastics and Composites* 33 (20):1879–1892. doi: 10.1177/0731684414548612.

Kumar, T.S.M., M. Chandrasekar, K. Senthilkumar, N. Ayrilmis, S. Siengchin, et al. 2021. "Utilization of bamboo fibres and their influence on the mechanical and thermal properties of polymer composites." In: Jawaid, M., Mavinkere Rangappa, S., Siengchin, S. (eds.) *Bamboo Fiber Composites*, pp. 81–96. Springer, Singapore.

Li, W., X. Qiao, K. Sun, and X. Chen. 2008. "Mechanical and viscoelastic properties of novel silk fibroin fiber/poly (ε-caprolactone) biocomposites." *Journal of Applied Polymer Science* 110(1):134–139. doi: 10.1002/app.28514.

Li, X., L.G. Tabil, and S. Panigrahi. 2007. "Chemical treatments of natural fiber for use in natural fiber-reinforced composites: A review." *Journal of Polymers and the Environment* 15(1):25–33. doi:10.1007/s10924-006-0042-3.

Manikandan Nair, K.C., S. Thomas, and G. Groeninckx. 2001. "Thermal and dynamic mechanical analysis of polystyrene composites reinforced with short sisal fibres." *Composites Science and Technology* 61(16):2519–2529. doi: 10.1016/s0266-3538(01)00170-1.

Martínez-Hernández, A.L., C. Velasco-Santos, M. de-Icaza, and V.M. Castaño. 2007. "Dynamical–mechanical and thermal analysis of polymeric composites reinforced with keratin biofibers from chicken feathers." *Composites Part B: Engineering* 38(3):405–410. doi: 10.1016/j.compositesb.2006.06.013.

Mittal, M., and R. Chaudhary. 2018. "Effect of fiber content on thermal behavior and visco-elastic properties of PALF/Epoxy and COIR/Epoxy composites." *Materials Research Express* 5(12). doi: 10.1088/2053-1591/aae274.

Mohammed, L., M.N.M. Ansari, G. Pua, M. Jawaid, and M.S. Islam. 2015. "A review on natural fiber reinforced polymer composite and its applications." *International Journal of Polymer Science* 2015:1–15. doi: 10.1155/2015/243947.

Mukhopadhyay, S., and R. Fangueiro. 2009. "Physical modification of natural fibers and thermoplastic films for composites — A review." *Journal of Thermoplastic Composite Materials* 22(2):135–162. doi: 10.1177/0892705708091860.

Pappu, A., V. Patil, S. Jain, A. Mahindrakar, R. Haque, and V.K. Thakur. 2015. "Advances in industrial prospective of cellulosic macromolecules enriched banana biofibre resources: A review." *International Journal of Biological Macromolecules* 79:449–458 doi: 10.1016/j.ijbiomac.2015.05.013.

Pothan, L.A., Z. Oommen, and S. Thomas. 2003. "Dynamic mechanical analysis of banana fiber reinforced polyester composites." *Composites Science and Technology* 63(2):283–293. doi: 10.1016/s0266-3538(02)00254-3.

Pothan, L.A., S. Thomas, and G. Groeninckx. 2006. "The role of fibre/matrix interactions on the dynamic mechanical properties of chemically modified banana fibre/polyester composites." *Composites Part A: Applied Science and Manufacturing* 37(9):1260–1269. doi: 10.1016/j.compositesa.2005.09.001.

Rana, A.K., B.C. Mitra, and A.N. Banerjee. 1999. "Short jute fiber-reinforced polypropylene composites: Dynamic mechanical study." *Journal of Applied Polymer Science* 71(4):531–539. doi: 10.1002/(sici)1097-4628(19990124)71:4<531::Aid-app2>3.0.Co;2-i.

Ray, D., B.K. Sarkar, S. Das, and A.K. Rana. 2002. "Dynamic mechanical and thermal analysis of vinylester-resin-matrix composites reinforced with untreated and alkali-treated jute fibres." *Composites Science and Technology* 62(7–8):911–917. doi: 10.1016/s0266-3538(02)00005-2.

Righetti, M., P. Cinelli, N. Mallegni, C. Massa, L. Aliotta, and A. Lazzeri. 2019. "Thermal, mechanical, viscoelastic and morphological properties of poly (lactic acid) based bio-composites with potato pulp powder treated with Waxes." *Materials* 12(6). doi: 10.3390/ma12060990.

Romanzini, D., A. Lavoratti, H.L. Ornaghi, S.C. Amico, and A.J. Zattera. 2013. "Influence of fiber content on the mechanical and dynamic mechanical properties of glass/ramie polymer composites." *Materials & Design* 47:9–15 doi: 10.1016/j.matdes.2012.12.029.

Saheb, D.N., and J.P. Jog. 1999. "Natural fiber polymer composites: A review." *Advances in Polymer Technology* 18(4):351–363. doi: 10.1002/(sici)1098-2329(199924)18:4<351::Aid-adv6>3.0.Co;2-x.

Sanjay, M.R., S. Siengchin, J. Parameswaranpillai, M. Jawaid, C.I. Pruncu, and A. Khan. 2019. "A comprehensive review of techniques for natural fibers as reinforcement in composites: Preparation, processing and characterization." *Carbohydrate Polymers* 207:108–121 doi: 10.1016/j.carbpol.2018.11.083.

Sathishkumar, T.P., P. Navaneethakrishnan, S. Shankar, R. Rajasekar, and N. Rajini. 2013. "Characterization of natural fiber and composites – A review." *Journal of Reinforced Plastics and Composites* 32(19):1457–1476. doi:10.1177/0731684413495322.

Saw, S.K., G. Sarkhel, and A. Choudhury. 2011. "Dynamic mechanical analysis of randomly oriented short bagasse/coir hybrid fibre-reinforced epoxy novolac composites." *Fibers and Polymers* 12(4):506–513. doi: 10.1007/s12221-011-0506-5.

Senthilkumar, K., N. Saba, N. Rajini, M. Chandrasekar, M. Jawaid, et al. 2018. "Mechanical properties evaluation of sisal fibre reinforced polymer composites: A review." *Construction and Building Materials*. doi: 10.1016/j.conbuildmat.2018.04.143.

Senthilkumar, K., N. Rajini, N. Saba, M. Chandrasekar, M. Jawaid, et al. 2019. "Effect of alkali treatment on mechanical and morphological properties of pineapple leaf fibre/polyester composites." *Journal of Polymers and the Environment* 27(6):1191–1201.

Senthilkumar, K., T. Ungtrakul, M. Chandrasekar, T.S.M. Kumar, N. Rajini, et al. 2020. "Performance of sisal/hemp bio-based epoxy composites under accelerated weathering." *Journal of Polymers and the Environment* 29:1–13.

Shahid, M., Shahid-ul-Islam, and F. Mohammad. 2013. "Recent advancements in natural dye applications: A review." *Journal of Cleaner Production* 53:310–331 doi:10.1016/j.jclepro.2013.03.031.

Shanmugam, D., and M. Thiruchitrambalam. 2013. "Static and dynamic mechanical properties of alkali treated unidirectional continuous Palmyra Palm Leaf Stalk Fiber/jute fiber reinforced hybrid polyester composites." *Materials & Design* 50:533–542. doi: 10.1016/j.matdes.2013.03.048.

Srinivas, K. 2017. "A review on chemical and mechanical properties of natural fiber reinforced polymer composites." *International Journal of Perform Ability Engineering*. doi: 10.23940/ijpe.17.02.p8.189200.

Thiagamani, S.M.K., S. Krishnasamy, and S. Siengchin. 2019. "Challenges of biodegradable polymers: An environmental perspective." *Applied Science and Engineering Progress* 12(3):149.

Varghese, S., B. Kuriakose, S. Thomas, and A.T. Koshy. 1994. "Mechanical and viscoelastic properties of short fiber reinforced natural rubber composites: Effects of interfacial adhesion, fiber loading, and orientation." *Journal of Adhesion Science and Technology* 8(3):235–248. doi: 10.1163/156856194x01086.

15 Influence of Compatibilizer on Viscoelastic Properties of the Thermoplastic Polymer-Based Biocomposites

S. Bolka and B. Nardin
Faculty of Polymer Technology

CONTENTS

DOI: 10.1201/9781003173625-17

15.1 INTRODUCTION: BACKGROUND AND DRIVING FORCES

A composite material consists of two or more physically distinct phases, the combination of which produces mechanical properties that are different from those of the individual constituents. Thermoplastic polymer-based biocomposites are of great importance due to their remarkable strength and stiffness combined with very low weight. When using natural fiber reinforcement, the main advantage is that they have almost no effect on the wear of processing equipment. With glass fiber-reinforced composites, the wear of processing equipment is a major disadvantage. The strength to weight and stiffness to weight ratios are many times higher than steel or aluminum and property combinations that cannot be achieved with metals, ceramics, or polymers alone [1]. Recently, not only different contents, particle sizes, and aspect ratios of natural fibers have been developed [2–4], but also hybrid (different (natural) fibers) thermoplastic polymer-based biocomposites [5–7]. The aim of the hybridization of natural fibers is to combine the maximum influence of advantages and the minimum influence of disadvantages of both natural fibers. Thermoplastic polymer-based biocomposites have gained importance in all fields of research due to their environmentally friendly properties. The main disadvantage of natural fibers is their thermal stability. This can be avoided by using a suitable thermoplastic matrix. Polypropylene (PP) and polyethylene (PE) are the most suitable due to their high processing temperature, as the thermal decomposition temperature of natural fibers is higher [8,9]. The hydrophilic nature of natural fibers and the hydrophobic nature of PP and PE pose a challenge to researchers. It can be solved in several ways: by thermal treatment of the natural fibers, by chemical modification of the natural fiber surface, or by the addition of compatibilizers. Modification by compatibilizers is often more commercially acceptable and can be carried out together with mixing and pelletizing of biocomposites based on thermoplastic polymers. Although twin screw extruders are suitable for processing, at least antioxidants should be added to prevent thermal degradation during processing. The addition of a compatibilizer improves the interfacial adhesion between natural fibers and the thermoplastic matrix. The improved adhesion leads to higher stiffness and strength. The addition of maleic anhydride-grafted polypropylene (PP-g-MA) in red pine fibers PP biocomposites [4] increased strength and stiffness, but had no significant effect on toughness. Researchers reported [4] higher strength, stiffness, and toughness by adding PP-g-MA in sawdust PP biocomposites. The addition of anhydride-modified PE (PE-g-MA) in softwood high-density polyethylene (PE-HD) biocomposites [10] increased the strength, stiffness, and elongation at break in tensile tests. PP-g-MA addition in pine fibers PE-HD biocomposites [11] increased bending and impact properties. Treatment of waste paper and various compatibilizers for biocomposites with waste paper have been reported [12,13]. Recycled PP (rPP) and waste paper compatibilized with PP-g-MA facilitate dispersion of waste paper in rPP matrix, strength and stiffness, improve surface bonding and toughness, and waste paper promotes rPP crystallization [13]. Maleic anhydride-grafted styrene-(ethylene-butene)-styrene triblock copolymer (SEBS-g-MA) improves polystyrene (PS)-cellulose interactions and elongation at break in tensile tests [14].

Biocomposites are characterized by traditional characterization methods and with the right combination, biocomposites can be characterized in detail. Dynamic mechanical analysis (DMA) is complementary to traditional mechanical (tensile and flexural tests) and thermal (differential scanning calorimetry (DSC) and thermogravimetric analysis (TGA)) methods. The dynamic nature of DMA results in measured quantities: storage modulus (E'), loss modulus (E''), and loss factor (tan δ).

They are all affected by the following DMA parameters: frequency, temperature, time, amplitude, and measurement mode. The dynamic mechanical behavior of the measured samples allowed us to study the adhesion of the natural fiber–thermoplastic matrix at the interface, the morphology, the glass transition, the damping behavior, and the elastic behavior. The viscoelastic properties of the thermoplastic polymer-based biocomposites can be evaluated by DMA and confirmed with additional tensile, flexural, DSC, TGA, and impact tests. E' can be related to the stiffness of the sample, E'' to the internal friction, and tan δ as a mechanical damping factor to the molecular motions and viscoelasticity including the contribution of defects [15]. The increase in E' during the incorporation of sandalwood PE-HD biocomposites [16] is higher at higher loading fractions and more pronounced at higher temperatures. At the same time, E' also increases with increasing weight fraction of natural fibers. Higher E' was associated with inhibition of matrix relaxation by fiber addition at about 55°C. Researchers elaborated [17] on the improvement of E' and the decrease of tan δ when PP-g-MA was added to the biocomposites of wood flour PP. The addition of the compatibilizer increases the interfacial adhesion between wood flour and PP matrix and consequently the stiffness. In biocomposites made of jute fibers-reinforced PP [18], a slight increase in E' was observed by the addition of PP-g-MA. The higher the fiber content, the higher E' and E'', but the lower tan δ. Two peaks in tan δ are associated with the glass transition temperatures of the amorphous part of PP (β-transition) and the α-transition of the rigid amorphous segments in the PP crystals. Similar results were reported in the study reported Ref. [19] for jute fibers PE-HD biocomposites compatibilized with PE grafted with maleic anhydride (PE-g-MA). Higher E' and E'' and lower tan δ were measured after incorporation of the compatibilizer. The second peak for the α-transition (around 80°C) is related to the reorientation of the defects in the PE-HD crystals. The "E" peak (α-transition) was shifted to a higher temperature and value after the addition of the compatibilizer, which is due to the decrease of the mobility of the matrix at the relaxation temperature caused by a better interfacial adhesion between fibers and matrix. Alkali treatment [20,21] of natural fibers (oil palm fibers and argan nutshells) also increases E' and E'' and decreases tan δ due to better adhesion of natural fibers to PE matrix.

In this chapter, the preparation of composites with thermoplastic polymer-based biocomposites with the addition of different biomasses and various commercial compatibilizers is described. Compatibilizers are widely used in composites and allow the use of standard processing equipment. In addition, an incompatible compatibilizer was chosen for the matrix PE-HD to test the viscoelastic properties and reinforcing effect for use in recycled thermoplastic materials where traces of PE are normally present. The viscoelastic behavior of these composites was investigated by DMA. The mechanical and thermal properties were investigated using universal testing machine (UTM), DSC, and impact tests.

15.2 EXPERIMENTAL

15.2.1 MATERIALS

PE-HD in the form of granules with a combination of various natural fibers and various commercial compatibilizers based on PE-g-MA, styrene-maleic anhydride (SMA), and PP-g-MA were used. Natural fibers (waste paper, cellulose fibers, Fiber-Plast 35E, miscanthus fibers, beech sawdust, and hemp husks) were used without any surface pretreatment despite drying to a moisture content below 5 wt.% and an attempt at NaOH surface modification together with the addition of compatibilizer PE-g-MA. Table 15.1 shows the composition of the biocomposites, samples. Sample

TABLE 15.1
Composition of Biocomposites

Sample[a]	PE-HD (wt.%)	Fiber[b,c]	Compatibilizator[d]	Compatibilizator (wt.%)
1	100	–	–	–
2-1	68.62	1	–	0
2-2	66.62	1	1	2
3-2	66.62	2	2	2
4-2	66.62	2	3	2
5-2	66.62	2	1	2
6-1	68.62	3	–	0
6-2	66.62	3	4	2
7-2	66.62	3	5	2
8-2	66.62	3	3	2
9-2	66.62	3	4	2
10-2	66.62	2	3	2
11-2	66.62	4	3	2
12-2	66.62	5	3	2
13-2	66.62	6	3	2
14-2	66.62	7	3	2
15-3	66.62	2	3	2
16-2	66.62	7	6	2
17-2	64.62	7	6	4
18-2	62.62	7	6	6
19-2	66.62	2	6	2
20-2	65.62	2	6	3
21-2	64.62	2	6	4
22-2	63.62	2	6	5

[a] In all biocomposites samples lubricant (1 wt.%) and antioxidant (0.38 wt.%) were added.

[b] In all biocomposites 30 wt.% fibers were added.

[c] 1 – beech sawdust, 2 – waste paper, 3 – Fiber-Plast 35E, 4 – cellulose fibers 6 mm, 5 – cellulose fibers 0,6 mm, 6 – hemp husk, 7 – miscanthus fibers.

[d] 1 – FG 1901 GT, 2 – E 265, 3 – PMD 353D, 4 – SMA, 5 – Lushan PR-3C, 6 – Exxelor PE 1040 K2.

TABLE 15.2

Collected Results of Tensile Tests of Pristine PE-HD and All Biocomposites

Sample	E_t (GPa)	STD	σ_m (MPa)	STD	ε_m (%)	STD	ε_{tb} (%)	STD
1	1.02	0.12	22.5	0.3	8.24	0.31	471.86	85.46
2-1	2.34	0.14	23.2	0.8	2.69	0.18	3.01	0.29
2-2	1.90	0.15	29.6	0.4	2.80	0.07	3.19	0.13
3-2	2.03	0.07	35.2	0.3	5.30	0.32	5.62	0.40
4-2	2.00	0.12	31.3	0.3	4.32	0.06	4.71	0.12
5-2	1.65	0.09	29.8	0.1	5.06	0.06	5.67	0.11
6-1	2.20	0.19	19.6	0.3	2.79	0.28	3.04	0.37
6-2	2.72	0.23	23.1	0.2	2.63	0.06	2.86	0.16
7-2	2.21	0.12	24.8	0.4	2.12	0.13	2.30	0.15
8-2	2.24	0.23	24.2	0.3	2.18	0.13	2.31	0.2
9-2	2.42	0.46	19.7	0.3	2.25	0.09	2.59	0.17
10-2	1.59	0.13	26.9	0.2	4.29	0.30	4.48	0.46
11-2	3.52	0.32	42.2	1.2	2.83	0.23	2.83	0.29
12-2	2.57	0.63	26.1	0.2	2.49	0.13	2.89	0.13
13-2	0.89	0.11	16.0	0.2	4.24	0.48	4.75	0.66
14-2	1.20	0.15	25.7	0.1	2.70	0.08	2.90	0.15
15-3	2.09	0.47	26.5	0.4	3.95	0.13	4.14	0.25
16-2	2.38	0.13	33.5	0.2	3.03	0.09	3.09	0.14
17-2	2.64	0.15	36.4	0.2	3.50	0.13	3.61	0.21
18-2	2.44	0.09	35.3	1.5	3.94	0.17	4.23	0.26
19-2	1.98	0.13	32.2	0.4	5.38	0.4	5.52	0.49
20-2	1.67	0.12	31.0	0.5	6.40	0.22	6.72	0.44
21-2	1.63	0.09	32.6	0.2	6.40	0.09	6.91	0.41
22-2	1.68	0.09	32.3	0.4	6.48	0.38	6.72	0.59

1 is a pristine PE-HD matrix for reference. Values of mechanical properties together with standard deviation values are presented in Table 15.2 for tensile and Table 15.3 for flexural test.

15.2.2 COMPOUNDING AND INJECTION MOLDING

For all experiments, the materials were mixed separately and extruded on a Labtech LTE 20-44 twin screw extruder. The screws had a diameter of 20 mm, an L/D ratio of 44:1, the screw speed was 400 rpm, and the temperature profile was increasing from the hopper (130°C) to the die (165°C). The temperatures were set as low as possible to prevent thermal degradation of the natural fibers during extrusion [13]. The composite material was extruded through the die with two round orifices with a diameter of 3 mm. After extrusion, the two filaments were cooled in a water bath at a temperature of 25°C and pelletized on the Scheer granulating machine to the approximate dimensions of 3 mm diameter and 4 mm length.

Due to the water bath cooling of the material after extrusion, the granules were cooled in an oven at 80°C for 4 hours to a moisture content of max. 0.1%. Injection

TABLE 15.3

Collected Results of Flexural Tests of Pristine PE-HD and All Biocomposites

Sample	E_f (GPa)	STD	σ_{fM} (MPa)	STD	ε_{fM} (%)	STD
1	0.73	0.02	18.9	0.3	7.36	0.12
2-1	2.37	0.05	32.7	0.2	4.37	0.11
2-2	2.33	0.01	37.8	0.1	4.49	0.11
3-2	1.88	0.02	36.1	0.1	6.47	0.11
4-2	1.81	0.02	34.4	0.1	6.02	0.04
5-2	1.58	0.02	31.3	0.1	6.32	0.07
6-1	2.08	0.03	27.7	0.2	4.70	0.09
6-2	1.97	0.02	28.0	0.1	4.32	0.11
7-2	2.25	0.08	34.4	0.3	4.08	0.13
8-2	1.92	0.17	32.4	1.0	4.42	0.11
9-2	2.04	0.17	28.3	0.5	3.87	0.21
10-2	1.56	0.13	29.7	0.2	6.31	0.12
11-2	2.36	0.05	39.4	0.8	4.00	0.10
12-2	2.27	0.06	30.7	0.2	3.27	0.02
13-2	0.76	0.06	17.8	0.1	6.59	0.10
14-2	1.72	0.03	30.0	0.1	5.25	0.10
15-3	1.67	0.06	32.0	0.5	5.95	0.11
16-2	2.15	0.03	39.0	0.3	5.60	0.07
17-2	2.16	0.02	40.5	0.1	5.96	0.12
18-2	2.05	0.02	38.9	0.1	6.02	0.08
19-2	1.59	0.03	30.7	0.1	6.75	0.07
20-2	1.57	0.01	30.8	0.1	6.76	0.09
21-2	1.60	0.04	31.6	0.1	6.76	0.05
22-2	1.49	0.07	30.6	0.2	6.85	0.06

molding was performed on the injection molding machine Krauss Maffei 50–180 CX with a screw diameter of 30 mm and a clamping force of 500 kN. The cold runner mold was used for the sample production, the mold had two cavities, one with dumbbell-shaped mold of type 1BA (ISO 527-1) and the second with cuboid-shaped mold (ISO 178/ ISO 179). The temperature profile was increasing from the hopper (145°C) to the mold (165°C), the injection speed was set to 60 mm/min, during plastication the backpressure was set to 195 bar and the screw speed to 175 rpm. The mold temperature was set to 40°C and the cooling time was set to 10 s.

15.2.3 METHODS FOR CHARACTERIZATION OF THE COMPOSITES

The thermomechanical properties were investigated using a dynamic mechanical analyzer Perkin Elmer DMA 8000. TT_DMA software, version 14310, was used to evaluate the results. The viscoelastic properties of the samples were analyzed by recording the E', E'', and tan δ as a function of temperature. For the analysis of the test samples, the DMA instrument was operated in dual cantilever mode. The viscoelastic analyses were performed on samples with dimensions of approximately 42 × 5 × 2 mm.

The samples were heated at 2°C/min from room temperature (22°C) to 120°C under air atmosphere. A frequency of 1 Hz and an amplitude of 20 μm were used.

Thermal measurements were performed using a differential scanning calorimeter (DSC 2, Mettler Toledo) under a nitrogen atmosphere (20 mL/min). The temperatures of the samples were raised from 0°C to 180°C at a heating rate of 10°C/min and kept in the molten state for 5 min to clear the thermal history. After cooling at 10°C/min, the samples were reheated to 180°C at 10°C/min. The crystallization temperature (T_c), melting temperature (T_m), and enthalpy of fusion (ΔH_m) were determined from the cooling scan and the second heating scan. Moreover, the crystallinity values (X_c) were calculated using Equation 15.1, where $H_m = 293$ J/g [22].

$$X_c = \left(\Delta H_m / \Delta H^0{}_m \right) \times \left(100\% / \text{wt.\%} PE \right) \tag{15.1}$$

where X_c (%) is the degree of crystallinity, wt.% PE is the weight % of PE-HD in the sample, ΔH_m (J/g) is the enthalpy of fusion of the sample, and ΔH^0_m (J/g) is the enthalpy of fusion of 100% crystallized PE-HD.

Impact tests were carried out on Pendel Dongguan Liyi Test Equipment, type LY-XJJD5 impact testing machine according to ISO 179. The specimens for impact test were injection molded according to ISO 179, with dimensions 80 × 10 × 4 mm.

The flexular tests were carried out according to ISO 178 on Shimadzu AG-X plus with a load cell of maximum 10 kN, at a crosshead speed of 2 mm/min and a span of 64 mm. TrapeziumX software, version 1.3.1, was used to evaluate the results. The specimens were injection molded according to ISO 178 with dimensions 80 × 10 × 4 mm. Five replicates for each sample were tested. All tests were carried out at 23°C.

Tensile tests were carried out according to ISO 527-1 on the Shimadzu AG-X plus equipped with a force transducer of maximum 10 kN, at a crosshead speed of 1 mm/min until the elongation of 0.25%, then at 50 mm/min until failure, measuring the elongation at mid-span of each specimen with a Shimadzu TRViewX optical extensometer. TrapeziumX software, version 1.3.1, was used to analyze the results. The specimens were injection molded on dumbbell-shaped type 1BA specimens according to ISO 527-2. Five replicates were tested for each specimen. All tests were conducted at 23°C.

15.2.4 Investigation of the Properties of the Composites

The investigation of the properties is presented in six sections. In the first, the influence of two different compatibilizers varied:

- Biocomposites with and without compatibilizer vs. untreated PE-HD matrix.
- Influence of the compatibilizer on the addition of different recycled paper batches in biocomposites.
- Influence of the compatibilizer on the addition of waste paper in biocomposites.
- Influence of compatibilizer amount on biocomposites with miscanthus fibers.

- Influence of compatibilizer amount on biocomposites with waste paper.
- Influence of different natural fibers on biocomposites.
- Influence of compatibilizer on the addition of Fiber-Plast 35E in biocomposites.

15.2.4.1 Biocomposites with and without Compatibilizer vs. Pristine PE-HD Matrix

The authors present the influences of different compatibilizers in biocomposites compared to the pristine PE-HD matrix. E' as a function of temperature is shown in Figure 15.1. It can be seen that there is a significant increase in E' of the original PE-HD matrix with the incorporation of beech sawdust and Fiber-Plast 35E. This is due to an increase in the stiffness of the matrix after the addition of natural fibers. Fiber-Plast 35E achieved a higher E' compared to the beech sawdust. It can be speculated that Fiber-Plast 35E allowed a higher degree of stress transfer at the fiber–matrix interface [23]. As expected, the addition of the compatibilizer SEBS-g-MA to the biocomposite of beech sawdust PE-HD lowered the E' throughout the temperature range. It can be concluded that the elastic segments in the compatibilizer lower E', but the maleic anhydride achieves a good interaction with the surface of the natural fibers. The difference decreases with increasing temperature, which is probably due to the amorphous structure of the compatibilizer. The addition of SMA in Fiber-Plast 35E PE-HD biocomposite has almost no effect on E'.

Figure 15.2 shows the results of the loss modulus. The peak of E' is shifted to higher values and higher temperature with the addition of natural fibers. The peak can be related to the α-relaxation temperature [24]. With the addition of natural fiber, the magnitude of loss modulus increases. For FG 1901 GT compatibilizer, the peak is at a lower temperature but has the same value as without compatibilizer. SMA compatibilizer has no discernible effect on E''. The compatibilizer FG 1901 GF prevents inhibition of the relaxation process when beech sawdust is added to the PE-HD matrix [25].

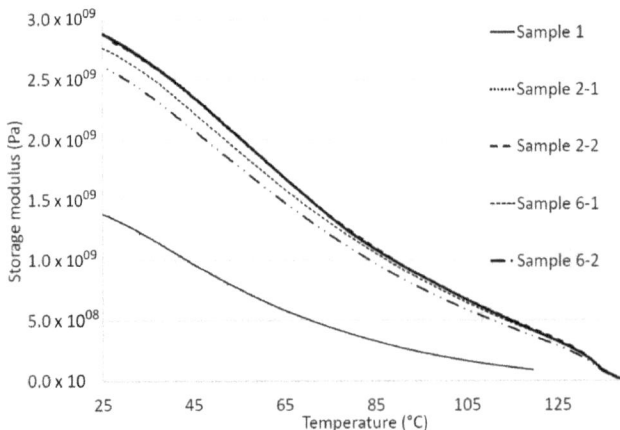

FIGURE 15.1   Storage modulus of the biocomposite samples with and without compatibilizer.

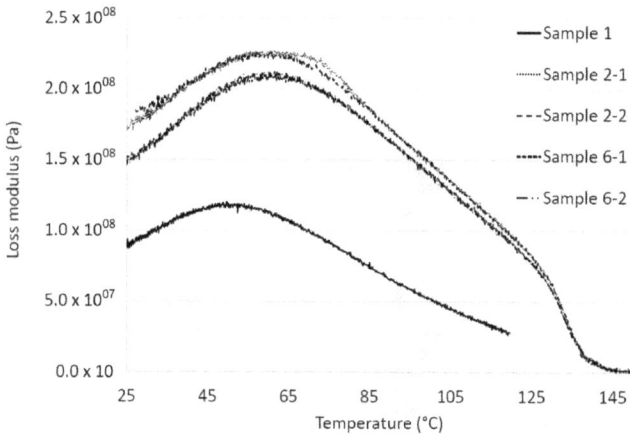

FIGURE 15.2 Loss modulus of the biocomposite samples with and without compatibilizer.

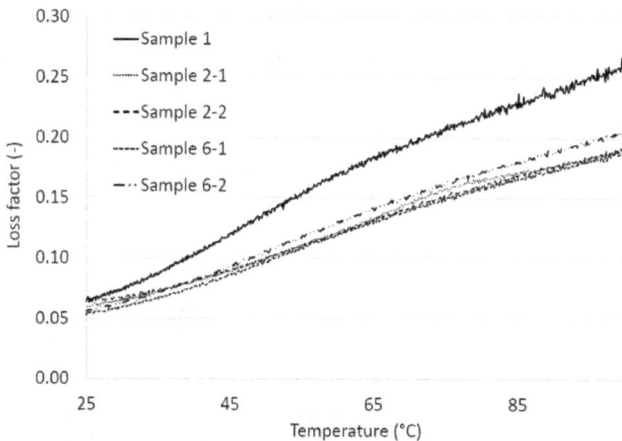

FIGURE 15.3 Loss factor of the biocomposite samples with and without compatibilizer.

The loss factor (Figure 15.3) is highest throughout the temperature range for the pristine PE-HD matrix. The addition of natural fibers lowers tan δ as most of the stress is taken by the fibers where the energy dissipation is much lower than in the matrix [26]. FG 1901 GF compatibilizer shows more favorable tan δ values than composite without compatibilizer. The difference is small but most pronounced in the α-relaxation region where the shoulder appears. As can be seen in Figure 15.3, the best elastic behavior has an unmodified sample with Fiber-Plast 35E. The best damping behavior of biocomposites with natural fibers has a sample with Fiber-Plast 35E modified with SMA. It can be concluded that SMA as a compatibilizer in biocomposite gives higher energy dissipation than FG 1901 GF compatibilizer.

The effect of compatibilizer on strength and stiffness (Tables 15.2 and 15.3) is shown in Figure 15.4 for tensile test and in Figure 15.5 for flexural test. The FG 1901 GF lowers the tensile and flexural stiffness and increases the tensile and flexural

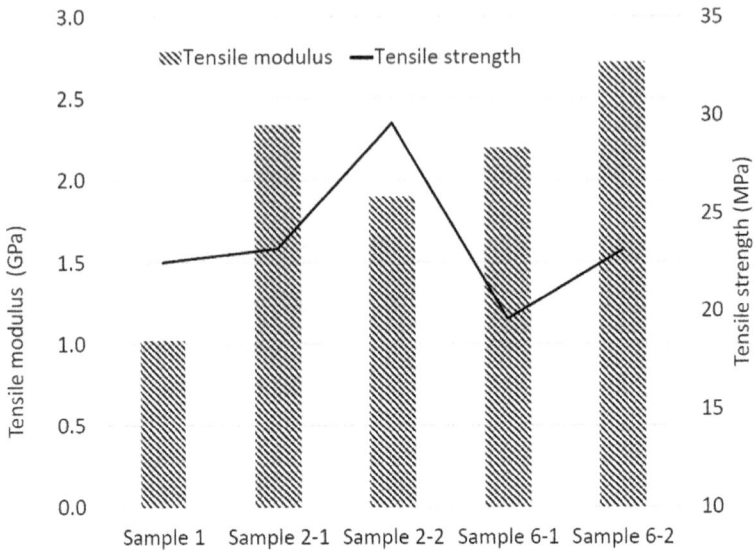

FIGURE 15.4 Tensile strength and stiffness of the biocomposite samples with and without compatibilizer.

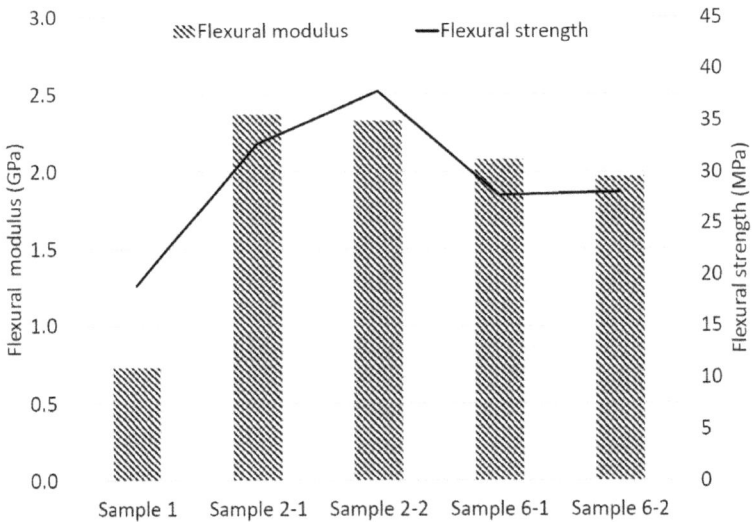

FIGURE 15.5 Flexural strength and stiffness of the biocomposite samples with and without compatibilizer.

strength. The PMD 353D increases the tensile and flexural strength and tensile stiffness and decreases the flexural stiffness. The results correlate well with the DMA results, as the higher strength indicates good compatibility of the natural fibers in the PE-HD matrix after the addition of the compatibilizer. For the beech sawdust, the impact strength test results confirm these results. The impact strength (Table 15.4) of

TABLE 15.4

Collected Results of Impact Tests of Pristine PE-HD and All Biocomposites

Sample	Impact Strength (kJ/m²)	STD
1	No break	–
2-1	8.0	0.2
2-2	8.5	0.3
3-2	18.4	1.6
4-2	13.9	1.0
5-2	16.2	1.4
6-1	8.0	0.7
6-2	6.1	0.6
7-2	7.9	0.7
8-2	8.2	0.9
9-2	6.4	0.7
10-2	11.8	1.1
11-2	11.8	1.3
12-2	8.5	0.5
13-2	9.5	1.7
14-2	8.9	1.3
15-3	11.2	0.9
16-2	11.0	0.7
17-2	14.5	1.2
18-2	15.4	1.3
19-2	15.2	1.3
20-2	18.8	1.4
21-2	19.4	1.2
22-2	19.7	1.8

the sample with compatibilizer FG 1901 GT is higher than that of the sample without compatibilizer. The impact strength of the sample with PMD 353D compatibilizer is lower than that of the sample without compatibilizer.

The compatibilizer SMA increases the crystallinity (Table 15.5) of the PE-HD matrix, but has no effect on the melting temperature of PE-HD. The addition of the compatibilizer lowered the crystallization temperature. Addition of Fiber-Plast 35D in PE-HD matrix reduced the crystallinity.

15.2.4.2 Influence of the Compatibilizer on the Addition of Different Waste Paper Batches in Biocomposites

In this part, the influence of compatibilizer on the mechanical properties of bio-composites is presented from the viewpoint of different batches of waste paper. It is illustrated in Figure 15.6 that the highest increase in stiffness was obtained with E 265 followed by PMD 353D addition in the waste paper PE-HD biocomposite. The lowest stiffness was obtained with the addition of Eccelor PE 1040 K2 into the

TABLE 15.5

Collected Results of DSC Tests of Pristine PE-HD and All Biocomposites

Sample	T_m (°C)	ΔH_m (J/g)	T_c (°C)	ΔH_c (J/G)	X_c (%)
1	130.7	203.1	116.7	201.2	69.3
2-1	130.8	138.9	116.9	136.8	69.1
2-2	130.9	135.1	116.8	134.7	69.2
3-2	131.2	138.8	117.9	130.1	71.1
4-2	131.0	135.6	117.8	123.6	69.5
5-2	131.2	131.2	118.1	131.0	67.2
6-1	130.2	137.8	118.0	137.6	68.5
6-2	130.7	137.7	115.9	133.0	70.5
7-2	130.6	133.1	118.4	133.9	68.2
8-2	129.9	137.9	117.4	128.9	70.6
9-2	130.2	131.8	116.9	123.8	67.5
10-2	131.6	140.7	117.2	140.7	72.1
11-2	130.8	142.7	115.7	138.0	73.1
12-2	131.1	135.4	117.6	125.6	69.4
13-2	130.9	126.2	113.8	123.1	64.7
14-2	130.9	147.0	114.9	145.0	75.3
15-3	131.0	135.4	116.0	130.9	69.4
16-2	131.1	132.4	116.3	131.8	67.8
17-2	132.1	125.9	115.4	111.0	66.5
18-2	131.6	122.2	115.9	118.8	66.6
19-2	130.9	140.4	116.9	130.1	71.9
20-2	131.2	133.0	116.7	127.4	69.2
21-2	131.0	125.8	116.9	114.0	66.4
22-2	132.2	136.1	115.8	131.8	73.0

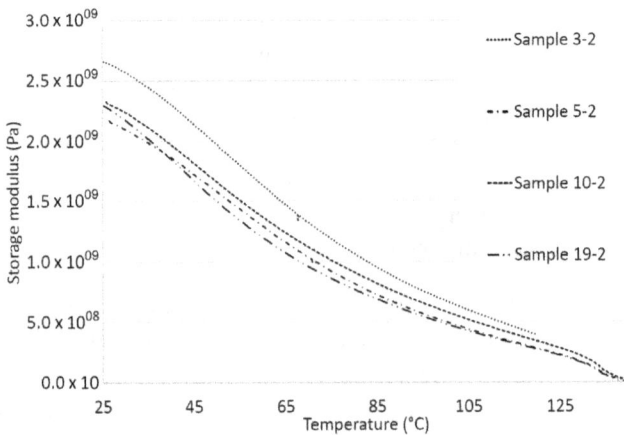

FIGURE 15.6 Storage modulus of the biocomposite samples with different batches of waste paper.

recycled paper PE-HD biocomposite. The addition of FG 1901 GT showed lower E' values at temperatures up to 40°C, and at higher temperatures the shape of the E' curve is like other compatibilizers. This behavior may be due to the elastomeric nature of the compatibilizer. From the results, it can be concluded that the best stress transfer is provided at the interface E 265, PMD 353D, FG 1901 GF, and Exxelor PE 1040 K2, respectively. It can be assumed that all compatibilizers achieve good interaction with the waste paper surface due to the high E' values up to the PE-HD melting range, although E 265 and PMD 353D have the highest values even in the melting range of the PE-HD matrix. This fact confirms that the addition of E 265 and PMD 353D to the biocomposite best binds the interface between the waste paper and PE-HD matrix.

Figure 15.7 presents the results of the loss modulus. The "E" curve for the compatibilizer Exxelor PE 1040 K2 in the biocomposite of waste paper PE-HD is different from the other curves. The E'' for Exxelor PE 1040 K2 reaches its maximum value at 39°C and the height is greater compared to other biocomposites (250 MPa). The compatibilizer Exxelor PE 1040 K2 together with the waste paper limits the mobility of the PE-HD matrix, although the difference is larger than expected. Also, the difference in the morphology of the PE-HD matrix could be caused by a larger energy dissipation due to the accelerated relaxation process in the crystal structure of the PE-HD matrix. The α-relaxation peak with other compatibilizers appears at 52°C, 57°C, and 62°C for PMD 353D, FG 1901 GT, and E 265, respectively. The biocomposite with compatibilizer E 265 reached the E'' peak with the highest amplitude. The biocomposite with the compatibilizer PMD 353D has the E'' peak with the lowest amplitude, which is probably due to the fact that the high melting point of the compatibilizer allows a lower dissipation of energy in the compatibilized matrix. It can be concluded that the best interactions between the waste paper surface and the PE-HD matrix were achieved with the E 265 compatibilizer.

The loss factor (Figure 15.8) is highest throughout the temperature range for the biocomposite with the compatibilizer Exxelor PE 1040 K2. From α-relaxation toward

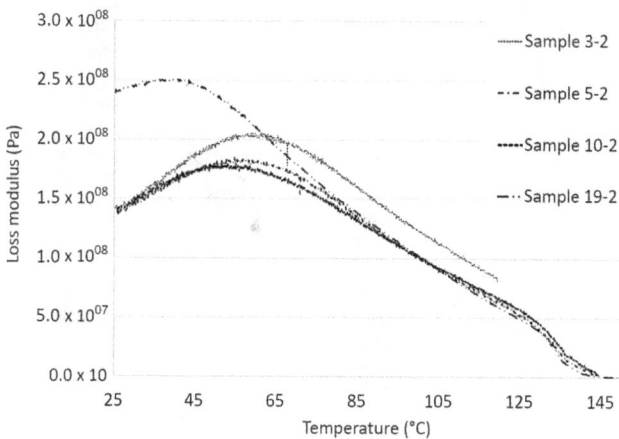

FIGURE 15.7 Loss modulus of the biocomposite samples with different batches of waste paper.

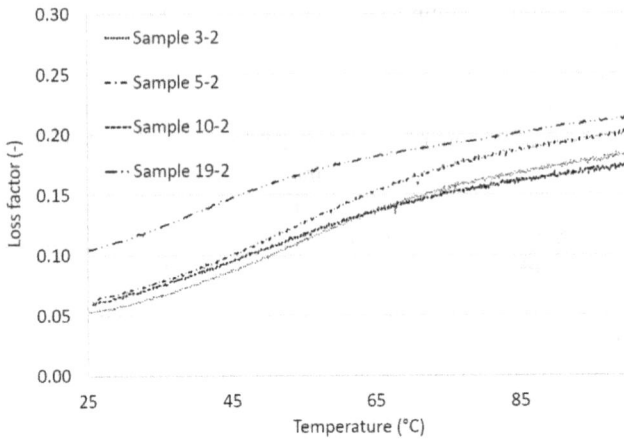

FIGURE 15.8 Loss factor of the biocomposite samples with different batches of waste paper.

higher temperatures, the difference with the other biocomposites is smaller. The lowest tan δ values up to the α-relaxation range were achieved by the biocomposite with E 265, and at elevated temperature with FG 1901 GF. Biocomposites with E 265 and FG 1901 GF have the best elastic behavior. The best damping behavior has biocomposite Exxelor PE 1040 K2. The conclusion can be drawn that Exxelor PE 1040 K2 as a compatibilizer in the biocomposite allows the highest energy dissipation.

The effect of compatibilizer on strength and stiffness (Tables 15.2 and 15.3) is shown in Figure 15.9 for tensile test and Figure 15.10 for flexural test. The highest strength and stiffness were obtained for the specimen with Compatibilizer E 265. This correlates well with the DMA results. The lowest strength and stiffness were

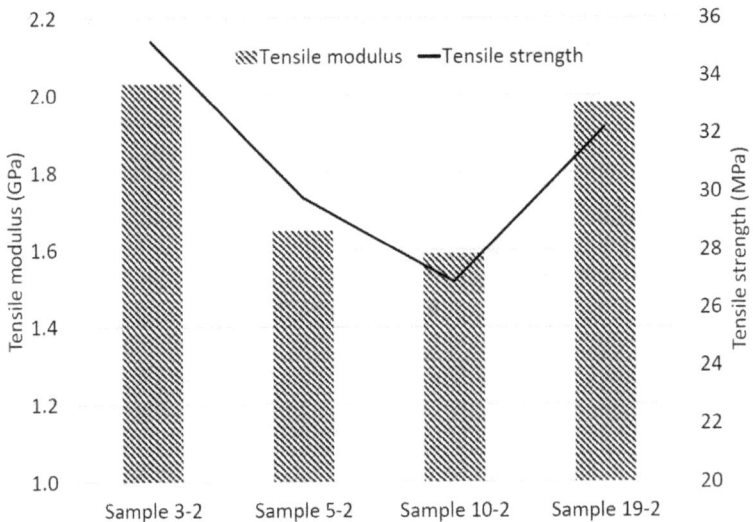

FIGURE 15.9 Tensile strength and stiffness of the biocomposite samples with different batches of waste paper.

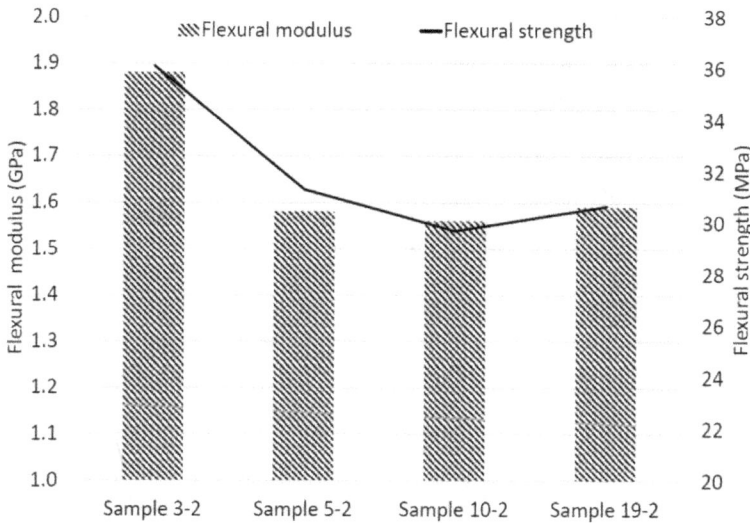

FIGURE 15.10 Flexural strength and stiffness of the biocomposite samples with different batches of waste paper.

achieved by the sample with PMD 353D. This may be due to the low compatibility between the PE-HD matrix and PP in the compatibilizer. The sample with Exxelor 1040 K2 compatibilizer achieved the second best strength and stiffness. This is not correlated with the DMA results where the E' was the lowest. The reason could be the different composition of the waste paper due to different batches of waste paper. The best impact strength (Table 15.4) was obtained for the sample with E 265 followed by FG 1901 GF, Exxelor PE 1040 K2, and PMD 353D. From the results of the tensile, flexural, and impact strength tests, it can be concluded that the waste paper in the sample with Exxelor PE 1040 K2 was probably larger than the other waste paper grades.

The compatibilizers PMD 353D and Exxelor PE 1040 K2 increase the crystallinity (Table 15.5) of PE-HD matrix but do not affect the melting and crystallization temperature of PE-HD. With the addition of the compatibilizer FG 1901 GT, the crystallinity was reduced.

15.2.4.3 Influence of the Compatibilizer on the Addition of Waste Paper in Biocomposites

In this part of the chapter, the influence of different compatibilizers on the addition of waste paper in biocomposites is presented from the viewpoint of the influence of different compatibilizers on the mechanical properties of the biocomposite. As shown in Figure 15.11, it can be seen that the highest increase in stiffness was obtained with E 265, followed by PMD 353D. As expected, the lowest stiffness was obtained with the addition of FG 1901 GT into the biocomposite of waste paper PE-HD. The shape of the E' curve is similar for all compatibilizers tested. As the temperature increases, the difference in E' becomes smaller. From the results, it can be concluded that the best stress transfer is enabled at the interface E 265, PMD 353D, or FG 1901 GF. The

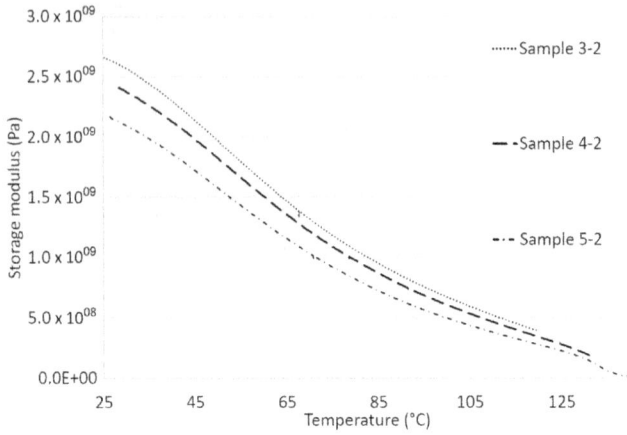

FIGURE 15.11 Storage modulus of the biocomposite samples with waste paper and different compatibilizer.

results are in agreement with the previous results. It can be confirmed the assumption that all compatibilizers achieve good interaction with the surface of the waste paper due to high E' values up to PE-HD melting range, although E 265 and PMD 353D showed the highest values even in the melting range of PE-HD matrix. This fact confirms that the addition of E 265 and PMD 353D to the biocomposite best bonded the interface between the waste paper and PE-HD matrix, as these two biocomposites achieved E' values greater than 0.5 GPa above 105°C.

Figure 15.12 shows the loss modulus results. E'' curves have the same shape. α-Relaxation peaks are at 57°C, 58°C, and 62°C for PMD 353D, FG 1901 GT, and E 265, respectively. The biocomposite with the compatibilizer E 265 reached the E'' peak with the highest amplitude. The biocomposite with the compatibilizer PMD 353D has the E'' peak with the lowest amplitude, which is probably due to the fact

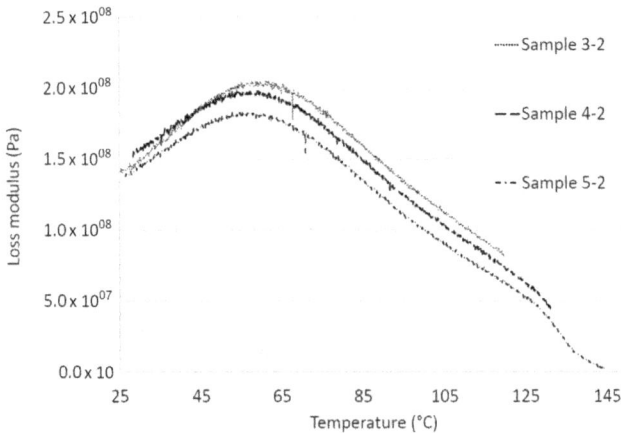

FIGURE 15.12 Loss modulus of the biocomposite samples with waste paper and different compatibilizer.

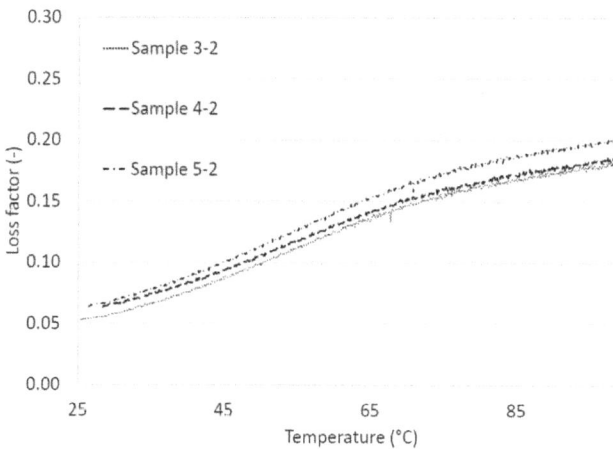

FIGURE 15.13   Loss factor of the biocomposite samples with waste paper and different compatibilizer.

that the high melting point of the compatibilizer allows a lower dissipation of energy in the compatibilized matrix. One can confirm the conclusion that the best interactions between the waste paper surface and the PE-HD matrix were achieved with the E 265 compatibilizer.

The loss factor (Figure 15.13) is the highest in the whole temperature range for the biocomposite with FG 1901 GF compatibilizer. From the α-relaxation toward higher temperatures, the difference with the other biocomposites is greater. The lowest tan δ values in the whole temperature range were obtained by the biocomposite with E 265. The biocomposite with E 265 has the best elastic behavior. The biocomposite with FG 1901 GF has the best damping behavior. It can be concluded that E 265 as a compatibilizer in the biocomposite gives the best energy dissipation. The addition of PMD 353D in the biocomposite enables elastic behavior at elevated temperatures.

The effect of compatibilizer on strength and stiffness (Tables 15.2 and 15.3) is shown in Figure 15.14 for tensile test and Figure 15.15 for flexural test. The highest strength and stiffness were obtained for the specimen with compatibilizer E 265. The lowest strength and stiffness were achieved by the specimen with FG 1901 GT. This is in good correlation with the DMA results. The best impact strength (Table 15.4) was obtained for the sample with E 265, followed by FG 1901 GF and PMD 353D. According to the DMA results, it can be concluded that the most suitable compatibilizer for waste paper is PE-HD biocomposite E 265.

The compatibilizer FG 1901 GT reduced the crystallinity (Table 15.5), the other two compatibilizers have no effect on the thermal properties of the biocomposites.

15.2.4.4   Influence of the Compatibilizer Amount on the Biocomposites with Miscanthus Fibers

In this part, the influence of the compatibilizer on the mechanical properties of the miscanthus fibers in the biocomposites is presented. E' as a function of temperature is shown in Figure 15.16. The addition of 2 and 4 wt.% of compatibilizer Eccelor PE

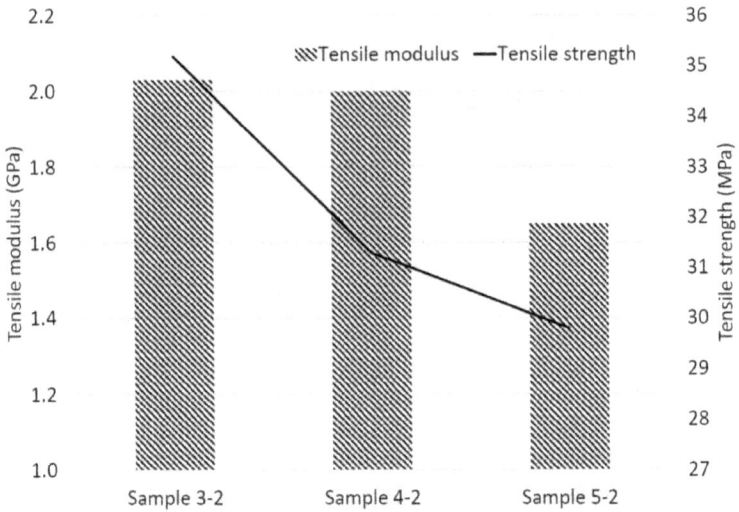

FIGURE 15.14 Tensile strength and stiffness of the biocomposite samples with waste paper and different compatibilizer.

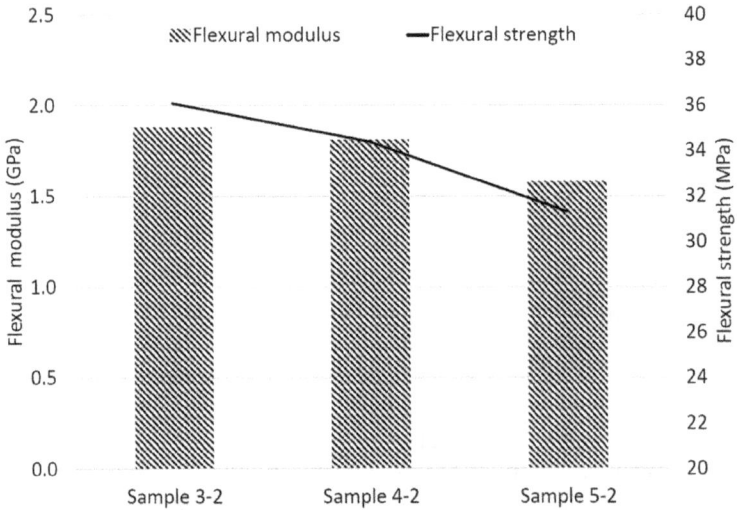

FIGURE 15.15 Flexural strength and stiffness of the biocomposite samples with waste paper and different compatibilizer.

1040 K2 into the miscanthus fibers PE-HD biocomposites showed the same values. The addition of 4 wt.% lowers the E' values over the entire temperature range. As the temperature increases, the difference decreases and in the melting range of the PE-HD matrix the values equalize. It can be seen that the compatibilizer amount between 2 and 4 wt.% is sufficient for compatibilizing the miscanthus fibers into the PE-HD matrix. Compatibilizer Exxelor PE 1040 K2 is suitable for compatibilizing the miscanthus fibers into the PE-HD matrix.

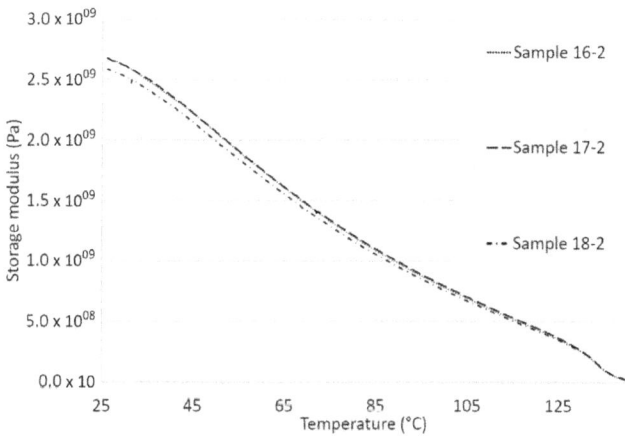

FIGURE 15.16 Storage modulus of the biocomposite samples with miscanthus fibers and different compatibilizer.

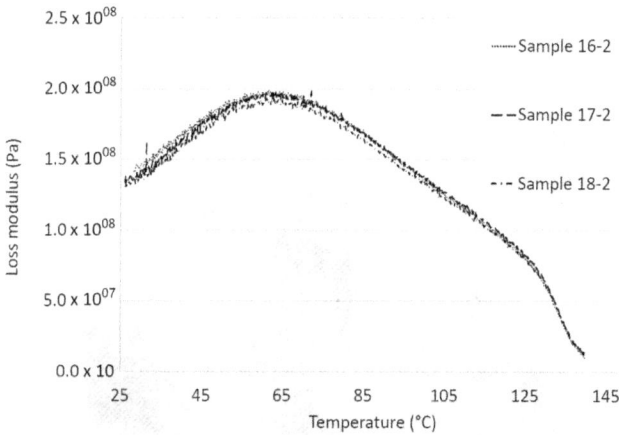

FIGURE 15.17 Loss modulus of the biocomposite samples with miscanthus fibers and different compatibilizer.

Figure 15.17 shows the results of the loss modulus. The lowest E'' in the whole temperature range is shown by the sample with 6 wt.% compatibilizers. The other two samples have the same curve shape. The peak temperature is in the same range for all samples with a slight shift to higher temperatures with increasing compatibilizer content −62°C. It can be concluded that the compatibilizer PE 1040 K2 prevents the inhibition of the relaxation process in the miscanthus fibers PE-HD biocomposites.

The loss factor (Figure 15.18) is practically the same for all amounts of the compatibilizer. It can be concluded that 2 wt.% of the compatibilizer is sufficient to allow the elastic behavior of the miscanthus fibers PE-HD biocomposite.

The effect of compatibilizer on strength and stiffness (Tables 15.2 and 15.3) is illustrated in Figure 15.19 for tensile test and in Figure 15.20 for flexural test. The

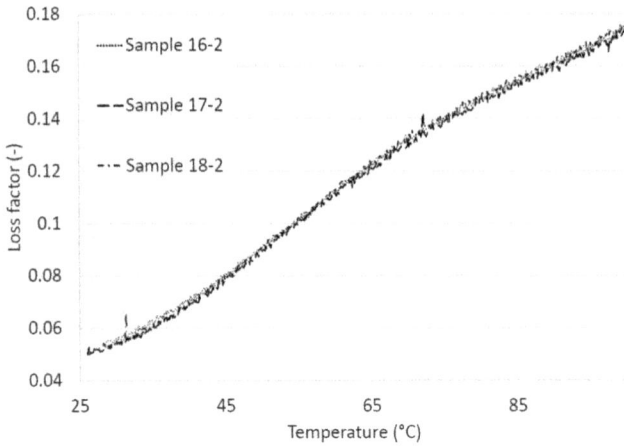

FIGURE 15.18 Loss factor of the biocomposite samples with miscanthus fibers and different compatibilizer.

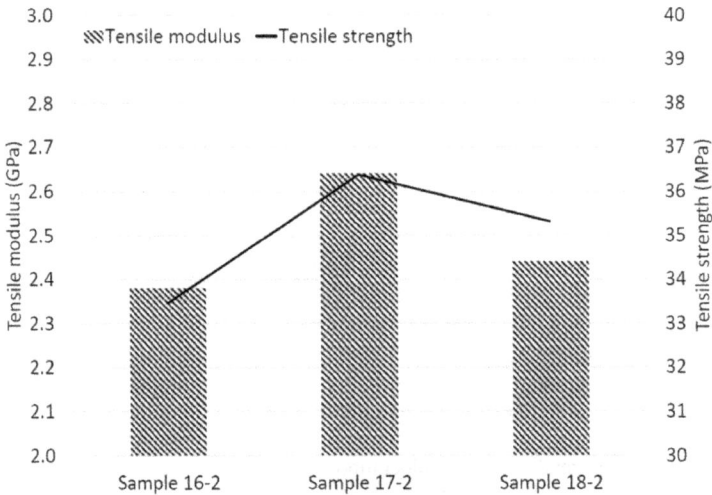

FIGURE 15.19 Tensile strength and stiffness of the biocomposite samples with miscanthus fibers and different compatibilizer.

highest strength and stiffness were obtained for the sample with the 4 wt.% compatibilizers Exxelor PE 1040 K2 in miscanthus fibers PE-HD biocomposite. The best impact strength (Table 15.4) was obtained for the sample with 6 wt.% followed by 4 and 2 wt.%. According to the DMA results, it can be concluded that the most suitable compatibilizer amount for miscanthus fiber PE-HD biocomposite is 4 wt.% Exxelor PE 1040 K2.

The compatibilizer Exxelor PE 1040 K2 lowers the crystallinity in miscanthus fiber PE-HD biocomposites (Table 15.5), but does not affect the melting and crystallization temperature of PE-HD. As the amount of compatibilizer Exxelor PE 1040 K2 increases, the crystallinity decreases.

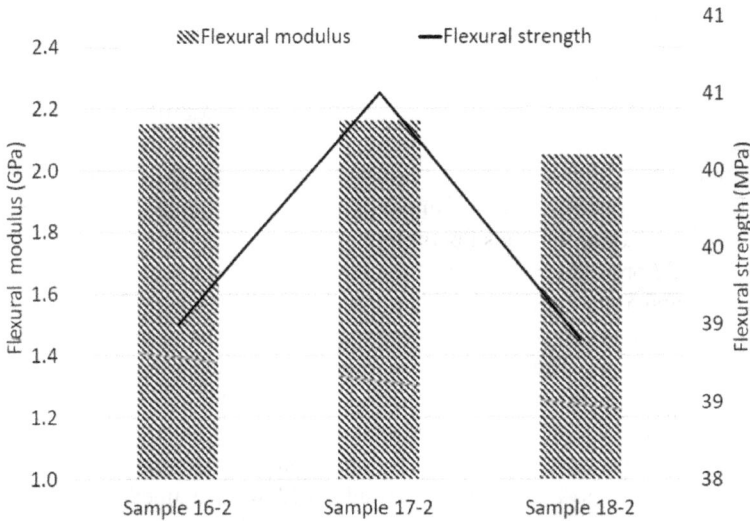

FIGURE 15.20 Flexural strength and stiffness of the biocomposite samples with miscan-thus fibers and different compatibilizer.

15.2.4.5 Influence of the Amount of Compatibilizer on the Biocomposites with Waste Paper

In this part, the influence of the amount of compatibilizer on the biocomposites with waste paper is presented. E' as a function of temperature is shown in Figure 15.21. The addition of 2 and 4 wt.% of the compatibilizer Eccelor PE 1040 K2 in waste paper PE-HD biocomposites showed the same values. Addition of 3 and 5 wt.% of compatibilizer Eccelor PE 1040 K2 in waste paper PE-HD biocomposites showed the same values, i.e., lower than samples with 2 and 4 wt.% compatibilizer in the whole

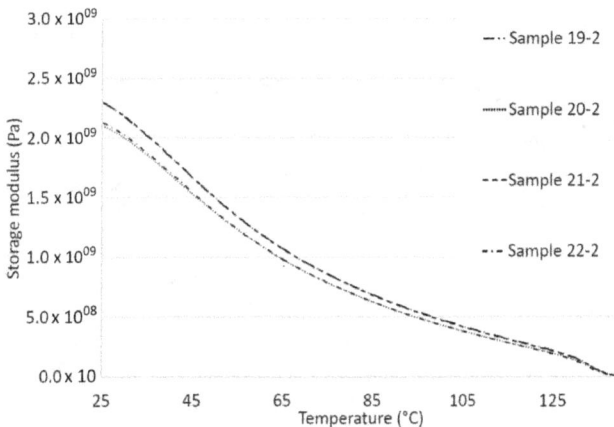

FIGURE 15.21 Storage modulus of the biocomposite samples with waste paper and differ-ent amount of compatibilizer.

temperature range. With increasing temperature, the difference becomes smaller and in the melting range of the PE-HD matrix the values equalize. It can be assumed that the inhomogeneity in the waste paper causes this difference, as shown in the previous tests with different batches of the waste paper. Compatibilizer Exxelor PE 1040 K2 is suitable for compatibilizing the waste paper in PE-HD matrix, although E′ is lower compared to the composites with the same compatibilizer with miscanthus fibers. From these results, it can be concluded that the waste paper biomass has wider inhomogeneity than miscanthus fibers.

Figure 15.22 presents the results of the E″. The same trend as for E′ is seen. The addition of 3 and 5 wt.% of compatibilizer Eccelor PE 1040 K2 in waste paper PE-HD biocomposites showed lower values compared to the samples with 2 and 4 wt.% addition of compatibilizer. The differences can be better seen from the E″ curves. The lowest E″ is exhibited by the sample with 5 wt.% compatibilizer. As with the addition of miscanthus fibers, a slight shift of the E″ peak toward higher temperatures with increasing compatibilizer amount can be seen with the addition of waste paper in PE-HD biocomposites. The highest amplitude of E″ was obtained with 2 wt.% compatibilizer and the lowest with 5 wt.%. It can be concluded that the compatibilizer PE 1040 K2 prevents the inhibition of the relaxation process in the waste paper fibers PE-HD biocomposites.

The loss factor (Figure 15.23) is close for all samples. The highest values in the whole temperature range are reached by the sample with 3 wt.% of the compatibilizer and the lowest with 4 wt.%. It can be concluded that 2 wt.% of compatibilizer is sufficient to allow elastic behavior of the recycled paper PE-HD biocomposite.

The effect of compatibilizer on strength and stiffness (Tables 15.2 and 15.3) is illustrated in Figure 15.24 for tensile test and Figure 15.25 for flexural test. The highest tensile stiffness was obtained by the specimen with 2 wt.% compatibilizer and the highest flexural stiffness was obtained by the specimen with 4 wt.% compatibilizer. The highest strength was achieved by the sample with 4 wt.% compatibilizer.

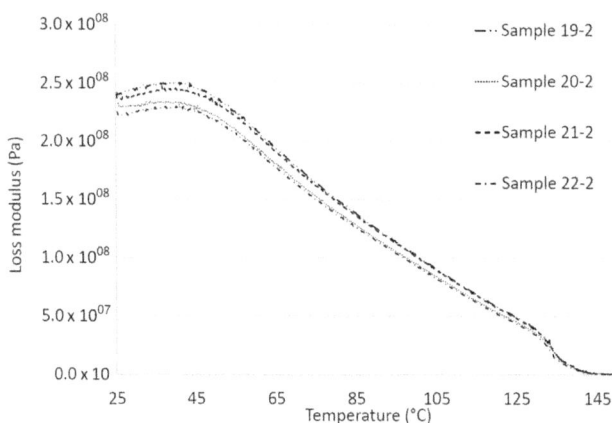

FIGURE 15.22 Loss modulus of the biocomposite samples with waste paper and different amount of compatibilizer.

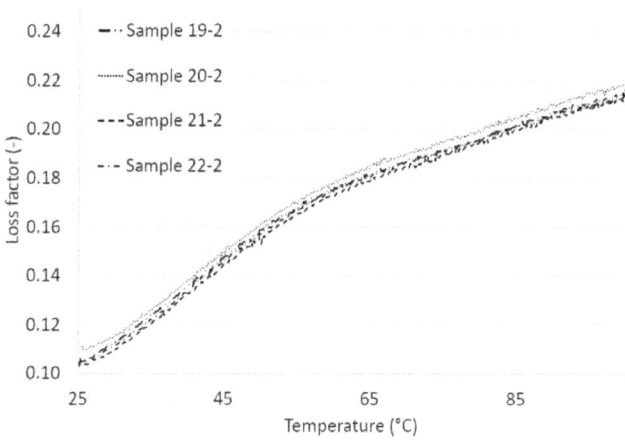

FIGURE 15.23 Loss factor of the biocomposite samples with waste paper and different amount of compatibilizer.

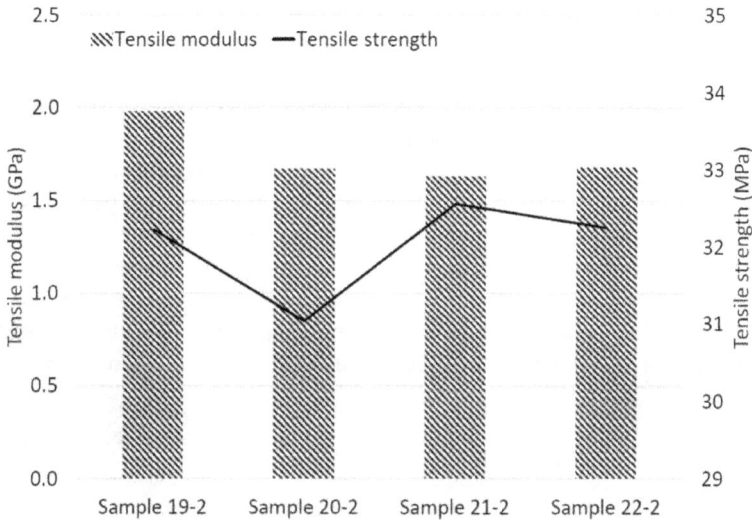

FIGURE 15.24 Tensile strength and stiffness of the biocomposite samples with waste paper and different amount of compatibilizer.

The best impact strength (Table 15.4) was obtained for the sample with 5 wt.% compatibilizer followed by 4, 3, and 2 wt.%. From the DMA results, it can be concluded that the most suitable compatibilizer amount for waste paper PE-HD biocomposite is 4 wt.% Exxelor PE 1040 K2. This correlates with the amount of compatibilizer for miscanthus fiber.

As the amount of Exxelor PE 1040 K2 increases, the melting temperature increases and the crystallization temperature decreases. No effect on crystallinity was observed (Table 15.5).

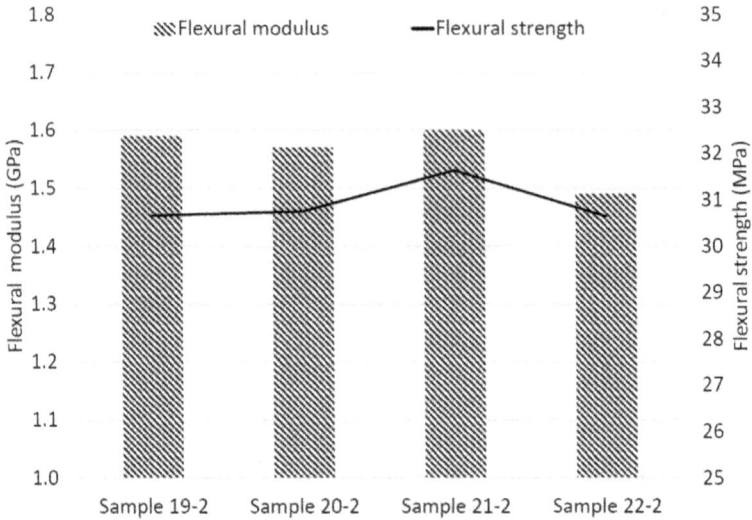

FIGURE 15.25 Flexural strength and stiffness of the biocomposite samples with waste paper and different amount of compatibilizer.

15.2.4.6 Influence of Different Natural Fibers on Biocomposites

In this chapter, the authors present how different natural fibers affect the properties of biocomposites. The following natural fibers were compared: waste paper, cellulose fibers 6 mm, cellulose fibers 0.6 mm, hemp husks and miscanthus fibers. E′ as a function of temperature is shown in Figure 15.26. As expected, a large scatter of values can be seen. With the exception of hemp husks, a large improvement in E′ is seen in all samples. Cellulose fiber 6 mm and cellulose fiber 0.6 mm achieved the highest improvement. Miscanthus fibers achieved significantly lower E′ values, waste paper even slightly less. It can be concluded that the stiffness of the biocomposites can be

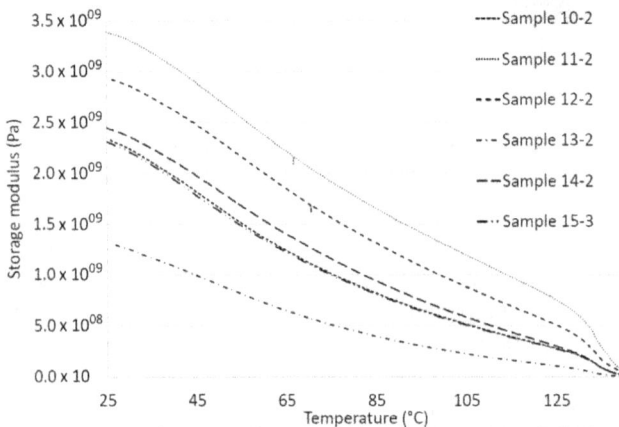

FIGURE 15.26 Storage modulus of the biocomposite samples with different natural fibers.

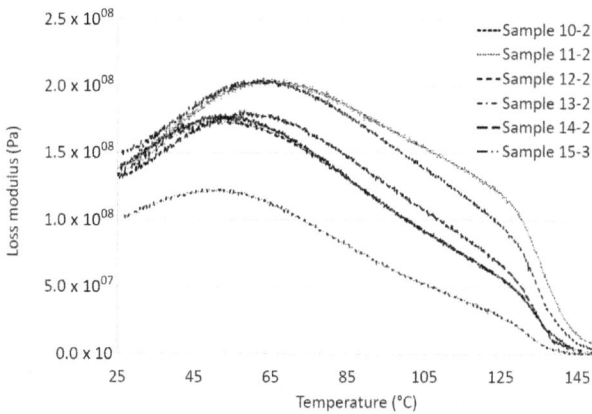

FIGURE 15.27 Loss modulus of the biocomposite samples with different natural fibers.

controlled by the addition of the selected natural fiber. The E′ range of 1.0–3.5 GPa can be achieved with the addition of 30 wt.% natural fibers.

Figure 15.27 presents the results of the E″. The same tendency as for E′ can be seen. Cellulose fibers reach the highest amplitude, 0.6 mm long fibers show a peak at a slightly higher temperature. It can be concluded that longer cellulose fibers inhibit the matrix relaxation process more than shorter cellulose fibers, but are still significantly more affected than all other natural fibers. Samples with hemp husks reached the most beautiful amplitude and peak temperature, which is probably due to the hemp oil residue contained in them. Miscanthus fibers inhibit matrix relaxation much more than waste paper.

The loss factor is shown in Figure 15.28. The highest values throughout the temperature range were achieved by samples with cellulose fibers 6 mm, followed by cellulose fibers 0.6 mm. The lowest values were achieved by samples with hemp husks. With increasing temperature, the difference of tan δ to the other samples also increases.

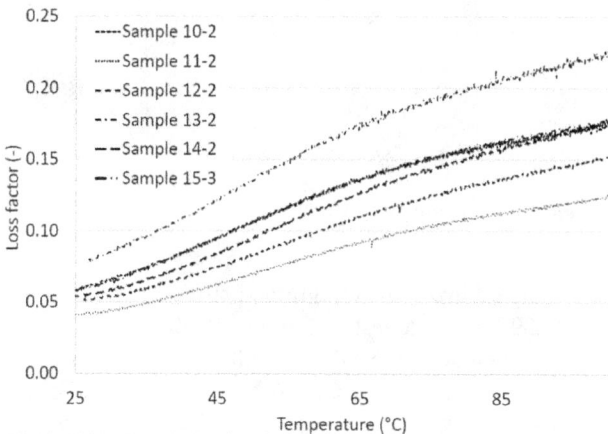

FIGURE 15.28 Loss factor of the biocomposite samples with different natural fibers.

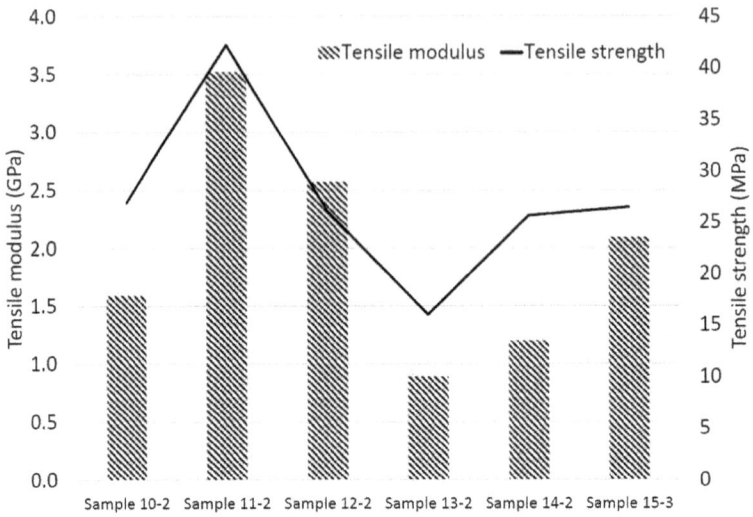

FIGURE 15.29 Tensile strength and stiffness of the biocomposite samples with different natural fibers.

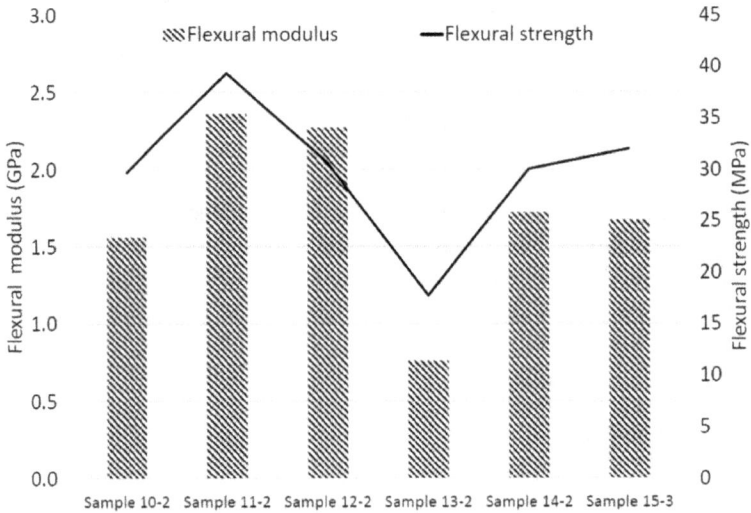

FIGURE 15.30 Flexural strength and stiffness of the biocomposite samples with different natural fibers.

The effect of compatibilizer on strength and stiffness (Tables 15.2 and 15.3) is shown in Figure 15.29 for tensile test and Figure 15.30 for flexural test. The results have excellent correlation with the DMA results. The highest strength and stiffness were achieved by the cellulose sample 6 mm, followed by the cellulose sample 0.6 mm, miscanthus fiber, waste paper and hemp husk. The best impact strength (Table 15.4) was obtained for the sample with waste paper, followed by cellulose fiber 6 mm, hemp husk and miscanthus fiber.

Hemp husk had the strongest effect on the thermal properties (Table 15.5). The crystallinity was reduced and the crystallization temperature also decreased.

15.2.4.7 Influence of the Compatibilizer on the Addition of Fiber-Plast 35E in Biocomposites

The authors present the influence of the addition of Fiber-Plast 35E to the compatibilizer in biocomposites. As shown in Figure 15.31, the highest E' was obtained with Lushan PR-3C. PMD 353D and SMA have lower values. In the matrix melting region, all samples have the same values. As the temperature increases, the difference between PMD 353D and SMA decreases.

Figure 15.32 shows the E″ results. E″ curves have the same peak temperature. The sample with PMD 353D reaches the lowest E″ value, while the sample with Luschan

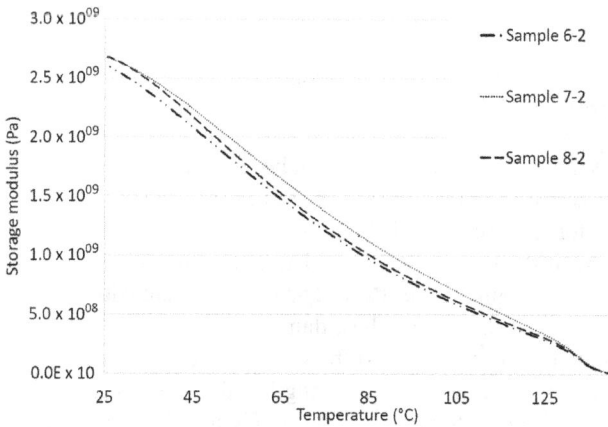

FIGURE 15.31 Storage modulus of the biocomposite samples with miscanthus fibers and different compatibilizer.

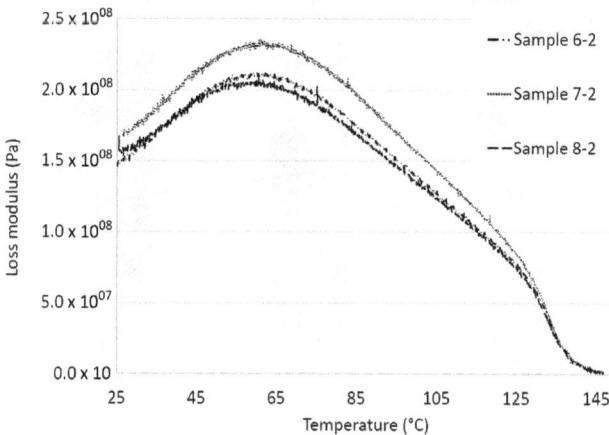

FIGURE 15.32 Loss modulus of the biocomposite samples with miscanthus fibers and different compatibilizer.

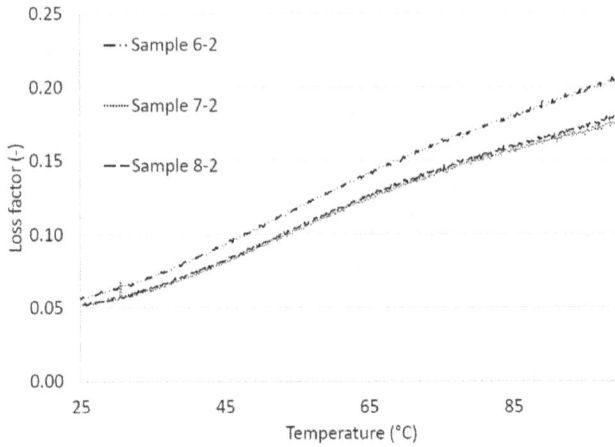

FIGURE 15.33 Loss factor of the biocomposite samples with miscanthus fibers and different compatibilizer.

PR-3C reaches the highest. Therefore, it can be concluded that the inhibition effect of all compatibilizers is the same for matrix relaxation.

The loss factor (Figure 15.33) is highest throughout the temperature range for the biocomposite with the compatibilizer SMA. The samples with Lushan PR-3C and PMD 353D show the same curve shape. The biocomposite with Fiber-Plast 35E with SMA as compatibilizer shows the best damping behavior. The biocomposites with PP-g-MA compatibilizer have almost the same elastic behavior.

The effect of compatibilizer on the strength and stiffness (Tables 15.2 and 15.3) is shown in Figure 15.34 for the tensile test and Figure 15.35 for the flexural test. The

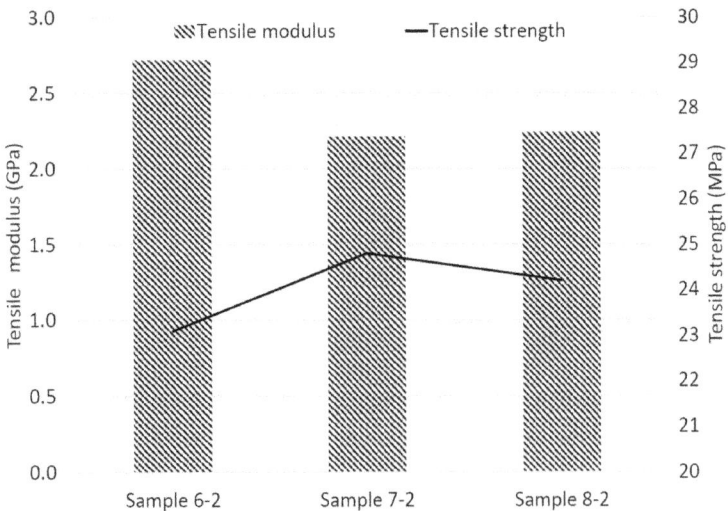

FIGURE 15.34 Tensile strength and stiffness of the biocomposite samples with miscanthus fibers and different compatibilizer.

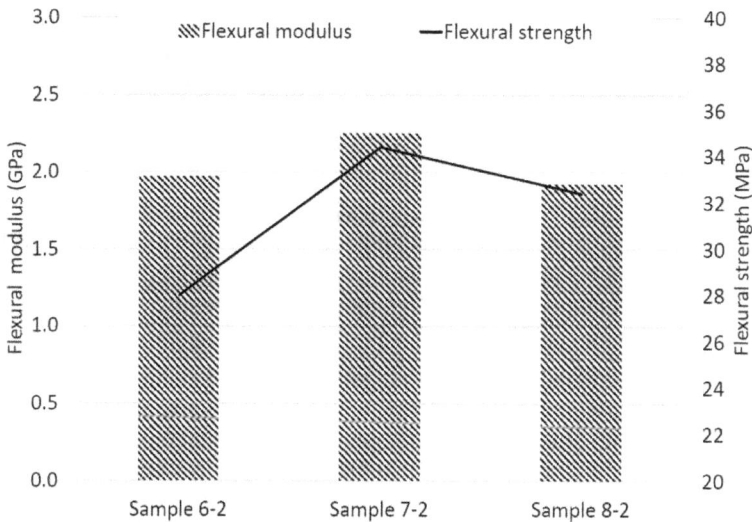

FIGURE 15.35 Flexural strength and stiffness of the biocomposite samples with miscanthus fibers and different compatibilizer.

results have excellent correlation with the DMA results. The sample with Lushan PR-3C achieved the highest strength and stiffness, except for tensile stiffness. The best impact strength (Table 15.4) was obtained for the sample with PMD 353D, followed by Lushan PR-3C and SMA.

No significant effect on the thermal properties was found for these specimens (Table 15.5).

15.3 CONCLUSIONS

The research work focused on different cases, forms and amounts of compatibilizers on the biocomposite materials. The greatest effect on the viscoelastic properties of the thermoplastic PE-HD-based biocomposites was measured with the PE-g-MA compatibilizer E 265. The compatibilizer has no effect on the thermal behavior of the biocomposite, but the highest strength and stiffness of the biocomposites were obtained. All biocomposites were prepared with 30 wt.% of different natural fibers and the thermoplastic matrix PE-HD. Except the sample with hemp husks, all the biocomposites have higher strength compared to the untreated PE-HD matrix, all the biocomposites including the sample with hemp husks have higher stiffness. The storage modulus was improved for all biocomposite samples, although "incompatible" compatibilizers with PE-HD matrix were also tested. These particular compatibilizers showed that the traces of the PE in the recycled PP, PS would not affect the compatibilization with natural fibers. Based on the results, it can be concluded that there is an excellent improvement in storage modulus, strength and stiffness with compatibilized cellulosic fibers. Miscanthus fibers in the biocomposites also exhibited good mechanical properties but caused brittleness. A good combination of mechanical properties and toughness is waste paper incorporated in PE-HD matrix with compatibilizer. For blends as matrix, SEBS-*g*-MA compatibilizer is the best option to improve the toughness of biocomposites with natural fibers.

REFERENCES

1. K.P. Ashik, R.S. Sharma, A review on mechanical properties of natural fiber reinforced hybrid polymer composites, *J. Miner. Mater. Charact. Eng.* 03 (2015) 420–426. https://doi.org/10.4236/jmmce.2015.35044.
2. A. Nourbakhsh, A. Karegarfard, A. Ashori, A. Nourbakhsh, Effects of particle size and coupling agent concentration on mechanical properties of particulate-filled polymer composites, *J. Thermoplast. Compos. Mater.* 23 (2010) 169–174. https://doi.org/10.1177/0892705709340962.
3. M. Chaharmahali, J. Mirbagheri, M. Tajvidi, S.K. Najafi, Y. Mirbagheri, Mechanical and physical properties of wood-plastic composite panels, *J. Reinf. Plast. Compos.* 29 (2010) 310–319. https://doi.org/10.1177/0731684408093877.
4. E. Pérez, L. Famá, S.G. Pardo, M.J. Abad, C. Bernal, Tensile and fracture behaviour of PP/wood flour composites, *Compos. Part B Eng.* 43 (2012) 2795–2800. https://doi.org/10.1016/j.compositesb.2012.04.041.
5. S.K. Nayak, S. Mohanty, S.K. Samal, Influence of short bamboo/glass fiber on the thermal, dynamic mechanical and rheological properties of polypropylene hybrid composites, *Mater. Sci. Eng. A.* 523 (2009) 32–38. https://doi.org/10.1016/j.msea.2009.06.020.
6. S.H. Aziz, M.P. Ansell, The effect of alkalization and fibre alignment on the mechanical and thermal properties of kenaf and hemp bast fibre composites: Part 2- cashew nut shell liquid matrix, *Compos. Sci. Technol.* 64 (2004) 1231–1238. https://doi.org/10.1016/j.compscitech.2003.10.002.
7. M. Tajvidi, R.H. Falk, J.C. Hermanson, Effect of natural fibers on thermal and mechanical properties of natural fiber polypropylene composites studied by dynamic mechanical analysis, *J. Appl. Polym. Sci.* 101 (2006) 4341–4349. https://doi.org/10.1002/app.24289.
8. J. Yu, L. Sun, C. Ma, Y. Qiao, H. Yao, Thermal degradation of PVC: A review, *Waste Manag.* 48 (2016) 300–314. https://doi.org/10.1016/j.wasman.2015.11.041.
9. M. Poletto, A.J. Zattera, M.M.C. Forte, R.M.C. Santana, Thermal decomposition of wood: Influence of wood components and cellulose crystallite size, *Bioresour. Technol.* 109 (2012) 148–153. https://doi.org/10.1016/j.biortech.2011.11.122.
10. I. Ghasemi, B. Kord, Long-term water absorption behaviour of polypropylene/wood flour/organoclay hybrid nanocomposite, *Iran. Polym. J. (English Ed.)* 18 (2009) 683–691.
11. J. Kajaks, K. Kalnins, A. Zagorska, J. Matvejs, Some exploitation properties of wood plastic composites based on recycled high density polyethylene and plywood production residues, *Solid State Phenom.* 267 SSP (2017) 76–81. https://doi.org/10.4028/www.scientific.net/SSP.267.76.
12. A. Scholten, D. Meiners, Use of recycled waste paper as fiber reinforcement for polypropylene-relationship of fiber extraction process and mechanical properties of the composites, *AIP Conf. Proc.* 2055 (2019). https://doi.org/10.1063/1.5084825.
13. M. Pracella, M.M.U. Haque, V. Alvarez, Functionalization, compatibilization and properties of polyolefin composites with natural fibers, *Polymers (Basel)* 2 (2010) 554–574. https://doi.org/10.3390/polym2040554.
14. Z. Yang, H. Peng, W. Wang, T. Liu, A compatibilized composite of recycled polypropylene filled with cellulosic fiber from recycled corrugated paper board: mechanical properties, morphology, and thermal behavior, *J. Appl. Polym. Sci.* 122 (2011) 2789–2797. https://doi.org/10.1002/app.34321.

15. Z. Zhang, P. Wang, J. Wu, Dynamic mechanical properties of EVA polymer-modified cement paste at early age, *Phys. Procedia.* 25 (2012) 305–310. https://doi.org/10.1016/j.phpro.2012.03.088.

16. M. Atagür, N. Kaya, T. Uysalman, C. Durmuşkahya, M. Sarikanat, K. Sever, Y. Seki, A detailed characterization of sandalwood-filled high-density polyethylene composites, *J. Thermoplast. Compos. Mater.* (2020) 1–18. https://doi.org/10.1177/0892705720939157.

17. G. Chui-gen, S. Yong-ming, W. Qing-wen, S. Chang-sheng, Dynamic-mechanical analysis and SEM morphology of wood flour/polypropylene composites, *J. For. Res.* 17 (2006) 315–318.

18. T.T.L. Doan, H. Brodowsky, E. Mäder, Jute fibre/polypropylene composites II. Thermal, hydrothermal and dynamic mechanical behaviour, *Compos. Sci. Technol.* 67 (2007) 2707–2714. https://doi.org/10.1016/j.compscitech.2007.02.011.

19. S. Mohanty, S.K. Verma, S.K. Nayak, Dynamic mechanical and thermal properties of MAPE treated jute/HDPE composites, *Compos. Sci. Technol.* 66 (2006) 538–547. https://doi.org/10.1016/j.compscitech.2005.06.014.

20. S. Shinoj, R. Visvanathan, S. Panigrahi, N. Varadharaju, Dynamic mechanical properties of oil palm fibre (OPF)-linear low density polyethylene (LLDPE) biocomposites and study of fibre-matrix interactions, *Biosyst. Eng.* 109 (2011) 99–107. https://doi.org/10.1016/j.biosystemseng.2011.02.006.

21. H. Essabir, M.E.I. Achaby, E.M. Hilali, R. Bouhfid, A.Ei. Qaiss, Morphological, structural, thermal and tensile properties of high density polyethylene composites reinforced with treated argan nut shell particles, *J. Bionic Eng.* 12 (2015) 129–141. https://doi.org/10.1016/S1672-6529(14)60107--.

22. M. Wagner, M. Wagner, Thermal Analysis in Practice, 2017. https://doi.org/10.3139/9781569906446.fm.

23. J.P. Siregar, M.S. Salit, M.Z. Ab Rahman, K.Z.H. Mohd Dahlan, Thermogravimetric analysis (TGA) and differential scanning calometric (DSC) analysis of pineapple leaf fibre (PALF) reinforced high impact polystyrene (HIPS) composites, *Pertanika J. Sci. Technol.* 19 (2011) 161–170.

24. B. John, K.T. Varughese, Z. Oommen, P. Pötschke, S. Thomas, Dynamic mechanical behavior of high-density polyethylene/ethylene vinyl acetate copolymer blends: The effects of the blend ratio, reactive compatibilization, and dynamic vulcanization, *J. Appl. Polym. Sci.* 87 (2003) 2083–2099. https://doi.org/10.1002/app.11458.

25. D. Romanzini, H.L. Ornaghi, S.C. Amico, A.J. Zattera, Influence of fiber hybridization on the dynamic mechanical properties of glass/ramie fiber-reinforced polyester composites, *J. Reinf. Plast. Compos.* 31 (2012) 1652–1661. https://doi.org/10.1177/0731684412459982.

26. J.M. Felix, P. Gatenholm, The nature of adhesion in composites of modified cellulose fibers and polypropylene, *J. Appl. Polym. Sci.* 42 (1991) 609–620. https://doi.org/10.1002/app.1991.070420307.

16 Interphase Damping Modification for Vibration Reduction and Control

Shengyu Li and Xiaoning Tang
Wuhan Textile University

CONTENTS

16.1 INTRODUCTION OF COMPOSITES INTERPHASE

With the development of modern industry, the importance of fiber-reinforced composites has gradually increased in various engineering fields. At present, large amounts of literatures have investigated the comprehensive properties of biocomposites, which mainly include mechanical strength, flexibility, thermal and vibration insulation, and chemical stability [1,2]. Generally, composites consist of at least three components: fiber, matrix, and interphase. Biocomposites are usually made of different natural fibers, binders, or matrices, which have been developed to meet the requirement of different end-use applications. Among various composites, biocomposites exhibit the advantage of good durable and sustainable properties, which have attracted more and more attention in various research fields for the past several decades. The natural fibrous resource was used in biocomposites products for a long history, which can be dated back to the beginning of human civilization [3,4]. For instance, the straw byproduct can be used to reinforce mud bricks, thus increasing the strength for building applications. In the world markets, the demand of consumers for such kinds of products has been rapidly increasing. The continued interest in biocomposites can be attributed to several advantages, as well as the potential low-carbon and environmentally friendly benefits [5]. For better engineering applications, the strength of the fiber is inherently higher than that of the bulk matrix, while the

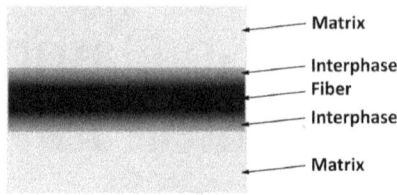

FIGURE 16.1   Schematic diagram of interphase region in fiber-reinforced composites.

matrix can effectively transfer the external forces to the fiber, thus well distributing the applied stresses. In addition, the matrix also plays the role of protecting fibrous components from damages due to a severe environment. Interphase is generally the adjoining region of the fiber surface, which possesses a considerable thickness. The schematic diagram of the interphase is shown in Figure 16.1. The properties of interphase are distinguishing from fibrous phase and bulk matrix.

The construction of interphase is essential to achieve the transformation of stress between fiber and matrix. Interphase can be summarized as the interlayer with different physical or chemical characteristics from both fiber and matrix [6]. For various composites, the modification of the interphase region is an effective approach to adjust the interaction conditions between fiber and matrix, thus achieving desired tensile strength, modulus, and damping. According to reported studies, there are several design approaches concerning interphase engineering. For instance, the strong bonding between fiber and matrix can generally achieve good stiffness and strength, while weak interfacial bonding conditions can enhance the energy absorption performance of composites [7–9]. It can be stated that the interphase plays an important factor in composite properties for engineering applications, which is gradually being a hot research topic in both academia and industry. The morphologies and characteristics of the interphase can be well controlled on the foundation of various structural parameters. In summary, the interphase of composites can be well designed and fabricated under optimized process conditions, thus achieving desired interfacial region. In this chapter, we have mainly introduced the modification of interphase for damping and vibration reduction applications.

16.2   INTERPHASE DAMPING

Compared to metallic and ceramic materials, biocomposites generally exhibit higher damping capacity, which is mainly due to the viscoelastic characteristics of polymeric base materials. Fiber-reinforced composites in engineering fields frequently suffered from the dynamic external force, and the vibration effects can result in unwanted noise, thus decreasing the lifetime of composite structures. In the past few decades, it is an appealing challenge to investigate the damping of composites for better vibration control applications. This performance can be enhanced by exploring the connatural characteristics to consume the forced mechanical energy. The energy absorption of fiber-reinforced composite is an important performance related to engineering, and dynamic mechanical analysis is an indispensable and effective approach for characterizing the damping behavior [10–12]. In detail, the storage

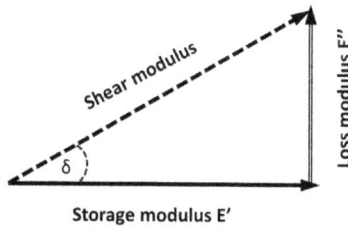

FIGURE 16.2 Relationship between storage modulus E', loss modulus E'', and mechanical damping factor tan δ.

modulus (E') typically means Young's modulus, and it has determined the stiffness of composite materials. In other words, E' can be used to represent the capability to store applied energy. Loss modulus (E'') refers to the viscous response behavior, which can effectively dissipate the applied energy. The mechanism of loss modulus might be explained by the internal friction of materials, and the process is sensitive to various molecular motions and microstructural morphologies. Generally, tan δ is a dimensionless number to express the mechanical damping factor, and it can be calculated by the proportion between two parameters of loss and storage modulus, the diagram is presented in Figure 16.2. In the practical manufacture process, composites damping can be well designed by choosing different constitutive parameters, including fiber component, matrix volume ratio, fiber stacking sequence, and configuration structures. The successful dynamic mechanical characterization of viscoelastic composite materials can well investigate the applied loading conditions, defects of microstructures, and interaction conditions of interphase.

The vibration attenuation of composite materials is entirely distinguishing from conventional metals and alloys, the sources of energy dissipation are primarily attributed to the firm viscosity nature of the matrix, fiber, and interphase region. Generally speaking, the major contribution of composite damping performance can be attributed to the matrix. Under large vibration effects with strong pressure force, thermoplastic composites exhibit obvious nonlinear damping caused by the presence of strain concentration phenomenon. The damping of damage mainly consists of the friction damping because of the slip in interphase and the energy dissipation effects in matrix cracks area [13]. The damping due to interphase has also possessed considerable effects on the comprehensive damping of composites, which has also accordingly affected the mechanical performance. In comparison with fiber and matrix damping, the damping of interphase is more complex and it can be well designed by different configurations in the interfacial region.

The interfacial region is a crucial factor for the properties of the fiber-reinforced composite, it can be strong or weak due to different processing techniques [14]. According to reported works of literature, fiber-reinforced composites with different damping conditions can be obtained by varying the microstructures of interphase. The damping control process in engineering typically requires efficient energy conversion and dissipation. Based on reported studies, the weak interface bonding is easy to absorb more mechanical energy. The debonding process can also effectively improve the contacted area of stick-slip effects with cyclic force, thus achieving a higher damping factor [15,16]. Considering the effects of high shear deformations on energy dissipation, it is an effective approach to design microstructural configurations at this region. To analyze the damping of the interfacial region, the interphase is generally assumed to exhibit the intermediate characteristic between fibers and matrix. The high shear strain of interphase can be mainly attributed to the mismatch between the elastic stiffness of fiber and matrix components. The surface modification of reinforced components is effective to optimize the mechanical properties, thus adjusting the damping behavior. In the following sections, this chapter has gathered the recent advances of different treatment methods for enhanced interphase damping. Based on the published works of literature, these methods can be mainly classified as chemical treatment, physical treatment, fiber coating, and microstructures design of interphase.

16.3 CHEMICAL TREATMENT

The performance of fibrous materials reinforced composites can be well controlled by adjusting the interphase adhesion conditions between fiber and matrix. This process can be achieved by modifying the surface morphologies of fiber, such as various treatment methods to tailor the bonding systems of interlayer in the composite. As for the fabrication of biocomposites, it suffers difficulty due to the inherent incompatibility between lignocellulosic fibrous and matrix. Therefore, the application of surface treatment of fiber is indispensable to produce hydrophilic natural fibrous bonded strongly with the hydrophobic polymeric matrix. At present, physical, chemical, or biological treatments are widely used in the surface modification of biocomposites [17]. In a moisture environment, it is generally unstable due to the absorption of water and the weak adhesion with the matrix in the interphase region. The wax and pectin in the cell have also inhibited the inter bonding with the matrix of grafting the fiber surface activated chemical groups [18]. Consequently, it is necessary to modify the natural fibrous surface, thus decreasing the hydrophilic efficiency and increasing the adhesion activity of reinforced fiber to matrix materials.

Among various treatment techniques, chemical treatment has been considered as the most widely developed approach to achieve the structural modification of the reinforced fibers, thus improving the interfacial bonding conditions. According to reported studies, natural fibrous materials have been successfully treated with different kinds of chemicals, including alkali, silane, acetic acid, and hydrochloric acid [19]. It has been reported that these chemical treatments can significantly improve mechanical properties. The mechanism can be explained by the modified crystalline structure of natural fibers, which can also remove the weak components from the fiber surface, mainly including hemicelluloses, impurities, and associated debris. The overall results of the modification can improve the interphase conditions between fiber and matrix, also some chemical modification methods have influenced the chemical composition and fiber–matrix adhesion strength.

Recently, chemical treatment has attracted more and more attention to modify the interphase of composites. To dissipate more energy, it is usually necessary to decrease the bonding strength of the interfacial region through surface modification. Geethamma et al. [20] studied the effects of reagent surface treatment on coir fiber-reinforced biocomposites, and dynamic mechanical performance was measured. It has been found that the weak bonding conditions in the interphase can effectively attribute to the relative motion of fibers, thus increasing the damping capability. As it is known to all, the surface pristine coir fiber surface exhibits an uneven surface due to the existence of globular protrusions. After alkali treatment, those globular protrusions on the surface were disappeared, while large amounts of voids were produced. On the other hand, the interaction with a polymeric solution in the following fabrication process will be improved due to the increased surface area of these fibers. In the reported work, the effect of coir fiber chemical treatment on damping was studied, the result has indicated that the biocomposites containing fibers subjected to bleaching exhibited a low dynamic mechanical factor under high-temperature region, thus indicating the good damping efficiency of elastomeric materials for engineering applications.

Furthermore, the influence of chemical modification and nanoclay addition of coconut sheath fiber on the free vibration performance was also investigated, both intrinsic frequency and standard damping efficiency were measured. Based on the work of Rajini et al. [21], modal analysis was used to study the effects of different surface-treated fibers, and hybrid composites were manufactured with different content of nanoclay particles. The improved energy and vibration control were achieved through the incorporation of slight nanoclay in control, alkali-treated, and silane-treated samples. The result has indicated that the glass transition temperature of the alkali-modified composite was slightly higher than the pristine composite without any treatment. The interfacial region is crucial for the bonding conditions of interphase, and it can be strong or weak due to adjusting the processing parameters. Different from chemical treatment, physical modification of fibers can effectively modify surface properties without changing the inherent structural compositions. For instance, plasma treatment was widely adopted to treat natural fiber surfaces. The mechanical strength of natural fibers with plasma treatment was also increased obviously. It is also effective to deposit abundant chemical groups on the fiber surface, thus forming strong covalent bonds with the matrix, thereby obtaining a strong fiber–matrix interface [22]. It has been found that the increased mechanical bonding of fiber and matrix can effectively improve composite configuration performance [23]. In addition, Cruz and Fangueiro [24] concluded the physical treatment to increase the mechanical strength of the fibers, is effective to tailor structural and topological properties. Plasma treatment has also induced surface etching, thus can increase the surface roughness and good mechanical interlocking of interphase.

Based on reported literatures, the vibration reduction and control properties can be well controlled by adjusting the interphase conditions, and these literatures of chemical modification methods are summarized in Table 16.1. It can be stated that poor interfacial bonding generally consumed energy in higher efficiency than strong bonding. To explain the damping mechanism in composites, the debonding process can improve the friction with periodic external forces, thus increasing energy consumption. The reduction of impact strength is also a result of modified fiber/matrix interphase adhesion. For the preparation of vibration reduction and control composites, high energy dissipating efficiency and large shear strain deformation are

TABLE 16.1

A Summary of Fiber Surface Chemical Treatment for Composites

No.	Chemical Treatments	Key Findings	References
1	Alkalization treatment of natural kenaf and hemp bast fibers	The storage modulus of surface-treated fiber-reinforced composites is higher than that of untreated samples	Aziz and Ansell [25]
2	Alkali and toluene diisocyanate treated coir fiber-reinforced natural rubber	The treatment has an influence on the surface morphologies hence determining the interfacial bonding conditions	Geethamma et al. [20]
3	Alkali-treated natural fiber with silane coupling	The fiber–matrix interaction can be modified by adjusting fiber surface, thus increasing the contact area and fiber wetting and impregnation	Herrera-Franco and Valadez-Gonzalez [26]
4	Dewaxing, alkali, and acetic acid-treated jute fiber-reinforced green composites	The surface modification can increase the roughness of fibers, hence increasing the tensile strength	Hossain et al. [27]
5	Alkali treatment agave fiber	Composites with poor interfacial bonding tend to consume more energy	Mylsamy and Rajendran [28]
6	Alkali treatment of flax- and linen-fabric	The decreased fiber/matrix bonding can improve the damping efficiency for better vibration control applications	Yan [29]
7	Alkali-treated Doum fibers with a coupling agent	The application of a coupling agent is an effective approach to improve rheological properties	Essabir et al. [30]
8	Alkali, $KMnO_4$, and formic acid-treated banana fiber	Alkali-treated composites exhibit high storage modulus and glass transition temperature due to the obvious surface characteristics after treatment	Indira et al. [31]
9	Sodium hydroxide-treated coconut sheath and sisal fiber	The damping factor was decreased due to the improved interphase bonding, and the mechanical properties were improved	Kumar et al. [32]
10	Alkali-treated jute-reinforced polypropylene	Alkali treatment can effectively improve the interphase bonding of jute fiber/polypropylene nonwoven composites	Karaduman et al. [33]

necessary. In a word, fiber surface chemical treatment is effective to increase energy dissipation efficiency through interfacial damping phenomenon, which is closely related to friction effects.

16.4 FIBER COATING

Coating is an easy and cost-effective method to deposit functional layers onto the surface of target materials, which mainly include metals, ceramic, plastic films, and fibers. It has been deeply developed in various industrial fields. The dip-coating process can be defined as the deposition of aqueous solutions onto the substrate surface [34,35]. In detail, the paint-coat materials were firstly dispersed in the aqueous solutions and coated on the substrate, followed by the drying of the sedimentary wet coating to generate a film. Fibers coated with functional materials exhibited the inherent advantages of providing functionality to manufacture textiles-related products. Moreover, considering the available various coating methods, many kinds of materials can be effectively incorporated into the fiber surface layer. The coating process also enables the incorporation of functional aids into fibrous materials, thus achieving the functional finishing of fibrous-based composites [35].

Among various surface treatment techniques, fiber coating is a powerful tool in adjusting the morphologies, thus determining the damping efficiency. The coating of viscoelastic materials on reinforced fibers is a widely studied method to increase the energy attenuation efficiency of composites. According to Lurie et al. [36], the ultra-thin coating layer in the composites can obviously increase the effective loss moduli. The damping of laminated composites was investigated through the generalized self-consistent method, where the corresponding hypothesis approach was used. It has been found that the viscoelastic process can be explained by the large shearing dissipation mechanism of vibration reduction and control.

On the other hand, Finegan and Gibson [37,38] studied the effects of accurately designed fiber coating layer on the modification of energy dissipation, the mechanism was also investigated at the microstructural scale under forced loading conditions. It has been found that the loss factor of copper fiber with polyvinyl chloride coating was gradually higher with the increasing content, and the stiffness strength was gradually decreased. In detail, the strain energy method was employed to analyze the damping behavior by finite element numerical simulations through closed-form elasticity theory. It has been found that coating treatment is an efficient factor influencing the damping behavior, which can be characterized by the analytical micromechanical model. Vazquez et al. [7] prepared the mono-epoxy and di-epoxy coated glass fiber and studied the relationship between damping behavior and coating of composites. The result indicated that the damping efficiency decreased with the increasing coating thickness. Gu et al. [39] fabricated an interphase layer of carbon fibers through the coating of pyro carbon on the surface, and a finite element method was applied to study the effects of the coating process on the damping in longitudinal direction effected by different layer thickness and layer characteristics. It has been found that the poor coating elastic modulus can obviously increase the energy absorption of fiber-reinforced composites. In addition, to increase the admissible lifetime expectancy of composites, reinforced textile was coated with Resorcinol–Formaldehyde–Latex. The coating treatment can obtain strong interphase of unidirectional textile–rubber composites, thus to be used as a sustainable adhesive in composites under fatigue loading conditions [40]. The shear deformation plays an important role in energy dissipation, and the interphase damping is effective to promote the enhancement caused by the highly stored shear strain energy in the interfacial area. It can be stated that fiber surface treatment is efficient in tailoring the mechanical performance and thus adjusting the vibration reduction and control process. The surface modification method of fiber coating has been widely studied, and it is effective to achieve desired damping performance for vibration applications.

Based on the coating of fiber, the interphase damping can also be controlled by the addition of nanofillers into the layer. According to reported studies, the integration of uniformly oriented nanoparticles into the coating layer has been increasing in the manufacture of various functional materials including photo-chemistry devices, smart electronics textiles, renewable energy devices, and damping composites [9]. The prepared composites incorporated with nanostructured zinc oxide nanowire arrays of the interphase region are effective to achieve high damping enhancement without the expense of stiffness [41]. The result has also indicated that the damping efficiency can be adjusted by changing the morphologies of nanowire-scaled zinc oxide. The nanowire-modified interphase can increase the energy dissipation efficiency, and the composite flexural rigidity was also increased. Furthermore, the ratio of zinc oxide nanowires also significantly determined the composites damping with nanoscale-tailored interphase conditions.

16.5   INCORPORATION OF FILLERS IN THE INTERPHASE

In the fabrication process of composites, both multiple components and interphase conditions between different components should be considered. The slight slippage in the interphase has provided an effective approach for the energy dissipation in the vibration process. In addition, the dissipation efficiency can also be obviously enhanced by increasing the interphase area in the composites. Among various modification methods, the incorporation of nanoparticles can effectively increase the amount of interphase area, thus improving energy dissipation capacity. According to Agrawal et al. [42], the interphase area near the incorporated fillers exhibited different characteristics to the matrix, it is a commonly used method to achieve the functional modification of composites. Remillat [43] fabricated elastic particles-reinforced polymer composites, and the damping mechanism was also studied to describe the microstructures of the interface region. The result indicated that the incorporation of exfoliated graphite is an effective approach to improve damping efficiency. The improvement of energy dissipation can be primarily attributed to the interphase de-bonding, and the mechanism can also be explained by the shearing deformation process under forced loading [44].

Coal ash is a byproduct of the coal power plant, and its use should be encouraged. The coal ash particle exhibited a millimeter size with special honeycomb-like construction, and it has a large specific surface area. The incorporation of coal ash into the composites has produced a poor interface, and it is important for tailoring the interphase of concrete composites [45]. Hollow glass microsphere was also used to reinforce synthetic foam, and the fracture surface in the interphase region can be observed. The nanoscale dispersion of second phase nanoclay in the composites can significantly increase the composites' damping efficiency [46]. The increase of damping can be achieved by the synergetic effects between viscoelastic and interphase contact areas, which are mainly attributed to the incorporation of mica powders [47]. Recently, the silanization of graphene nanoplatelet was used as nanoreinforcement in composites, thus improving damping factor and static energy dissipation capability. It has been found that the superior energy dissipation capacity was obtained with 3.0 wt.% content of graphene nanosheet, where the damping factor is approximately three times of the control sample [48].

Apart from graphene, the interest concerning the application of carbon nano-tubes fiber-reinforced composites for energy absorption is also rapidly increased. In detail, the high damping ratio of carbon nanotubes-reinforced composites can be obtained by the interphase friction between the nanofillers and matrix. The result indicated that this combination structure of carbon nanotubes and composite exhibited high damping capacity, which is promising for potential vibration applications [49]. Gardea et al. [50] studied the energy absorption mechanisms of carbon nano-tubes composites. It has been found that the friction effects of carbon nanotubes have obviously contributed to the shear deformation process, thus improving energy dissipation, and the effects were enhanced at large strain conditions. The high specific area of carbon nanotubes exhibits a unique advantage in damping applications. The matrix plasticity and tearing process can be attributed to the misorientation of carbon nanotubes, and it is generally a primary method for the attenuation of vibration energy. Furthermore, the damping capacity of carbon nanotube in carbon fiber-reinforced composites was also experimentally investigated, and the results indicated that the addition of carbon nanotubes can effectively improve the damping in composites [51]. Khan et al. [52] indicated that the increase of carbon nanotubes loading in the composites can increase the loss modulus, which can be explained by the sliding at the interfacial region with the hypothesis theory method. Epoxy resin reinforced with carbon nanotubes can simultaneously exhibit increased storage modulus and loss modulus, and this performance is valuable for engineering applications. In addition, the composites treated with carbon nanotubes exhibited a higher tan δ than control materials in a wide temperature range, which can be used in vibration control applications. The difference between the two components has been obviously increased with the incorporation of carbon nanotubes, and it can be used to explain the experimental result of dynamic mechanical analysis. It has provided an innovative approach to achieve the enhancement of damping capacity without decreasing the storage modulus.

The interaction condition between the fiber phase and nanoparticles is important for the bonding strength of the composite system. For instance, the strength of the interphase of arranged zinc oxide and carbon fiber can be observed through atomic force microscopy [53]. Low-temperature synthesis method was used to deposit carbon nanotubes on the fibrous surface [54]. It can significantly enhance the energy attenuation efficiency of 56% than the control sample, and the interphase damping was mainly caused by the frictional sliding at the grown interphase of different layers, respectively. Gardea et al. [55] prepared carbon nanotubes-reinforced polystyrene composites, where the highly aligned carbon nanotubes were obtained by twin-screw extrusion followed by a hot and drawing process. The interphase damping mechanism of strain energy attenuation was studied, the result has shown that carbon nanotube interphase slip exhibits a promising trend in engineering configuration vibration reduction structures.

Considering the above-mentioned literatures, the incorporation of auxiliary nanofillers into the matrix is an effective approach to improve the energy attenuation of composites. Currently, various functional additions have been reported, including graphite nanoplates, carbon nanotubes, and inorganic nano powders. The interphase damping can also be greatly improved by the incorporation of nanoparticles into the interphase by fiber coating technique. The mechanism can be explained by interphase stick-sliding and the increased interphase area. In a word, the addition of nanofillers is a powerful method for improving vibration reduction and control applications, it has a unique advantage in future vibration control engineering applications.

16.6 SUMMARY

It is a valuable topic to investigate the vibration control application of biocomposites, and the key factor should be focused on the design and fabrication process. In this chapter, we have gathered the recent progress on the damping behavior of biocomposites. Among various approaches, the modification of interphase is an effective method to obtain desired damping capacity. The chemical treatment of fibrous materials can obviously determine the surface morphologies, thus tailoring the dynamic mechanical properties. The enhancement of damping can also be achieved by fiber coating with a strong elastic layer, followed by the incorporation into the interphase region. Furthermore, the incorporation of nanofillers into the composites can significantly increase the amount of interphase region, thus improving the energy dissipation efficiency under shear deformation effects. In summary, interphase damping is a promising approach to achieve high energy attenuation efficiency, therefore for better application in the fields of vibration reduction and control engineering.

REFERENCES

1. Faruk O, Bledzki AK, Fink HP, Sain M. Progress report on natural fibre reinforced composites. *Macromolecular Materials and Engineering*. 2014;299(1):9–26.
2. Tang X, Yan X. A review on the damping properties of fibre reinforced polymer composites. *Journal of Industrial Textiles*. 2020;49(6):693–721.
3. Gurunathan T, Mohanty S, Nayak SK. A review of the recent developments in biocomposites based on natural fibres and their application perspectives. *Composites Part A: Applied Science and Manufacturing*. 2015;77:1–25.
4. La Mantia FP, Morreale M. Green composites: A brief review. *Composites Part A: Applied Science and Manufacturing*. 2011;42(6):579–88.
5. Lotfi A, Li H, Dao DV, Prusty G. Natural fibre–reinforced composites: A review on material, manufacturing, and machinability. *Journal of Thermoplastic Composite Materials*. 2019. doi: 10.1177/0892705719844454.
6. Olmos D, González-Benito J. Visualization of the morphology at the interphase of glass fibre reinforced epoxy-thermoplastic polymer composites. *European Polymer Journal*. 2007;43(4):1487–500.
7. Vazquez A, Ambrustolo M, Moschiar SM, Reboredo MM, Gerard JF. Interphase modification in unidirectional glass-fibre epoxy composites. *Composites Science and Technology*. 1998;58(3–4):549–58.
8. Kumar P, Chandra R, Singh SP. Interphase effect on fibre-reinforced polymer composites. *Composite Interfaces*. 2010;17(1):15–35.
9. Karger-Kocsis J, Mahmood H, Pegoretti A. Recent advances in fibre/matrix interphase engineering for polymer composites. *Progress in Materials Science*. 2015;73:1–43.
10. Vantomme J. A parametric study of material damping in fibre-reinforced plastics. *Composites*. 1995;26:147–153.

11. Tsai JL, Chi YK. Effect of fibre array on damping behaviors of fibre composites. *Composites Part B-Engineering.* 2008;39(7–8):1196–204.

12. Martone A, Giordano M, Antonucci V, Zarrelli M. Enhancing damping features of advanced polymer composites by micromechanical hybridization. *Composites Part A: Applied Science and Manufacturing.* 2011;42(11):1663–72.

13. Chandra R, Singh SP, Gupta K. Damping studies in fibre-reinforced composites: A review. *Composite Structures.* 1999;46(1):41–51.

14. Liu L, Jia CY, He JM, Zhao F, Fan DP, Xing LX, et al. Interfacial characterization, control and modification of carbon fibre reinforced polymer composites. *Composites Science and Technology.* 2015;121:56–72.

15. Saba N, Jawaid M, Alothman OY, Paridah MT. A review on dynamic mechanical properties of natural fibre reinforced polymer composites. *Construction and Building Materials.* 2016;106:149–59.

16. Zheng C, Wang S, Liang S. Interface bonding mechanisms of co-cured damping carbon fibre reinforced epoxy matrix composites. *Journal of Alloys and Compounds.* 2020;822:153739.

17. Varghese AM, Mittal V. Surface modification of natural fibres. In Shimpi, N.G. (Ed.) *Biodegradable and Biocompatible Polymer Composites: Processing, Properties and Applications.* Woodhead Publishing: Sawston, 2017, pp. 115–55.

18. Amiandamhen SO, Meincken M, Tyhoda L. Natural fibre modification and its influence on fibre-matrix interfacial properties in biocomposite materials. *Fibres and Polymers.* 2020;21(4):677–89.

19. Rajini N, Jappes JTW, Jeyaraj P, Rajakarunakaran S, Bennet C. Effect of montmorillonite nanoclay on temperature dependence mechanical properties of naturally woven coconut sheath/polyester composite. *Journal of Reinforced Plastics and Composites.* 2013;32(11):811–22.

20. Geethamma VG, Kalaprasad G, Groeninckx G, Thomas S. Dynamic mechanical behavior of short coir fibre reinforced natural rubber composites. *Composites Part A-Applied Science and Manufacturing.* 2005;36(11):1499–506.

21. Rajini N, Jappes JTW, Rajakarunakaran S, Jeyaraj P. Dynamic mechanical analysis and free vibration behavior in chemical modifications of coconut sheath/nano-clay reinforced hybrid polyester composite. *Journal of Composite Materials.* 2013;47(24):3105.

22. Oliveira FR, Erkens L, Fangueiro R, Souto AP. Surface modification of banana fibres by DBD plasma treatment. *Plasma Chemistry and Plasma Processing.* 2012;32(2):259–73.

23. Abdelmouleh M, Boufi S, Belgacem MN, Dufresne A. Short natural-fibre reinforced polyethylene and natural rubber composites: Effect of silane coupling agents and fibres loading. *Composites Science and Technology.* 2007;67(7–8):1627–39.

24. Cruz J, Fangueiro R. Surface modification of natural fibres: A review. *Procedia Engineering* 2016;155:285–8.

25. Aziz SH, Ansell MP. The effect of alkalization and fibre alignment on the mechanical and thermal properties of kenaf and hemp bast fibre composites: Part 1- polyester resin matrix. *Composites Science and Technology.* 2004;64(9):1219–30.

26. Herrera-Franco PJ, Valadez-Gonzalez A. A study of the mechanical properties of short natural-fibre reinforced composites. *Composites Part B-Engineering.* 2005;36(8):597–608.

27. Hossain MK, Dewan MW, Hosur M, Jeelani S. Mechanical performances of surface modified jute fibre reinforced biopol nanophased green composites. *Composites Part B-Engineering.* 2011;42(6):1701–7.

28. Mylsamy K, Rajendran I. The mechanical properties, deformation and thermomechanical properties of alkali treated and untreated Agave continuous fibre reinforced epoxy composites. *Materials & Design.* 2011;32(5):3076–84.

29. Yan L. Effect of alkali treatment on vibration characteristics and mechanical properties of natural fabric reinforced composites. *Journal of Reinforced Plastics and Composites.* 2012;31(13):887–96.

30. Essabir H, Elkhaoulani A, Benmoussa K, Bouhfid R, Arrakhiz FZ, Qaiss A. Dynamic mechanical thermal behavior analysis of doum fibres reinforced polypropylene composites. *Materials & Design.* 2013;51:780–8.

31. Indira KN, Jyotishkumar P, Thomas S. Viscoelastic behaviour of untreated and chemically treated banana fibre/PF composites. *Fibres and Polymers.* 2014;15(1):91–100.

32. Kumar KS, Siva I, Rajini N, Jeyaraj P, Jappes JTW. Tensile, impact, and vibration properties of coconut sheath/sisal hybrid composites: Effect of stacking sequence. *Journal of Reinforced Plastics and Composites.* 2014;33(19):1802–12.

33. Karaduman Y, Sayeed MMA, Onal L, Rawal A. Viscoelastic properties of surface modified jute fibre/polypropylene nonwoven composites. *Composites Part B-Engineering.* 2014;67:111–8.

34. Tang X, Kong D, Yan X. Facile dip-coating method to prepare micro-perforated fabric acoustic absorber. *Applied Acoustics.* 2018;130:133–9.

35. Chatterjee K, Tabor J, Ghosh TK. Electrically conductive coatings for fibre-based E-textiles. *Fibres.* 2019;7(6):51.

36. Lurie S, Minhat M, Tuchkova N, Soliaev J. On remarkable loss amplification mechanism in fibre reinforced laminated composite materials. *Applied Composite Materials.* 2014;21(1):179–96.

37. Finegan IC, Gibson RF. Improvement of damping at the micromechanical level in polymer composite materials under transverse normal loading by the use of special fibre coatings. *Journal of Vibration and Acoustics-Transactions of the ASME.* 1998;120(2):623–7.

38. Finegan IC, Gibson RF. Analytical modeling of damping at micromechanical level in polymer composites reinforced with coated fibres. *Composites Science and Technology.* 2000;60(7):1077–84.

39. Gu JH, Zhang XN, Gu MY. Effect of fibre coating on the longitudinal damping capacity of fibre-reinforced metal matrix composites. *Materials Letters.* 2005;59(2–3):180–4.

40. Valantin C, Lacroix F, Deffarges M-P, Morcel J, Hocine NA. Interfacial damage on fatigue-loaded textile-rubber composites. *Journal of Applied Polymer Science.* 2015;132(4):41346.

41. Malakooti MH, Hwang HS, Sodano HA. Morphology-controlled ZnO nanowire arrays for tailored hybrid composites with high damping. *ACS Applied Materials & Interfaces.* 2015;7(1):332–9.

42. Agrawal R, Nieto A, Chen H, Mora M, Agarwal A. Nanoscale damping characteristics of boron nitride nanotubes and carbon nanotubes reinforced polymer composites. *ACS Applied Materials & Interfaces.* 2013;5(22):12052–7.

43. Remillat C. Damping mechanism of polymers filled with elastic particles. *Mechanics of Materials.* 2007;39(6):525–37.

44. Han SJ, Chung DDL. Mechanical energy dissipation using carbon fibre polymer-matrix structural composites with filler incorporation. *Journal of Materials Science.* 2012;47(5):2434–53.

45. Wang T, Zhang JH, Zhang Y. Forming process and damping properties of carbon fibre-reinforced polymer concrete. *Journal of Reinforced Plastics and Composites.* 2014;33(1):93–100.

46. Chandradass J, Kumar MR, Velmurugan R. Effect of nanoclay addition on vibration properties of glass fibre reinforced vinyl ester composites. *Materials Letters.* 2007;61(22):4385–8.

47. Wang Y, Zhan M, Li Y, Shi M, Huang Z. Mechanical and damping properties of glass fibre and mica-reinforced epoxy composites. *Polymer-Plastics Technology and Engineering.* 2012;51(8):840–4.

48. Li B, Olson E, Perugini A, Zhong WH. Simultaneous enhancements in damping and static dissipation capability of polyetherimide composites with organosilane surface modified graphene nanoplatelets. *Polymer.* 2011;52(24):5606–14.
49. Zhou X, Shin E, Wang KW, Bakis CE. Interfacial damping characteristics of carbon nanotube-based composites. *Composites Science and Technology.* 2004;64(15):2425–37.
50. Gardea F, Glaz B, Riddick J, Lagoudas DC, Naraghi M. Identification of energy dissipation mechanisms in CNT-reinforced nanocomposites. *Nanotechnology.* 2016;27(10):105707.
51. DeValve C, Pitchumani R. Experimental investigation of the damping enhancement in fibre-reinforced composites with carbon nanotubes. *Carbon.* 2013;63:71–83.
52. Khan SU, Li CY, Siddiqui NA, Kim J-K. Vibration damping characteristics of carbon fibre-reinforced composites containing multi-walled carbon nanotubes. *Composites Science and Technology.* 2011;71(12):1486–94.
53. Patterson BA, Galan U, Sodano HA. Adhesive force measurement between HOPG and zinc oxide as an indicator for interfacial bonding of carbon fibre composites. *ACS Applied Materials & Interfaces.* 2015;7(28):15380–7.
54. Tehrani M, Safdari M, Boroujeni AY, Razavi Z, Case SW, Dahmen K, et al. Hybrid carbon fibre/carbon nanotube composites for structural damping applications. *Nanotechnology.* 2013;24(15):155704.
55. Gardea F, Glaz B, Riddick J, Lagoudas DC, Naraghi M. Energy dissipation due to interfacial slip in nanocomposites reinforced with aligned carbon nanotubes. *ACS Applied Materials & Interfaces.* 2015;7(18):9725–35.

17 Nanomaterials-Based Natural Fiber Composites
Dynamic Mechanical Analysis

Theivasanthi Thirugnanasambandan and
Senthil Muthu Kumar Thiagamani
Kalasalingam Academy of Research and Education

CONTENTS

17.1 INTRODUCTION

The natural fibers are abundant in nature, environmentally friendly, highly economical, and can be processed easily to make a composite. The difficulties in the adhesion of the fibers with the polymers render poor mechanical properties. The natural fibers are more hydrophilic in nature since they have more hydroxyl groups on their surface. This leads to their poor adhesion with the polymers because polymer matrices are hydrophobic. As a result, natural fibers are blended with the fibers such as cotton fibers or wool fibers to bring durable composites (Thiagamani et al. 2019a; Thiagamani, Krishnasamy, and Siengchin 2019b; Senthilkumar et al. 2019c). The interfacial adhesion of natural fibers with the polymers can be improved by functionalizing the natural fibers by means of chemical treatments. The functionalization process meets some difficulties during processing because of the complicated chemical compositions present in the structure of the natural fibers. Therefore, adding nanoscale materials in the composite can be the best alternate way to increase

DOI: 10.1201/9781003173625-19

the mechanical properties and interfacial adhesion (Chandrasekar et al. 2020; Senthilkumar et al. 2021).

Nanofillers are added in natural fiber-based polymer composites to improve the interfacial bonding between the natural fibers and the polymer matrix. The properties such as thermal stability, strength, water absorption, and flame retardancy are greatly enhanced even for very low loading of the nanofiller because these nanofillers are having more surface area or a high aspect ratio (Jose et al. 2021). The nanofillers can be easily added in powder form that will not affect the density of the polymer composites. The performance of the composite depends on the compatibility between the natural fiber and the polymer matrix (Shahroze et al. 2021; Thiagamani et al. 2021; Nasimudeen et al. 2021). Even though natural fibers possess many advantages such as low cost, lightweight, and high strength, they lack compatibility because natural fibers are hydrophilic, whereas polymer matrices are hydrophobic. This will result in the degradation of the composite. So, natural fibers are modified by various chemical processes such as alkylation, acetylation, and maleated coupling (Senthilkumar et al. 2019; Krishnasamy et al. 2019; Shahroze et al. 2019; Senthilkumar et al. 2021b). Nanofillers such as nanoclay, graphene, graphene oxide, carbon nanotubes, silicon carbide, nanosilica, and nanocellulose are mostly reported to add with natural fibers (Senthilkumar et al. 2018). Nanoclay reduces the water absorption in sisal fiber-based composites. Carbon nanotubes participate in the improvement of the mechanical properties of ramie and bamboo fibers (Devnani and Sinha 2019).

In dynamic mechanical analysis (DMA), the dynamic variables such as storage modulus (E'), loss modulus (E''), and damping factor (tan $\delta = E''/E$) are analyzed as a function of temperature. These variables give information about the bonding between the natural fibers and the polymer matrices in a composite material. The storage modulus is nothing but the elastic constant of the material that gives the data about the ability of the material when the load is applied. Thus, the stiffness behavior of the material can be understood. The loss modulus corresponds to the scattered energy inside the material which in turn provides the results regarding the viscous nature of the sample. It is also the amount of energy dissipated as heat energy during the application of load. The value of the transition temperature of the composite can be obtained directly from the loss modulus curve peak (Chandrasekar et al. 2019; Krishnasamy et al. 2019; Thiagamani et al. a2019a, 2020b). The damping phenomenon in a material is explained from the ratio of the loss modulus to the storage modulus, i.e., the loss factor. The damping factor can be resolved into two components. One is the horizontal component E Cos δ that corresponds to the storage modulus. The other one is the vertical component E Sin δ that corresponds to the loss modulus. DMA also displays the sine waves of stress and strain curves with respect to time when a vibrating force is applied to the sample (Ashok et al. 2019; Alothman et al. 2020).

17.2 NANOCLAY

A reduction in water mass uptake behavior is observed when nanoclays are added from 1 wt.% to 5 wt.% with epoxy/sisal fiber composites. Smectic clays for example montmorillonite have a more dispersing ability with the natural fibers. Montmorillonite is nothing but alumina/silicate layers where alumina sheets are placed in between two silicate sheets. Nanoclays are highly exfoliated when added to the polymer matrix. This happens because the hydrophilic nanoclays become organophilic through alkyl-ammonium cations. The storage modulus and the glass transition temperature are increased for the composites with nanoclays. These values are reduced after placing the samples in a water medium. The penetration of water inside the composites results in more plasticization in the polymer (Mohan and Kanny 2011).

The nanoclay (1 wt.%)/bamboo/woven kenaf/epoxy composites are fabricated. The commonly used layered silicate materials used as reinforcements are montmorillonite, organoclay, saponite, and halloysite nanotubes because of their easy processing nature. Nanoclays are also available at a low cost with superior properties. The DMA not only gives information about the bonding between natural fibers and the polymer matrix, but also discusses about the constituents of the composite, cross-linking density, and phase changes. The movement of the polymer chains is prevented by the toughening effect offered by the nanoclay thus producing a very good bonding between the natural fibers and the polymer matrix. If the tan delta value is high for a sample, the sample has more nonelastic strain components. The height of the tan delta peak for montmorillonite/bamboo/woven kenaf/epoxy composites is greatly reduced thus showing a low dissipation of energy from the composite. The addition of nanoclay affects the values of the storage modulus of the composites, which is shown in the storage modulus versus temperature curve in Figure 17.1. The storage modulus is found to be increased both in the glassy and rubbery regions when nanoclay is added with the fibers (Chee et al. 2021).

Figure 17.2 shows that the height of the loss modulus peak is raised when nanoclay is mixed with the fibers. The peak gives the value of glass transition temperature. Thus, the composite exhibits remarkable damping that is able to decrease the damaging forces (Chee et al. 2021).

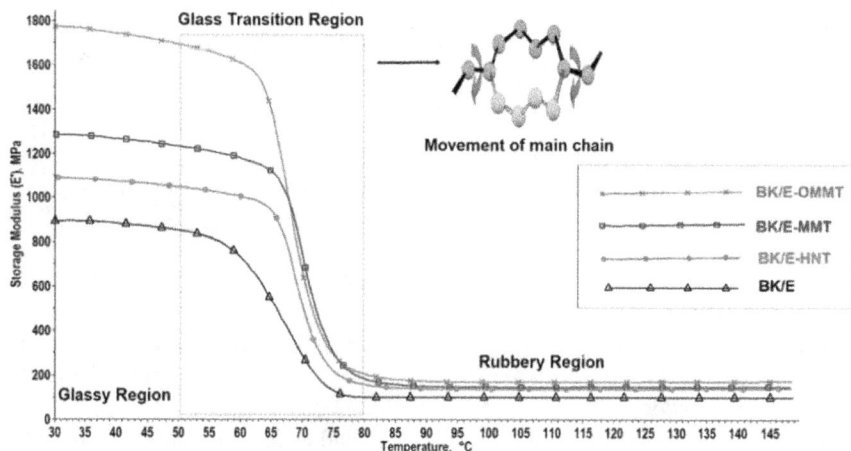

FIGURE 17.1 Changes in storage modulus by nanoclay addition (Chee et al. 2021) (Open Access).

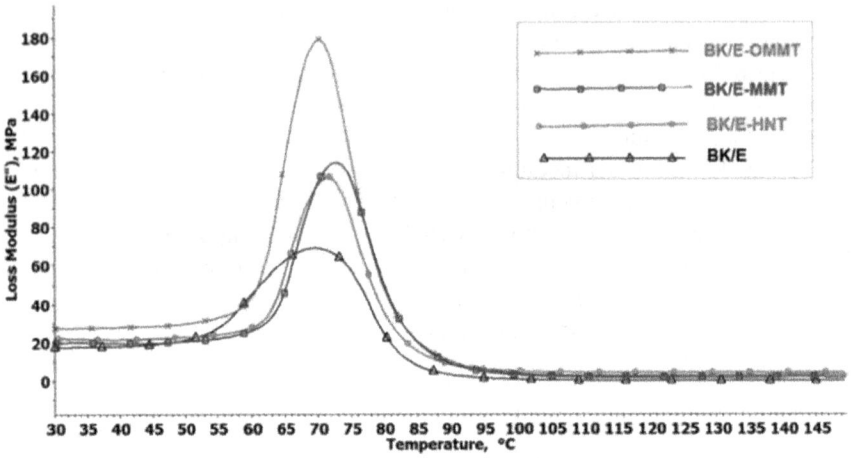

FIGURE 17.2 Changes in loss modulus by nanoclay addition (Chee et al. 2021) (Open Access).

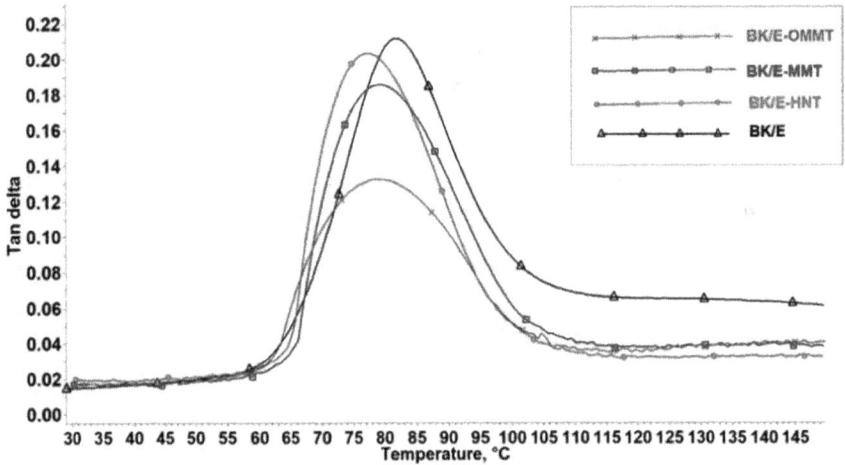

FIGURE 17.3 Changes in tan δ by nanoclay addition (Chee et al. 2021) (Open Access).

The curve for loss factor, i.e., the ratio of storage and loss modulus as a function of temperature is given in Figure 17.3. The peak height is reduced as nanoclay is added to the composite. This confirms that the movement of the polymer chain is restricted. The bonding of fibers with the epoxy is good and the damping of energy is less in the case of nanoclay added composites (Chee et al. 2021).

Nanoclay is functionalized using stearic acid and added with natural rubber/jute fiber composites. The stearic acid-modified nanoclay supports very good dispersion of jute fibers in natural rubber. DMA shows that the storage modulus value is increased and the loss factor peak height is reduced (Roy et al. 2018). The composites of 5 wt.% montmorillonite nanoclay/25 wt.% chemically modified jute fiber/polyester resin possess a glass transition temperature of up to 115°C (Arulmurugan and Venkateshwaran 2020).

17.3 GRAPHENE

The performance of jute fibers can be enhanced by coating with graphene-based materials such as graphene oxide and graphene flakes. The bonding or mechanical interlocking of these materials with natural fibers increases both interfacial shear strength and tensile strength. The composites made from this kind of natural fibers are superior to the conventional synthetic composites and can have applications in the automobile and aerospace industries. The adhesion of graphene oxide is more compatible with natural fibers as it possesses more reactive functional groups on its surface (Sarker et al. 2018).

An enhancement in the dynamic mechanical properties of kevlar/*Cocos nucifera* sheath (CS)/epoxy composites is demonstrated when graphene nanoplatelets are added as a filler. The stress transfer rate is increased for 50/50 of kevlar/CS with 0.75 wt.% of graphene nanoplatelets. This is similar to the effect obtained from kevlar/epoxy composite without natural fibers. This is due to the effective cross-linking between the natural fiber and epoxy that is offered by the graphene nano-platelets. The conclusion is that synthetic fibers such as Kevlar can be replaced by natural fibers with the aid of nanomaterials and is suitable for structural applications. The nanomaterials having applications in this area are graphene, graphene oxide, carbon nanotubes, carbon fibers, nanoclay, silicon carbide, and nanosilica particles. These nanomaterials are having a high surface area for two-dimensional materials such as graphene or a high aspect ratio for one-dimensional materials such as a carbon nanotube. Even though carbon nanotubes can bring high strength and stiffness to the composite, they require a large production cost. This makes them not suitable for their commercial applications. Graphene has a large surface area of 2600 m^2/g is responsible for the efficient transfer of stress. So, the load-carrying capability of the composite is enhanced. In DMA, the addition of graphene to the polymer/natural fiber composite increases the storage modulus and the loss modulus. The fabrication process of including graphene nanoplatelets in kevlar/CS/epoxy composites is shown in Figure 17.4. The graphene nanoplatelets are sonicated in acetone solution to get a very good dispersion. It is then added with epoxy by a magnetic stirrer. Finally, the laminates are produced by the hand lay-up method. (Jesuarockiam et al. 2019).

The thermal behavior of the piassava fiber/graphene oxide/epoxy composites is analyzed by DMA. The graphene oxide coating increases the viscous stiffness and damping capacity of the composite (Filho et al. 2020). The natural fiber is functionalized with graphene oxide from 20 to 40 vol.% and produces changes in viscous stiffness and damping capacity of the composite (Filho et al. 2021). A composite made of L-12 epoxy resin/hemp fibers 0.7 wt.%/graphene 0.3 wt.% exhibit superior mechanical performance (Hallad et al. 2018).

Graphene is having high electrical conductivity than metals. Multifunctional composites are made from graphene-coated jute fibers for electromagnetic interference

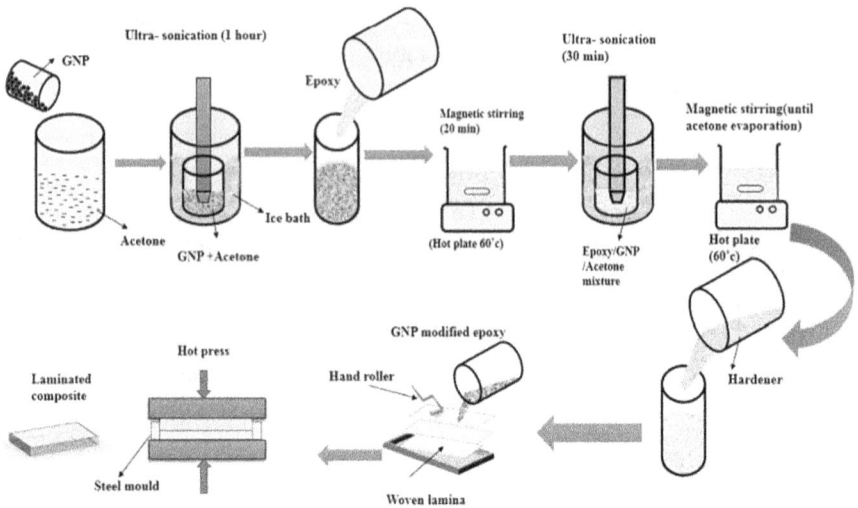

FIGURE 17.4 Fabrication of graphene-coated natural fiber/epoxy composites (Jesuarockiam et al. 2019) (Open Access).

shielding applications (Karim et al. 2021). Natural fibers can be the best alternate to synthetic fibers in polymer composites. To overcome the challenges such as damage from heat, they are treated with nanomaterials. Graphene oxide functionalized epoxy composites show improved performance as demonstrated from the DMA (Filho et al. 2021).

17.4 CARBON NANOTUBES

Carbon nanotubes enhance the mechanical properties such as Young's modulus, tensile strength, and fracture resistance when used as a reinforcement (Shen et al. 2014). If the mechanical properties and the thermal stability of natural fiber-based biodegradable composites are improved using advanced nanomaterials, they can have applications in automotive industries, construction areas and to make the interior parts of aircraft. Nanoclay also referred to as layered silicates disperse uniformly over the composites resulting in very high strength. A composite based on polylactic acid/ramie fiber/carbon nanotube/montmorillonite is developed. Carbon nanotubes possess remarkable electrical and mechanical properties. When this composite is exposed to elevated temperatures, the DMA data give information regarding the stiffness and energy dissipation of the composite. The properties of the composite depend on the uniform dispersion of the filler, volume fraction, and load transfer from the filler to the matrix. Montmorillonite shows a very good reinforcing effect than carbon nanotubes. The storage modulus is increased enormously due to the presence of montmorillonite and carbon nanotubes by rendering a large transfer of stress from the polymer to the ramie fiber that gives a high stiffness to the composite. In DMA, the damping peak will be high for poor bonding between the fiber and the matrix because of high-energy dissipation. In many composites, the damping peak becomes high when nanomaterials are added. The nanomaterials increase the movement of polymer chains. The glass transition temperature is also decreased with the addition of nanomaterials.

Multi-walled carbon nanotubes (MWCNT) are dispersed in Chinese ink to mix with kenaf fibers. Chinese ink is prepared using soot and animal glue that assists in enhancing the adhesion of MWCNT with kenaf fibers. The MWCNT/glass/kenaf/ epoxy- and MWCNT/carbon/kenaf/epoxy-laminated composites are fabricated in which the laminates show a more hierarchical structure. This behavior is suitable to obtain superior flexural properties and is available as more lightweight laminates. These composites show an increase in storage modulus and glass transition temperature. Composite laminates are having more applications in various industries. Even though they possess high fatigue strength, they are subjected to delamination, i.e., the layers in these laminates are getting separated. The delamination occurs by the brittle nature of the resin. Nanomaterials are having the capacity to form hierarchical composites. Hierarchical composites are nothing but multiscale composites that can offer multifunctional properties. Chinese ink as a dispersing agent for CNTs will bring the CNT/natural fiber-based composites for large-scale applications in industries (Hidayah et al. 2019).

17.5 SILICON CARBIDE

The lightweight composites are developed by adding nanoparticles with natural fibers. These composites are having superior physical and thermomechanical properties. The needle punch nonwoven jute fibers are functionalized with 15 wt.% of silicon carbide and added in epoxy composite. The composite is made by hand-layup techniques. The storage modulus and loss modulus are largely increased in the resulting composite. The composite is found to be long-lasting and reusable similar to synthetic composites since jute offers very good strength, low extensibility, and more breathability (Patnaik et al. 2018).

Epoxy composites are fabricated by hand lay-up method in which *Luffa cylindrica* fibers reinforced with silicon carbide particles are incorporated. DMA is performed at temperatures from 26°C to 160°C for a frequency of 1 Hz. A high storage modulus of 2300 MPa and a low loss modulus of 282.35 MPa are obtained at temperature between 30°C and 70°C for 3.5 wt.% Silicon Carbide particles. The glass transition temperature of the composite is found to be 80°C. The silicon carbide particles are able to reduce the damping factor and can increase the thermal stability of the composite up to 95.62°C (Chethan et al. 2019).

17.6 NANOSILICA

The nanosilica/areca fiber/neem oil/epoxy composites are developed after silanization by amino silane 3-aminopropyltriethoxysilane (APTES). Thus, the nanosilica-coated natural fibers are uniformly distributed in the epoxy matrix. The storage modulus reaches a value of 9.8 GPa for 2 vol.% of nanosilica particles in 30 vol.% of areca/epoxy-neem composite. Pure biocompatible polymer composites lack the most essential properties such as thermomechanical, physical, chemically inactive, and insulation to electricity. The loss factor becomes very much lower for these composites. Nanosilica supports filling the voids in the epoxy and also holds the molecules of epoxy. The storage modulus is increased as the crosslinking density is enhanced by preventing the formation of interpenetrating polymeric structure in the composite. Since nanosilica is added to the composite, the amount of heat energy to rotate the C–C primary chains is large with a high glass transition temperature up to 78°C. So, the movement of polymer chains is prevented by this method. Thus, a highly reacted phase of nanoparticles incorporated natural fiber with the epoxy matrix is achieved (Samuel et al. 2021).

The development of hybrid composites is able to overcome the challenges in natural fiber-reinforced composites. The strength of the composites is decided by various reasons such as the amount of natural fiber content, their arrangement in the polymer matrix, and the interfacial bonding between the natural fiber and the polymer matrix. Nanofillers are additional support to these factors in strengthening the composites. Shear blending, Mechanical stirring, and Ultra sonicating are the methods adopted for mixing the nanofillers in a natural fiber-reinforced composite. The mechanical properties of sisal/25 wt.% glass fiber/polyester composites are largely enhanced with 3 wt.% of nanosilica (Chowdary et al. 2020). Nanoparticles are also reported with synthetic fiber-reinforced polymer composites. For example, a high glass transition temperature and storage modulus is shown in nanosilica/aramid fiber/epoxy composite (Ravi et al. 2021).

17.7 NANOCELLULOSE

Nanocellulose is extracted from waste jute fibers by the acid hydrolysis process. A composite is made by coating 5 wt.% of cellulose nanocrystals over jute fabric and mixing in epoxy matrix. The nanocellulose colloidal solution is ultrasonicated using an ultrasonic probe and is coated on jute fabric by roller padding. The composite exhibits a high storage modulus and the height of the tan delta peak is reduced (Jabbar et al. 2017). Figure 17.5 shows the extraction of nanocellulose from plant materials by the acid hydrolysis method (Ferreira et al. 2019). The disordered regions in cellulose chains are hydrolyzed by strong acids such as sulfuric acid in this process. This process is able to modify the structure of the cellulose. Thus, obtained nanocellulose can be used as filler material to produce functional composites.

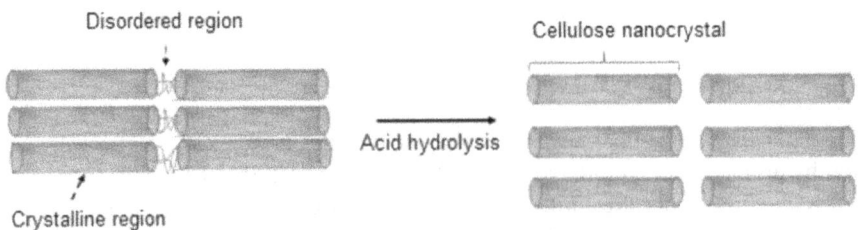

FIGURE 17.5 Extraction of nanocellulose by acid hydrolysis method (Ferreira et al. 2019) (Open Access).

Cellulose nanoparticles are more attractive nowadays among researchers since they offer a high strength to the composite than the conventional fillers such as carbon fibers, steel fibers, and glass fibers. They are biocompatible and biodegradable and produced from plant materials. Cellulose nanoparticles are also called nanocellulose and cellulose nanocrystals. Polylactic acid composites are prepared using cellulose nanoparticles that are extracted from kenaf fibers as fillers with a loading of 1–5 wt.%. Dynamic variables such as storage modulus, loss modulus, and tan δ are measured at temperatures from 30°C to 120°C at a rate of 3°C/min. The dimensions of the samples are the length of 10 mm, and thickness of 2.5 mm and a frequency of 1 Hz is fixed in the instrument. Since the cellulose nanoparticles are dispersed uniformly in the composite, the stress is easily transferred from the polymer matrix to the nanoparticles, which leads to an increase in the stiffness of the composite. Thus, the movement of the polymer chains and their deformation is greatly reduced. The storage modulus of the polylactic acid composite is found to be increasing with the increase in the cellulose nanoparticles content. The loss factor measurements show that the glass transition temperature shows a transition from 70.6°C to 67°C. The results thus exhibit the ability of the composite material to store and dissipate energy (Ketabchi et al. 2016).

Making high-performance composites from renewable resources becomes a current trend among the research community. Such composites are not producing any harmful effects regarding the health of human beings, as they are easily biodegradable. Natural fibers like sisal, hemp, kenaf, jute, coir, oil palm fiber are added with both thermosets and thermoplastics to make automobile parts. The dynamic mechanical properties are increased using hybrid composites. The addition of nanocellulose in hemp fiber-reinforced cellulose-filled epoxy composite softens the matrix to a large extent at higher temperatures. When the amount of nanocellulose is increased, the storage modulus also increases. The height of the loss modulus peak is more with a high glass transition temperature as the amount of the filler is increased (Palanivel et al. 2017).

17.8 OTHER NANOPARTICLES

NiZn ferrite nanoparticles are incorporated in the thermoplastic natural rubber matrix. The storage modulus is found to be increasing as the amount of filler increases. This result confirms that the stiffness of the composite is improved because of the effective interactions between the magnetic nanoparticles and the thermoplastic natural rubber matrix. More transporting channels are created which lead to a high stress transfer. When a polymer sample is analyzed at various temperatures, two transition peaks are observed. The first peak provides the value of the glass transition temperature of the polymer. The polymers possess both amorphous and crystalline components in its structure. The movement of polymer chains in the amorphous nature of the sample is clearly observed in the amorphous nature of the sample at the glass transition temperature. The second peak refers to a temperature at which the relaxation of the polymer chains in the amorphous portion of the polymers occurs. NiZn ferrite nanoparticles incorporated thermoplastic natural rubber matrix shows only one peak at 60°C. The glass transition temperature increases as the amount of filler increases. The loss modulus decreases as the amount of the filler increases which states that the sample exhibits high damping. The nanoparticles offer resistance to the movement of polymer chains. This process is indicated by a shift of the loss modulus peak toward higher temperatures as the filler content increases. Thus, it is confirmed that the nanoparticles have the capacity to enhance the thermal stability of the polymer composites. The relation between the elastic and viscous nature of the composite is provided by the tan delta curve. Most of the damping peaks for the composite are hidden by the addition of nanoparticles with the polymer matrix. As the amount of the filler increases, only less volume is available for the dissipation of the vibrational energy (Flaifel et al. 2013).

The mechanical properties such as the tensile strength and the interfacial shear strength are improved in a flax fiber/epoxy composite by grafting the flax fiber yarn with 2.34 wt.% of TiO_2 nanoparticles (Wang et al. 2015). Nanofillers such as calcium carbonate, polyhedral oligomeric silsesquioxane (coir nanofiller and carbon black are also reported in these natural fiber-based polymer composites (Saba et al. 2014)).

17.9 CONCLUSIONS

DMA is a method that is used to analyze the viscoelastic behavior of the polymer composites. The thermal stability of the polymer composites is tested by varying the temperatures. By performing various characterization techniques such as DMA for the composites, it is possible to construct composites of lightweight, high stiffness, and excellent thermomechanical properties. When nanomaterials are incorporated in natural fiber composites, the storage modulus is increased and the loss modulus is decreased. The glass transition temperature is shifted toward higher temperatures. Nanofillers such as nanoclay and graphene can be used to bring commercial natural fiber-based composites since they provide not only higher performance, but also these materials are more economical.

REFERENCES

Alothman, O.Y., M. Jawaid, K. Senthilkumar, M. Chandrasekar, B.A. Alshammari, H. Fouad, M. Hashem, and S. Siengchin. "Thermal Characterization of Date Palm/Epoxy Composites with Fillers from Different Parts of the Tree." Journal of Materials Research and Technology 9, no. 6 (2020): 15537–15546.

Arulmurugan, S., and Venkateshwaran, N. "Effect of Nanoclay Addition and Chemical Treatment on Static and Dynamic Mechanical Analysis of Jute Fibre Composites." Polímeros 29 (2020): 1–8.

Ashok, R.B., C.V. Srinivasa, and B. Basavaraju. "Dynamic Mechanical Properties of Natural Fiber Composites—A Review." Advanced Composites and Hybrid Materials 2, no. 4 (2019): 586–607.

Chandrasekar, M., R.M. Shahroze, M.R. Ishak, N. Saba, M. Jawaid, K. Senthilkumar, S.M.K. Thiagamani, and S. Siengchin. "Flax and Sugar Palm Reinforced Epoxy Composites: Effect of Hybridization on Physical, Mechanical, Morphological and Dynamic Mechanical Properties." Materials Research Express 6, no. 10 (2019): 105331.

Chandrasekar, M., S.M.K. Thiagamani, K. Senthilkumar, N.M. Nurazzi, M.R. Sanjay, N. Rajini, and S. Siengchin. "Inorganic Nanofillers-Based Thermoplastic and Thermosetting Composites." In Anjay Mavinkere Rangappa, Jyotishkumar Parameswaranpillai, Suchart Siengchin, Lothar Kroll (Eds.), *Lightweight Polymer Composite Structures*, 309–330. Boca Raton: CRC Press, 2020.

Chee, S.S., M. Jawaid, O.Y. Alothman, and H. Fouad. "Effects of Nanoclay on Mechanical and Dynamic Mechanical Properties of Bamboo/Kenaf Reinforced Epoxy Hybrid Composites." Polymers 13, no. 3 (2021): 395.

Chethan, S., S. Suresha, and G. Goud. "Influence of SiC Particulates on Dynamic Mechanical Response of Treated Luffa Cylindrica Epoxy Composites." Indian Journal of Science and Technology 12 (2019): 27.

Chowdary, M.S., G. Raghavendra, M.S.R. Niranjan Kumar, S. Ojha, and V. Boggarapu. "Influence of Nano-Silica on Enhancing the Mechanical Properties of Sisal/Kevlar Fiber Reinforced Polyester Hybrid Composites." Silicon (2020): 1–8.

Devnani, G.L., and S. Sinha. "Effect of Nanofillers on the Properties of Natural Fiber Reinforced Polymer Composites." Materials Today: Proceedings 18 (2019): 647–654.

Ferreira, Filipe V., Ivanei F. Pinheiro, Sivoney F. de Souza, Lucia H.I. Mei, and Liliane M.F. Lona. "Polymer Composites Reinforced with Natural Fibers and Nanocellulose in the Automotive Industry: A Short Review." Journal of Composites Science 3, no. 2 (2019): 51.

Flaifel, Moayad Husein, Sahrim Hj Ahmad, Aziz Hassan, Shamsul Bahri, A. Tarawneh Mou'ad, and Lih-Jiun Yu. "Thermal Conductivity and Dynamic Mechanical Analysis of NiZn Ferrite Nanoparticles Filled Thermoplastic Natural Rubber Nanocomposite." Composites Part B: Engineering 52 (2013): 334–339.

Garcia Filho, Fabio Da Costa, Fernanda Santos da Luz, Michelle Souza Oliveira, A.C. Pereira, U.O. Costa, and Sergio Neves Monteiro. "Thermal Behavior of Graphene Oxide-Coated Piassava Fiber and their Epoxy Composites." Journal of Materials Research and Technology 9, no. 3 (2020): 5343–5351.

Garcia Filho, Fabio da Costa, Michelle Souza Oliveira, Fernanda Santos da Luz, and Sergio Neves Monteiro. "Influence of Graphene Oxide Functionalization Strategy on the Dynamic Mechanical Response of Natural Fiber Reinforced Polymer Matrix Composites." *Characterization of Minerals, Metals, and Materials* 2021 (2021): 29.

Hallad, Shankar A., N. R. Banapurmath, Vishal Patil, Vivek S. Ajarekar, Arun Patil, Malatesh T. Godi, and Ashok S. Shettar. "Graphene Reinforced Natural Fiber Nanocomposites for Structural Applications." *IOP Conference Series: Materials Science and Engineering* 376, no. 1 (2018): 012072. Iop Publishing.

Hidayah, I. Nurul, D. Nuur Syuhada, H.P.S. Abdul Khalil, Z.A.M. Ishak, and M. Mariatti. "Enhanced Performance of Lightweight Kenaf-Based Hierarchical Composite Laminates with Embedded Carbon Nanotubes." Materials & Design 171 (2019): 107710.

Jabbar, Abdul, Jiří Militký, Jakub Wiener, Bandu Madhukar Kale, Usman Ali, and Samson Rwawiire. "Nanocellulose Coated Woven Jute/Green Epoxy Composites: Characterization of Mechanical and Dynamic Mechanical Behavior." Composite Structures 161 (2017): 340–349.

Jesuarockiam, Naveen, Mohammad Jawaid, Edi Syams Zainudin, Mohamed Thariq Hameed Sultan, and Ridwan Yahaya. "Enhanced Thermal and Dynamic Mechanical Properties of Synthetic/Natural Hybrid Composites with Graphene Nanoplateletes." Polymers 11, no. 7 (2019): 1085.

Jose, S., V.K. Smitha, S.M. Rangappa, S. Krishnasamy, D. Nandi, S. Siengchin, and J. Parameswaranpillai. "Micro-and Nanoscale Structure Formation in Epoxy-Clay Nanocomposites." In Jyotishkumar Parameswaranpillai, Harikrishnan Pulikkalparambil, Sanjay M. Rangappa and Suchart Siengchin (Eds.), *Epoxy Composites: Fabrication, Characterization and Applications*, 61–82. 2021. Wiley VCH, Germany.

Karim, Nazmul, Forkan Sarker, Shaila Afroj, Minglonghai Zhang, Prasad Potluri, and Kostya S. Novoselov. "Sustainable and Multifunctional Composites of Graphene-Based Natural Jute Fibers." Advanced Sustainable Systems 5, no. 3 (2021): 2000228.

Ketabchi, Mohammad Reza, Mohammad Khalid, Chantara Thevy Ratnam, and Rashmi Walvekar. "Mechanical and Thermal Properties of Polylactic Acid Composites Reinforced with Cellulose Nanoparticles Extracted from Kenaf Fibre." Materials Research Express 3, no. 12 (2016): 125301.

Krishnasamy, S., C. Muthukumar, R. Nagarajan, S.M.K. Thiagamani, N. Saba, M. Jawaid, S. Siengchin, and N. Ayrilmis. "Effect of Fibre Loading and Ca(OH)^2 Treatment on Thermal, Mechanical, and Physical Properties of Pineapple Leaf Fibre/Polyester Reinforced Composites." Materials Research Express 6, no. 8 (2021): 1–28

Krishnasamy, Senthilkumar, S.M.K. Thiagamani, C. Muthukumar, J. Tengsuthiwat, R. Nagarajan, S. Siengchin, S.O. Ismail, and N.C. Brintha. "Effects of Stacking Sequences on Static, Dynamic Mechanical and Thermal Properties of Completely Biodegradable Green Epoxy Hybrid Composites." Materials Research Express 6, no. 10 (2019): 105351.

Mohan, T. P., and K. Kanny. "Water Barrier Properties of Nano Clay Filled Sisal Fiber Reinforced Epoxy Composites." Composites Part A: Applied Science and Manufacturing 42, no. 4 (2011): 385–393.

Nasimudeen, N.A., S. Karounamourthy, J. Selvarathinam, S.M.K. Thiagamani, H. Pulikkalparambil, S. Krishnasamy, and C. Muthukumar. "Mechanical, Absorption and Swelling Properties of Vinyl Ester Based Natural Fibre Hybrid Composites." Young 355, no. 700 (2021): 295.

Palanivel, Anand, Anbumalar Veerabathiran, Rajesh Duruvasalu, Saranraj Iyyanar, and Ramesh Velumayil. "Dynamic Mechanical Analysis and Crystalline Analysis of Hemp Fiber Reinforced Cellulose Filled Epoxy Composite." Polímeros 27 (2017): 309–319.

Patnaik, Tapan Kumar, and S.S. Nayak. "Development of Silicon Carbide Reinforced Jute Epoxy Composites: Physical, Mechanical and Thermo-mechanical Characterizations." Silicon 10, no. 1 (2018): 137–145.

Ravi, S., K. Saravanan, D. Jayabalakrishnan, P. Prabhu, Vijayananth Suyamburajan, V. Jayaseelan, and A.V. Mayakkannan. "Silane Grafted Nanosilica and Aramid Fibre-reinforced Epoxy Composite: dma, Fatigue and Dynamic Loading Behaviour." Silicon (2021): 1–9.

Roy, Kumarjyoti, Subhas Chandra Debnath, Amit Das, Gert Heinrich, and Pranut Potiyaraj. "Exploring the Synergistic Effect of Short Jute Fiber and Nanoclay on the Mechanical, Dynamic Mechanical and Thermal Properties of Natural Rubber Composites." Polymer Testing 67 (2018): 487–493.

Saba, Naheed, Paridah Md Tahir, and Mohammad Jawaid. "A Review on Potentiality of Nano Filler/Natural Fiber Filled Polymer Hybrid Composites." Polymers 6, no. 8 (2014): 2247–2273.

Samuel, J. Ben, S. Julyes Jaisingh, K. Sivakumar, A. V. Mayakannan, and V. R. Arunprakash. "Visco-elastic, Thermal, Antimicrobial and Dielectric Behaviour of Areca Fibre-reinforced nano-Silica and Neem Oil-toughened Epoxy Resin Bio Composite." Silicon 13, no. 6 (2021): 1703–1712.

Sarker, Forkan, Nazmul Karim, Shaila Afroj, Vivek Koncherry, Kostya S. Novoselov, and Prasad Potluri. "High-performance Graphene-based Natural Fiber Composites." ACS Applied Materials & Interfaces 10, no. 40 (2018): 34502–34512.

Senthilkumar, K., I. Siva, S. Karthikeyan, H. Pulikkalparambil, J. Parameswaranpillai, M.R. Sanjay, and S. Siengchin. "Mechanical, Structural, Thermal and Tribological Properties of Nanoclay Based Phenolic Composites." In Mohammad Jawaid and Mohammad Asim (Eds.), Phenolic Polymers Based Composite Materials, 123–138. Singapore: Springer, 2021a.

Senthilkumar, K., N. Rajini, N. Saba, M. Chandrasekar, M. Jawaid, and S. Siengchin. "Effect of Alkali Treatment on Mechanical and Morphological Properties of Pineapple Leaf Fibre/Polyester Composites." Journal of Polymers and the Environment 27, no. 6 (2019): 1191–1201. http://dx.doi.org/10.1007/s10924-019-01418-x.

Senthilkumar, K., N. Saba, M. Chandrasekar, M. Jawaid, N. Rajini, S. Siengchin, N. Ayrilmis, F. Mohammad, and H.A. Al-Lohedan. "Compressive, Dynamic and Thermo-Mechanical Properties of Cellulosic Pineapple Leaf Fibre/Polyester Composites: Influence of Alkali Treatment on Adhesion." International Journal of Adhesion and Adhesives 106 (2021b): 102823.

Senthilkumar, K., N. Saba, N. Rajini, M. Chandrasekar, M. Jawaid, S. Siengchin, and O.Y. Alotman. "Mechanical Properties Evaluation of Sisal Fibre Reinforced Polymer Composites: A Review." Construction and Building Materials 174 (2018): 713–729.

Senthilkumar, K., S.M.K. Thiagamani, M. Chandrasekar, N. Rajini, R.M. Shahroze, S. Siengchin, S.O. Ismail, and M.P.I. Devi. "Recent Advances in Thermal Properties of Hybrid Cellulosic Fiber Reinforced Polymer Composites." International Journal of Biological Macromolecules 149 (2019): 1–13.

Shahroze, R.M., M. Chandrasekar, K. Senthilkumar, S.M.K. Thiagamani, M.R. Ishak, N. Rajini, S. Siengchin, and S.O. Ismail. "Mechanical, Interfacial and Thermal Properties of Silica Aerogel-Infused Flax/Epoxy Composites." International Polymer Processing 36, no. 1 (2021): 53–59.

Shahroze, R.M., M. Chandrasekar, K. Senthilkumar, T. Senthilmuthukumar, M.R. Ishak, and M.R.M. Asyraf. "A Review on the Various Fibre Treatment Techniques Used for the Fibre Surface Modification of the Sugar Palm Fibres." In Proceedings of the Seminar Enau Kebangsaan, 48–52. 2019.

Shen, Xi, Jingjing Jia, Chaozhong Chen, Yan Li, and Jang-Kyo Kim. "Enhancement of Mechanical Properties of Natural Fiber Composites via Carbon Nanotube Addition." Journal of Materials Science 49, no. 8 (2014): 3225–3233.

Thiagamani, Senthil Muthu Kumar, K. Senthilkumar, M. Chandrasekar, N. Rajini, S. Siengchin, and A. Varada Rajulu. "Characterization, Thermal and Dynamic Mechanical Properties of Poly(Propylene Carbonate) Lignocellulosic Cocos Nucifera Shell Particulate Biocomposites." Materials Research Express 6, no. 9 (2019a): 096426.

Thiagamani, Senthil Muthu Kumar, K. Senthilkumar, M. Chandrasekar, S. Karthikeyan, N. Ayrilmis, N. Rajini, and S. Siengchin. "Mechanical, Thermal, Tribological, and Dielectric Properties of Biobased Composites." In Anish Khan, Sanjay M. Rangappa, Suchart Siengchin and Abdullah M. Asiri (Eds.), Biobased Composites: Processing, Characterization, Properties, and Applications, 53–73. 2021. Wiley VCH.

Thiagamani, Senthil Muthu Kumar, K. Senthilkumar, M. Chandrasekar, S. Subramaniam, S.M. Rangappa, S. Siengchin, and N. Rajini. "Influence of Fillers on the Thermal and Mechanical Properties of Biocomposites: An Overview." In Anish Khan, Sanjay Mavinkere Rangappa, Suchart Siengchin and Abdullah M. Asiri (Eds.), Biofibers and Biopolymers for Biocomposites, 111–133. Switzerland: Springer, 2020.

Thiagamani, Senthil Muthu Kumar, M. Chandrasekar, K. Senthilkumar, N. Ayrilmis, S. Siengchin, and N. Rajini. "Utilization of Bamboo Fibres and Their Influence on the Mechanical and Thermal Properties of Polymer Composites." In Mohammad Jawaid, Sanjay Mavinkere Rangappa and Suchart Siengchin (Eds.), Bamboo Fiber Composites, 81–96. Singapore: Springer, 2020.

Thiagamani, Senthil Muthu Kumar, S. Krishnasamy, C. Muthukumar, J. Tengsuthiwat, R. Nagarajan, S. Siengchin, and S.O. Ismail. "Investigation into Mechanical, Absorption and Swelling Behaviour of Hemp/Sisal Fibre Reinforced Bioepoxy Hybrid Composites: Effects of Stacking Sequences." International Journal of Biological Macromolecules 140 (2019b): 637–646.

Thiagamani, Senthil Muthu Kumar, S. Krishnasamy, and S. Siengchin. "Challenges of Biodegradable Polymers: An Environmental Perspective." Applied Science and Engineering Progress 12, no. 3 (2019c): 149.

Wang, Hongguang, Guijun Xian, and Hui Li. "Grafting of Nano-TiO2 Onto Flax Fibers and the Enhancement of the Mechanical Properties of the Flax Fiber and Flax Fiber/ Epoxy Composite." Composites Part A: Applied Science and Manufacturing 76 (2015): 172–180.

18 Recent Developments in Computational Modeling of Viscoelastic Properties of Biocomposites

Renuka Sahu and Athul Joseph
Indian Institute of Science, Bangalore

Vishwas Mahesh
Siddaganga Institute of Technology
and
Indian Institute of Science, Bangalore

Vinyas Mahesh
National Institute of Technology Silchar

Dineshkumar Harursampath
Indian Institute of Science, Bangalore

CONTENTS

DOI: 10.1201/9781003173625-20

18.1   INTRODUCTION

Biomaterials are capturing the market due to their sustainability and ability to replace environment damaging polymeric, synthetic fibers, and matrix materials [1,2]. Biofibers such as flax, hemp, ramie, and silk are now being used for packaging [3,4]; structural applications such as turbines [5], furniture [6], and even aerospace [7,8], and automotive [9,10] applications. The ongoing call for sustainability and energy efficiency has expanded the scope of biomaterials and their usage in different industries. Biomaterials offer comparable property dimensions to artificially sourced materials. They are known to have high specific strength [11] and toughness [12] in addition to the added advantage of being lightweight, easily sourced, and economical [13]. For the composite case, biofibers are readily available and are easy to weave. Similarly, bio-sourced matrix materials are easily molded and are safe to handle. These properties give bio-based materials a significant upper hand and are more than enough to overcast their shortcomings such as hydrophilicity, low heat resistance, and low dimensional stability [2]. Moreover, studies are in place to remove or reduce the effect of these shortcomings on the overall performance of the biocomposites [14,15].

Naturally sourced materials have a wide property range than their synthetic counterparts. In addition, the material behavior is not significantly linear which offers a huge challenge to modeling and subsequent application of these materials. One such aspect of biocomposites is their viscoelastic nature. Viscoelasticity indicates that a material exhibits mechanical behavior in both the elastic regime such as linearly elastic solid and also as a viscous fluid upon application of a certain kind of force. Usually, upon application of fast loading, the material will behave as an elastic material whereas if it is loaded slowly the behavior will be predominantly viscous. This time/rate-dependent behavior is defined by the property of viscoelasticity. This effect of viscoelasticity on creep behavior and stress relaxation of biocomposites can be determined experimentally using dynamic mechanical analysis (DMA) [16]. Viscoelasticity studies have proven helpful in studying cellulose nanocrystals used in 3D printers [17]. These studies find similar applications in various damping and vibration problems [18–20]. Studies of creep and time-dependent failure and damage also make use of the viscoelastic characteristic of the material [21–23].

Numerical modeling techniques such as finite element method (FEM), micromechanics, viscoelastic constitutive theories, and micromechanical theories are some of the widely used tools. FEM has been successful in modeling and analyzing the mechanical behavior of biocomposites such as tensile, bending, buckling, and even fracture and damage behavior. Moreover, thermal and hygral loading conditions have also been studied using FEM. Similarly, biocomposites application to the structural field has been analyzed using micromechanics, continuum mechanics, and also machine learning (ML) and data analysis tools. Further, vibration and damping analyses have showcased the dexterity of existing modeling tools in handling complex mechanical phenomena and materials. Issue of material and geometric nonlinearity can also be easily addressed. This showcases the flexibility of FEM and other computational tools. Modeling of the viscoelastic nature of natural fiber composites (NFCs) through computational models is becoming easier as the technology is evolving.

Modeling of viscoelasticity is challenging as one has to consider two different property laws and governing conditions. Moreover, the viscoelastic behavior not only affects the normal mechanical properties of general loading such as tension, compression, and twist, but is also affected by external environmental factors such as temperature and moisture. This modeling is also dependent upon the experimental property data available for the different types of biomaterials. Hence, it is important to conduct studies on various naturally sourced fiber and matrix materials to create an extensive dataset dealing with such properties which can then be further used to study the behavior of these materials numerically at different conditions.

This chapter aims to highlight the various types of computational models used to study the viscoelastic nature of biocomposites. The first half of the chapter deals with viscoelasticity, its causes, and a few analytical models and theories that are in place for several years, to study viscoelasticity. Emphasis on evolving tools such as micromechanical models, FEM, and highly complicated tools such as MD and ANN has also been given. The focus is to introduce different numerical techniques and show the drawbacks and shortcomings as well. This will then pave the way for future developments in the numerical modeling of the viscoelastic nature of the biocomposites.

18.2 VISCOELASTICITY IN BIOMATERIALS

Viscoelasticity is the material behavior that encompasses both deformation and flow phenomena exhibited by the material. In nature, most of the materials exhibit nonlinearity in material characteristics and exhibit viscoelasticity. The viscoelasticity of biological tissue is incorporated by the presence of viscoelastic constituents such as cells, structural proteins, and extracellular matrices. Most biological tissues are viscoelastic and hence modeling this aspect of material nature becomes significant for proper application and study of these materials [24]. Many biological materials such as nacre, spider web consisting of both soft and hard elements also exhibit viscoelastic behavior [25]. Thus it is essential to study the viscoelastic parameters of these materials and also create mathematical models.

Rheology deals with material flow and finds application in various fields such as pharmacy, medicine, vibration analysis, acoustics, biotechnology, chemistry, and so on. Elastic materials are modeled using the classical elasticity relations given by Hooke's law which states the direct proportionality of stress to strain. There is no dependence of deformations on the strain rate [26]. While for the case of fluids, Newton's law of viscosity [26] is used to characterize the stress as given by Equation 18.1.

$$\tau = \mu \nabla v \tag{18.1}$$

Here, the stress (τ) is directly proportional to the rate of strain and is not dependent on the strain. When it comes to viscoelastic materials, both of these phenomena of elasticity and viscosity occur simultaneously. This behavior then becomes time-dependent and the rate determines the predominant nature of the material. Usually, for a slow loading rate, i.e., small strain rate, the behavior is dominated by elastic nature, whereas for higher strain rates the material behavior is dominated by viscous laws. For viscoelastic materials also, the behavior can be linear or nonlinear.

For linear viscoelastic materials, stress is linearly proportional to the strain history. Linear viscoelasticity is observed for low stresses, lower rate, and deformations, whereas nonlinear viscoelasticity is exhibited for the case of large deformations [27–34].

Rheology provides the main tools to analyze the viscoelastic properties of the polymers. Rheology is used to study the interconnection between the molecular structure and viscoelastic properties of the material and also to study the relationship between time, stress, and deformation [35]. Viscoelastic constitutive models are represented mathematically by either differential-based models or integral formulation-based models. It is evident that handling of differential form is easier than the integral form, but the integral form predicts the time dependence more efficiently [36]. Since viscoelasticity is the behavior in between the elastic and viscous limits, the two ends are modeled using linear elastic spring element for elastic behavior and linear viscous dashpot is used to model the purely viscous behavior. Some of the mechanical/rheological models widely used are given in the following sections.

18.2.1 Maxwell Model

Maxwell model (Figure 18.1) consists of purely elastic spring and purely viscous dashpot connected in series. This model simulates constant strain and constant stress conditions. It predicts the exponential decay of stress with time. The following equation (Equation 18.2) represents the model:

$$\sigma + \frac{\eta}{E}\dot{\sigma} = \eta\dot{\epsilon} \tag{18.2}$$

18.2.2 Kelvin (Voigt) Model

This model is described by linear viscous dashpot and linear elastic spring connected in parallel (Figure 18.2) and is used to model creep behavior in polymers. It is used to represent solid under reversible viscoelastic strain. The constitutive relation for this model is given by Equation 18.3:

$$\sigma = E\epsilon + \eta\dot{\epsilon} \tag{18.3}$$

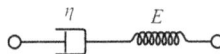

FIGURE 18.1   Representation of Maxwell Model [37].

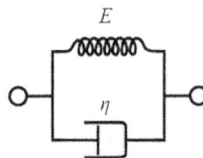

FIGURE 18.2   Representation of Kelvin-Voigt Model [37].

18.2.3 Standard Linear Solid Model

Another name given to this model is the Zener model. It consists of two springs and a dashpot (Figure 18.3) and can describe both creep and relaxation of the viscoelastic material.

18.2.4 Generalized Maxwell Model

This model is also known as the Wiechert model and is the most general form to represent linear viscoelasticity. It is a more complex model than other linear models and uses more elements (i.e., springs and dashpot). This model consists of N number of spring-dashpot Maxwell elements in parallel (Figure 18.4) and accounts that relaxation does not occur at a particular time but over the distribution of time.

Apart from the above-mentioned models, there are various other linear viscoelastic models such as the Burger model, three-element models, and so on. Also, there are nonlinear viscoelastic mathematical models to depict the nonlinearity of viscoelastic behavior [36,38,39]. Using the above-mentioned models, the viscoelastic nature of several biomaterials and nanoscale materials has been characterized [40,41]. Addressing the nonlinearity of viscoelastic behavior, bioinspired engineering models have been developed such as modeling snail's locomotion using large oscillatory amplitude shear [42]. Maxwell models have been used to compare the

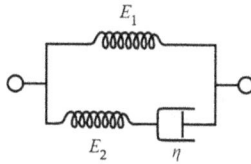

FIGURE 18.3 Representation of standard linear solid model [37].

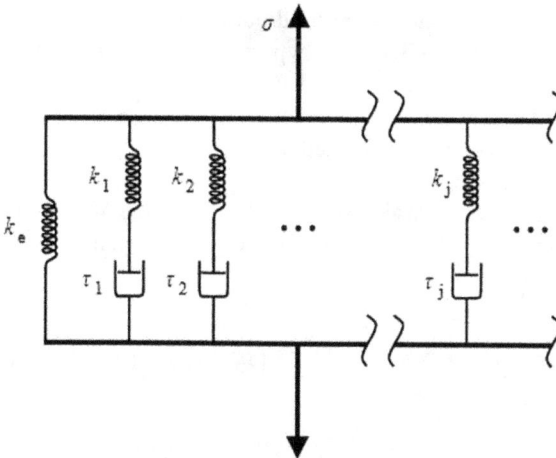

FIGURE 18.4 Representation of generalized Maxwell model [37].

rheological behavior of bio-based epoxy and synthetic epoxy to ensure the applicability of bio-based epoxies in ecocomposite [43]. Zhang et al. [44] carried out a full-fledged mathematical analysis to compare the behavior of various nonlinear viscoelastic models. Maxwell, Kelvin-Voigt, Zener models and their fractional forms had been studied. The fractional Zener model was found to be the most robust to achieve realistic responses for a wide range of loadings such as cyclic, creep, and uniaxial stress cycles. Thus, it is suggested that the end parameters must be decided to effectively choose the viscoelastic model according to the problem requirements.

18.3 MODELING OVERVIEW

18.3.1 FINITE ELEMENT MODELS

Using the viscoelastic theory given by Lodge, the viscoelastic behavior of biocomposites made up of high-density polyethylene (HDPE) matrix reinforced by wood particles was studied. This wood particle composite (WPC) is prepared using the process of thermoforming in which the WPC sheet is spread over the die and heated to its softening point. The Lodge's integral viscoelastic model was used for determining the mechanical behavior of WPC. Then, nonlinear dynamic finite element approach was adopted to simulate the inflation of the sheet. The lagrangian description was adopted and large deformation with finite strain was considered. The continuum sheet was modeled using isoparametric linear triangular elements and the assumption of plane stress was made as the sheet's thickness was less compared to other dimensions. Simulating an airflow load using the van der Waals gas equation, the formability of the WPC composite was studied. It was found that on increasing the wood particle content in the WPC, the elastic behavior becomes predominant making forming difficult. Stress-induced in the sheet was directly proportional to the wood particle content going as high as 8–9 times the stress generated in pure HDPE for WPC weight fraction 50% (shown in Figure 18.5) [45].

Another thermoforming simulation for jute woven fabric reinforced polybutylene succinate (PBS) biocomposite was carried out using FEM. To model the viscoelasticity of the biocomposites, a 3D visco-hyperelastic constitutive model that accounts for anisotropy and temperature dependence of the biocomposites was developed. The woven jute fiber was modeled as anisotropic hyperelastic material and the PBS matrix in the melted form was modeled as a viscoelastic material. The viscoelastic model is shown in Figure 18.6.

To characterize the viscoelastic nature of the matrix, the Mooney-Rivlin model was adopted. The temperature-dependent matrix energy function as given by Equation 18.4 was used.

$$\varphi_m\left(C_e, T\right) = \left[C_{10}^e\left(\overline{I_1^e} - 3\right) + C_{01}^e\left(\overline{I_2^e} - 3\right) + \frac{1}{D}(J-1)^2 \right] f_m\left(T\right) \qquad (18.4)$$

where φ_m is a function of the elastic strain of matrix, $f_m\left(T\right)$ is the temperature scaling function, $\overline{I_1^e}$, $\overline{I_2^e}$ are the invariants of deviatoric right Cauchy-Green strain tensor, and C_{10}^e, C_{01}^e, D are Mooney-Rivlin material parameters. The model parameters are

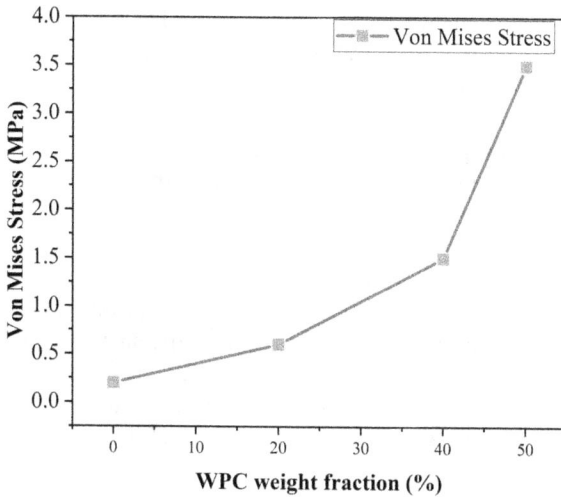

FIGURE 18.5 Variation of stress-induced in WPC/HDPE biocomposites with WPC weight fraction during thermoforming [45].

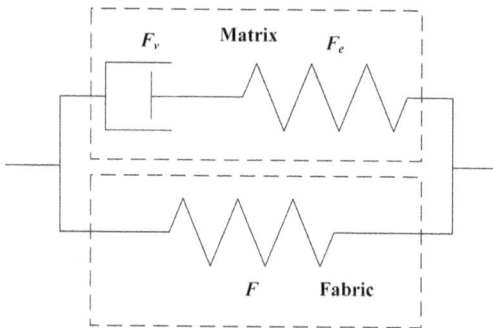

FIGURE 18.6 The rheological model used for PBS biocomposites [46] (License number: 5125281115719).

obtained by carrying out simulations of compression, uniaxial and biaxial tension, and compared with experimental results. The thermoforming process was then simulated by defining the constitutive model using user subroutine Vectorized User MATerial interface (VUMAT). The composite sheet was meshed using a solid element, C3D8R. The temperature dependency was also addressed. The maximum stamping force was found to be 555.67N and numerical results agreed well with their experimental counterpart [46].

Zhang et al. [47] came up with a novel finite element framework using Shamnskii-Newton-Raphson (SNR) method. The viscoelastic nature of biomaterials and bio-medical devices has been successfully represented using hyperelastic or fractional viscoelastic constitutive models for computation purposes. The drawback of these viscoelastic models in terms of higher computational cost, higher storage space was found to be subsided by the use of the finite element framework proposed. The above model was used to solve the Caputo fractional derivative. The proposed approximating technique led to a reduction of storage cost making it constant irrespective of the number of incrementation steps. The computational time also was now varying linearly with the number of time steps. Problems dealing with nonlinear viscoelasticity can be easily solved using this algorithm. The proposed model was further supported by applying it to the biomechanics problem of the liver [47].

Using the Prony series Maxwell model with damage function and Parallel Rheological framework (PRF), to model viscoelasticity in the oil palm fibers, Talib et al. [48] analyzed the stress cycle and damage in the oil palm fibers. For the Prony model, using user material (UMAT) subroutine in ABAQUS®, the damage function, damage stress, and damaged material tangent modulus were obtained. Similarly, the PRF model was available in the ABAQUS library and was used to obtain the plasticity response. It must be pointed out that the Prony model represented linear viscoelasticity, whereas the PRF model dealt with nonlinear viscoelastic behavior. 3D single element model (C3D8H) with eight integration points and eight node elements was used for the loading–unloading cyclic case. The major findings were that the PRF model is better suited for applications involving large deformation of the biocomposites whereas the Prony series model worked better for small deformation cases such as in biodegradation of the composite [48]. Natural fibers being viscoelastic due to the presence of components such as cellulose and lignin is beneficial for damping applications. Such damping behavior of flax fiber-reinforced unidirectional composites was investigated using finite element models. In the finite element framework, the total dissipated energy as a sum of different energy constituents such as strain energy along x, y, z directions were used. The damping loss factors for each of the three directions were evaluated. To model viscoelastic behavior, a viscoelastic layer of natural rubber was introduced in between the laminates. The effect of the viscoelastic layer's Young's modulus and thickness on vibration properties were determined (shown in Figure 18.7) [49].

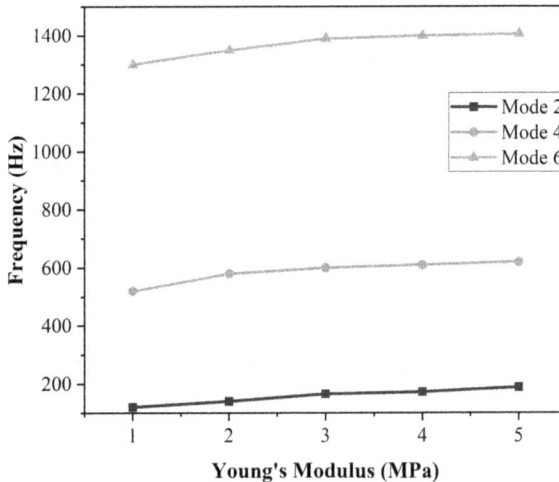

FIGURE 18.7 Variation of the frequency with Young's modulus of viscoelastic layer [49].

It was observed that on increasing the thickness of the viscoelastic layer, the loss factor increased for some values and remained mostly constant on further increasing thickness for all modes of vibration.

Guessasma et al. [50] used a finite element model to carry out nanoindentation analysis on starch–zein composite material and viscoelastic parameters were obtained. The indenter was meshed using 10 noded irregular tetrahedral 3D structural SOLID92 elements in ANSYS®. Each of these nodes had 3 degrees of freedom. To mesh the composite material specimen, the SOLID185 element in ANSYS®, which is an 8 noded 3D structural solid element, was used. Viscoelastic behavior was associated with these solid elements using the following equation (Equation 18.5):

$$\sigma = \int_0^t 2G(t-\tau)\frac{de}{d\tau}d\tau + I\int_0^t K(t-\tau)\frac{d\Delta}{d\tau}d\tau \qquad (18.5)$$

Here, σ is the Cauchy stress, I is the identity tensor, $G(t)$, $K(t)$ is the shear and bulk relaxation kernel functions, respectively, e is deviatoric strain, and Δ is the volumetric strain. τ and t are the past and present times, respectively. The contact was simulated by introducing an 8 noded surface contact element CONTA147. The viscoelastic parameters obtained for the case of starch were 6.72 GPa initial Young's modulus and 2.38 GPa final Young's modulus, whereas for zein it was 7.96 and 2.25 GPa, respectively, for initial and final Young's modulus [50]. Utilizing biocomposites as a protective gear, the viscoelastic nature of cork was studied to analyze its impact-absorbing behavior. Fernandes et al. [51] used FEM to model and design helmets used while driving a motorcycle to analyze its protectivity in case of a crash. Cork as a substitute to artificial liners was modeled using a hyperfoam model to incorporate its nonlinear behavior. To model, the impact, the human brain was modeled using coupled visco-hyperelastic mechanical model and Mooney-Rivlin strain energy was used. The viscoelastic behavior was characterized by the Prony series. Using different parametric studies, optimization of this newly developed helmet was carried out for double impact loading cases. The study was successful in proving the effectiveness and high performance of agglomerated cork liners to the existing synthetic liners. Hence, it can be said that biocomposites find effective application in protective gears as is proven by their ability to withstand several kinds of impact loads [51].

The nonlinear tensile behavior of hemp fiber composites was investigated by using a 3D viscoelastic model in a finite element framework. Constitutive law defining the viscoelastic nature of the fiber is implemented. Given by Equation 18.6 is the viscoelastic flow rule,

$$\dot{\varepsilon}^{ve} = \sum_{i=1}^n \dot{\xi}_i = \sum_{i=1}^n \frac{1}{\tau_i}\left(\mu_i \underline{\underline{S}}^{ve}\sigma - \underline{\xi}_i\right) \qquad (18.6)$$

Here, $\underline{\sigma}$ is the Cauchy true stress tensor, $\underline{\underline{S}}^{ve}$ is the viscoelastic compliance tensor, ε^{ve} is the viscoelastic strain tensor, μ_i represents relative weight at any i, and τ_i gives the corresponding release time. Eight node 3D elements were used to mesh the hemp fiber and stress–strain behavior was obtained. Apparent modulus–strain and apparent tangent modulus–strain behavior were also estimated as shown in Figure 18.8. Viscoelastic behavior helped determine severe nonlinearity in the stress–strain behavior of the fiber [52].

Wood is a natural composite that exhibits time-dependent behavior showing slow deformation under constant moisture conditions. This viscoelastic creep behavior was modeled using the Kelvin-Voigt element. Wood being a widely used natural material in construction, transportation, and other sectors, finds importance in numerical studies. Combining the hygroscopic nature of wood with the 3D ortho-tropic, viscoelastic material behavior, the material constants were taken as a function of moisture. This was done to capture the degradation of these characteristics over time. The finite element (FEM) approach, using the total strain in additive form, was implemented to carry out the numerical integration model. Quadratic brick elements were used to mesh the geometry of the wood specimen. Application of the prescribed model was done using a user-defined UMAT subroutine for different wood types. These wood types were then loaded such as to include both hygral and mechani-cal loading in various combinations. Derivation of a tangent operator for the com-plete model ensured that the asymptotic convergence occurred at a quadratic rate. Comparison with experimental findings was drawn out to establish the validity of the numerical model [53]. Vidal Salle et al. developed an ABAQUS® algorithm using UMAT subroutine to model the mechanical and mechanosorptive creep phenomena in wood. Wood was modeled using 3D nonlinear orthotropic viscoelastic constitutive law. The Maxwell model was used to describe the rheological characteristics for vari-ous elements such as hydral, thermal, elastic, and inelastic strains. In addition, the nonlinearities associated with creep were addressed using one element with a nonlin-ear dashpot. Applying the model for a wooden board, the stresses (radial, hoop, and shear) at a point in the board and their evolution with time were observed.

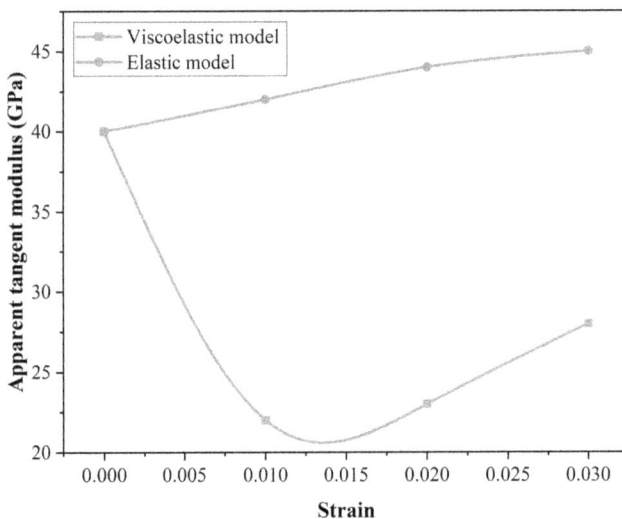

FIGURE 18.8 Apparent tangent modulus–strain behavior for elastic and viscoelastic model [52].

The viscoelastic and elastic stresses were closely correlated with just a 2%–3% difference in the two values [54]. Lv et al. [55] used ANSYS® Polyflow software to simulate the flow field and ultimately determine the rheological properties of wood–plastic composites (WPCs). The results showed that power law was for the characterization of fluids was used to predict the behavior of WPC reinforced with different concentrations of the wood fiber. It was shown that for a higher quantity of reinforced wood fiber, the pseudo-plasticity and shear thinning of the WPC increased. Carrying out a computational fluid dynamics (CFD) analysis using the commercial software ANSYS® Polyflow, extrusion of WPC and various properties such as viscosity, shear rate, velocity, and pressure was predicted. This analysis was carried out for different fiber contents and its effect on the WPCs flow field was found out. The results indicated that the shear rate along the axial position in the z-axis increased from 25 1/s to 32 1/s when the WPC weight fraction was increased from 40% to 80%, respectively. In addition, the viscosity increases 10^4 times along z-axis centerline on increasing WPC concentration from 40% to 80% [55].

Tagiltsev et al. [56] adopted a geometrically exact approach to take into account the presence of residual stresses in a composite structure while modeling under viscoelastic mechanical behavior. For this, a special field F_o was used to approximate the stress field in the structure. This field described the kinematics of the structure for two different reference configurations: stress-free state and load-free state of each material particle. This method can easily represent both elastic and inelastic deformations. This approach was successful in predicting the results of other widely used approaches "opening angle." Moreover, FEM simulations of cutting an artery were done by simulating the artery as a double-layered viscoelastic composite tube, where the material model is implemented using MSC code. Three-dimensional Hex20 brick elements were used and pressure loading was applied. At the final deformed configuration, an opening angle of 118±3 degrees was registered which was close to 120° as predicted by the semi-analytical approach.

To predict the nonlinear material behavior upon damage of flax fiber-reinforced composites, numerical and experimental studies were carried out. Due to the inherent structural built of the flax fibers, the mechanical behavior of composite laminates was highly on the nonlinear side. The omega sandwich panels exhibiting high strength find use in aeronautics and hence their damage was analyzed. A material model describing the four phases of nonlinear behavior was implemented to get a predictive finite element model. These four phases include linear elastic behavior, hardening behavior in a table damage development process, softening behavior at failure, and residual strain at loading–unloading. UMAT user-defined routine was used and single fully integrated solid elements C3D8 in ABAQUS® were used to simulate the three-point bend experimental test. The finite element model including the four-phase behavior up to damage and failure reproduced the crack locations and buckling zones and amplitudes of the omega panel under tension and three points bending. The deviation of finite element results was 9.3% for stiffness, 7.6% for failure force, and 6% for displacement at failure [57].

Zobeiry et al. [58] postulated a differential formulation of viscoelasticity using the generalized Maxwell model. This method was found to represent the viscoelastic behavior of unidirectional fiber-reinforced composites. Using finite difference scheme, the viscoelastic constitutive law was embedded in the FEM software by writing in a user-defined code in FORTRAN. The equation is given as (Equation 18.7):

$$\dot{\underline{\sigma}}_{pi} = P_i\left(\dot{\underline{\varepsilon}}_p - \dot{\underline{\varepsilon}}_{pf}\right) - \frac{1}{\tau_{pi}}\underline{\sigma}_{pi} \qquad (18.7)$$

The FEM code was then applied to the case of the viscoelastic cylinder and stresses were evaluated. The differential form solutions were in close correlation with the integral form and analytical solutions. The thermal stresses were also calculated and differential form was in complete agreement with the analytical solutions showing the efficacy of the proposed method [58].

To investigate the damping behavior of viscoelastic damping structures, a finite element-based optimization method was presented. Using first-order shear deformation theory, the layerwise finite element model was created. Nonstoring genetic algorithm (NSGA-II) was used and implemented in MATLAB®. It was observed that at a higher modal loss factor the modal frequency was lowest. The low elastic modulus of viscoelastic model can be attributed to this finding. To ensure the flexibility of the structure, a viscoelastic damping layer was introduced. The optimal position for introducing the viscoelastic layer was found to be near the middle of the skin plate [59]. Implementing viscoelastic behavior by generalized Maxwell model to pulp fiber, which is in turn idealized as a complex compound bar, a numerical model, was developed. The presence of microfibrils creates anisotropy in the material behavior of the pulp fiber. The large deformations were modeled by considering the finite strain theory. The pulp fiber was simulated in the finite element framework using UMAT subroutine in ABAQUS®. The mesh was developed using linear eight-noded hybrid brick element C3D8RH. The boundary conditions implemented were clamped at one end and tensile stress at the other end. It was observed that with softening due to its viscoelastic nature, the deformation of the fiber increases with time [60]. Yahyaei-Moayyed et al. [61] used the FEM model along with experimental simulation to describe the creep performance of the unidirectional aramid fiber-reinforced polymer sheet (AFRP). Southern yellow pine (SYP) and Douglas-fir (DF) timber beams were strengthened by AFRP to improve their mechanical performance and nonlinear finite element analysis was carried out on these timber beams. Creep was modeled using Norton's power law. C3D8 brick element in ABAQUS® was used to model the 3D nonlinear viscoelastic behavior of the beam in short- and long-term simulations [61]. The creep behavior depicted by power law is shown in Figure 18.9.

Modeling 3D viscoelastic-based finite element simulation of plant fibers (here hemp fiber), the effects of their structure, morphology, and organization on various mechanical parameters were determined. Tensile test simulations were performed by taking the cross-section as elliptical. The strong effect of cross-section nonlinearity, which in this case is the ellipticity, on the tensile behavior of the fiber was exhibited (shown in Figure 18.10). Apart from the ellipticity, other physical and structural parameters such as viscoelasticity of the fiber wall and microfibril angle of cellulose also influence the nonlinear response and are in fact coupled to each other. These are

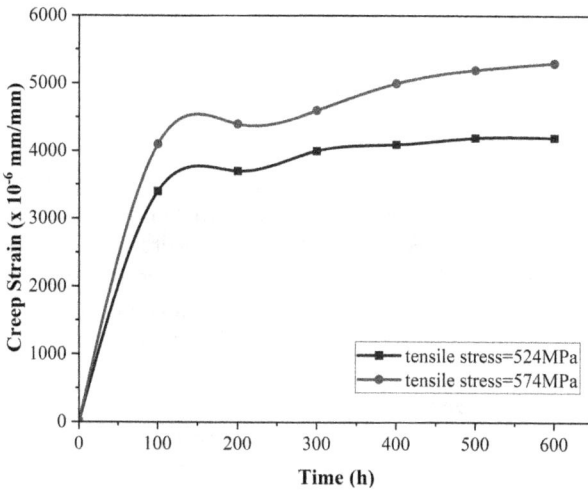

FIGURE 18.9   Creep behavior predicted by the power-law model for AFRP [61].

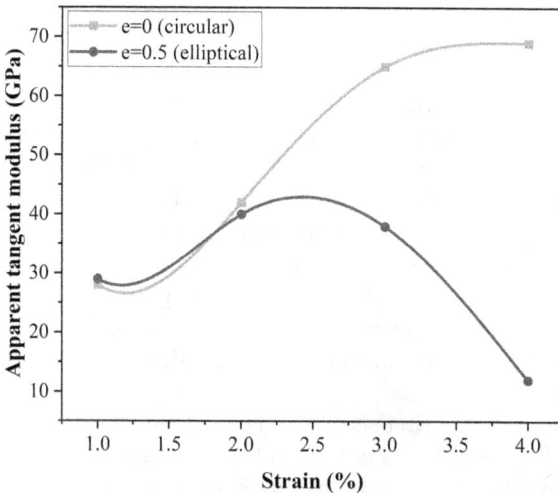

FIGURE 18.10   Apparent tangent modulus versus strain predicted by viscoelastic constitutive law for the circular and elliptical cross-section of fiber [62].

some of the factors that must be considered while modeling biocomposites to ensure accurate estimation of properties [62].

18.3.2   ANALYTICAL MODELS

Cherizol et al. [36] presented a detailed overview of all the analytical models in literature used to study the rheological behavior and viscoelastic characteristics of natural fiber-reinforced thermoplastic polymer biocomposites. The Phan-Thien-Tanner,

Giesekus-Leonov model, the White-Metzner model, the Oldroyd-B model, the K-BKZ model, the Upper Convected Maxwell (UCM) model, and the Giesekus-Leonov model models were some of the models that were investigated. These models were found to be most suitable for high shear rate type deformation in hybrid and green biomaterials. Some other models such as the Oldroyd-B model, Giesekus model, and Phan-Thien Tanner model were found to be more suitable for predicting shear thickening and thinning characteristics. In addition, the Giesekus, K-BKZ, Phan-Thien Tanner, Oldroyd-B, and UCM models were found to predict the transient first normal stress coefficient and steady shear viscosity better than the White-Metzner model [36]. To develop and optimize the biocomposites of Raffia vinifera fibers, an experimental study to determine the rheological model suited for these biocomposites was done first. Later, the nonlinear viscoelastic model of Schapery was implemented to analyze the effect on various rheological parameters with variation in stress for creep-recovery behavior. The Burger model consisting of eight elements was better suited to describe that the experimental points and the obtained parameters vary according to the stress. It was also found to better describe the creep-recovery behavior of Raffia vinifera fibers and viscoelastic parameters were obtained for the model. For cyclic loading, the energy dissipated in the first and second cycles was important and becomes negligible for further cycles. The deformation curves were found to achieve stability in the interval of 35–50 cycles [63]. Wu et al. [64] used a linear viscoelastic shear lag model (SLM) to simulate the mechanical behavior of biocomposites when being subjected to a triangular pulse load. This viscoelastic SLM is an extension to the existing SLM by just substituting the elastic or elastoplastic behavior with viscoelastic nature. The Kelvin and Maxwell model were also analytically and numerically solved, respectively, to determine the stress transmission behavior between the neighboring reinforcements in the biocomposites. The effect of fiber overlap length on the stress transfer was studied and it was found out that for longer fibril overlap length the relative sliding between the fibrils is smaller. The relative sliding for the Kelvin model was found to be the smallest and normal stress at the end was the largest, whereas for the Maxwell model this deformation was the largest [64].

Ji et al. [65] used the tension-shear chain (TSC) model to describe the mechanical properties of nanostructures of natural biological materials. Natural biological materials such as bone, teeth, and nacre are nanocomposites of protein and minerals with superior strength. The TSC model characterizes the nanostructure as 1D series of springs depicting the proteins and minerals. It was found that the high stiffness value of biomaterials was evidently due to the staggered alignment and large aspect ratio of the mineral platelets. Thereby using the TSC model, it can be concluded that to optimize the strength of biomaterials it is important to optimize the tensile strength of these mineral crystals. These mineral crystals achieve the strength of atomic bonds upon reaching the nano length scale due to the absence of sensitivity to flaws. By optimizing the mineral crystals' tensile strength, the fracture toughness of biocomposites is improved. This is so because a large amount of fracture energy can now be dissipated via shear in the proteins. Further, by incorporating the viscoelastic nature of proteins, the toughness of the biocomposites was further improved [65].

Giesekus model along with certain additional parameters was used to model the rheological nature of viscoelastic high-yield pulp (HYP) reinforced nylon11 (Ployamide11) (PA11) biocomposites (HYP/PA11). This model was derived from the Maxwell element and considers the shear stress, relaxation time behavior, and viscosity of the polymer. Here, not only the polymer but fiber and interphase phases were also included in the viscoelastic analysis. Rheological characteristics of the biocomposite such as the shear rate and viscosity affect the development of resultant microstructures, which then leads to changes in the mechanical characteristics and behavior of the composite as a whole. It was shown that upon increasing shear rate, the apparent viscosity of HYP/PA11 decreases. This is the shear thinning behavior and has been validated by both analytical and experimental results as depicted in Figure 18.11. The rheological behavior of HYP/PA11 using an analytical nonlinear model was experimentally validated for temperature of 200°C and shear rate of 5,000 per second. Strong correlation between theoretical and experimental results was found [66].

Hanipah et al. [67] modeled the nonlinear mechanical behavior of the oil palm mesocarp fibre (OPMF) using a large strain viscoelastic model. Due to the presence of silica bodies on the surface and cellular structure within the cross-section, modeling for these particular fibers becomes a difficult task. Considering a large strain energy potential, Van der Waals's large strain model given by Equation 18.8 the stress was defined. Coupled with stress softening functions, the Prony series consisting of a series of Maxwell springs and dashpot was used.

$$\sigma_o(\lambda) = \mu\lambda\left(\lambda - \frac{1}{\lambda^2}\right)\left[\frac{\sqrt{\lambda_m^2 - 3}}{\sqrt{\lambda_m^2 - 3} - \sqrt{\lambda^2 + 2\lambda^{-1} - 3}} - a\sqrt{\frac{\lambda^2 + 2\lambda^{-1} - 3}{2}}\right] \quad (18.8)$$

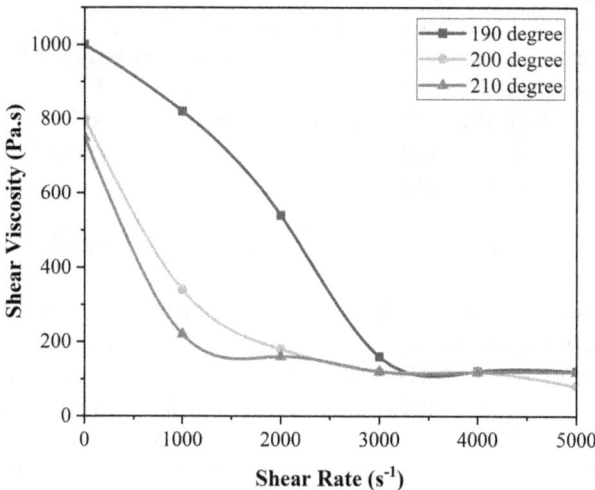

FIGURE 18.11   Variation of shear viscosity with shear rate at different temperatures [66].

Here, λ_m is the locking stretch constant which defines the stretch caused due to unfolding and lock of microstructure. a is the global interaction parameter and μ is the instantaneous initial shear modulus. The model was found to agree with the tensile test for stress relaxation and loading–unloading curve. Moreover, to model the biocomposites at the macro level, i.e., around 0.18 mm, the single fiber model gave good results. Similarly, for modeling at micro-scale, i.e., around 50 μm, the mechanical behavior of biocomposite was found to be significantly improved by the use of silica body [67]. Tazi et al. [68] analyzed the effect of wood filler on the visco-elastic and thermophysical behavior of the WPCs. The Lodge constitutive model was applied for predicting the viscoelastic deformation. The material's thermophysical properties were characterized using regression analysis. It was found that the loss modulus of the wood composite changed from around 0.1 MPa for pure HDPE to around 1 MPa on the addition of wood filler by 50% weight fraction. The heat capacity was also reduced from 20 J/°C to around 5 J/°C for the same content of wood filler addition [68]. Using constitutive nonlinear elasticity relations, the mechanical behavior of three types of NFCs-flax, jute and cotton, in polylactic acid (PLA) matrix was predicted. These bio-based composites are characterized by nonlinear elasticity, viscous effects, and plastic strains before failure. The analytical model predicted a viscoelastic relaxation final stress as 54 MPa for flax/PLA, 26 MPa for cotton/PLA, and 34 MPa for Jute/PLA composite. On validating with the experimental model, the deviation was found to be in the range of 2.6% to 5% [69].

Kontou et al. [70] studied two types of biodegradable polymer composites which were Ecovio®reinforced by 30 wt.% wood-flour, and flax/starch. Theoretically modeling the nonlinear viscoelastic behavior, the stress relaxation, creep-recovery and monotonic loading, stress relaxation, and creep-recovery cases were studied. The analytical model assumed the loading behavior to follow the viscoelastic path and unloading to follow the viscoplasticity. The viscoelastic and elastic parameters for the model were evaluated. The difference in the behavior of the material for the recovery stage suggested that these materials are highly sensitive to the loading sign and hence are structurally unstable. The behavioral change is exerted by the fact that the viscoelastic parameter for the recovery stage differed by 1.5 to 2 times the parameter for the loading stage [70].

Using the rheological equation of Klein for viscosity description, the extrusion modeling of the wood polymer composite made up of polypropylene matrix was done. The Klein equation is given by Equation 18.9:

$$\ln \eta = a_o + a_1 \ln \dot{\gamma} + a_{11} ln^2 \dot{\gamma} + a_{12} \ln \dot{\gamma} T + a_2 T + a_{22} T^2 \qquad (18.9)$$

Here, T is the temperature, η is the viscosity, $\dot{\gamma}$ is the shear rate, and model parameters are the terms a_o, a_1, a_{11}, a_2, a_{12}, a_{22}. Wood polymer composites are materials with pseudoplastic and viscoelastic properties. To model the slip phenomena during flow, the rheological model of Klein and power law was used. A novel global computer model of WPC extrusion with slip effects has been developed, and process simulations were performed to compute the extrusion parameters (throughput, power consumption, pressure, temperature, etc.), and to study the effect of the material rheological characteristics on the process flow. Simulations were validated experimentally and were discussed concerning both rheological and process modeling aspects. It was concluded that the location of the operating point of the extrusion process, which defines the thermo-mechanical process conditions, is fundamentally dependent on the characteristics of the rheological material, including slip effects [71].

Soft materials such as hydrogels, sponges, and xerogels are multiphase systems consisting of a three-dimensional polymeric architecture with lots of fluid in them. This multiphase composition enables them to undergo large deformation before undergoing failure. Having tremendous biological applications, the modeling of the nonlinear mechanical behavior of these materials becomes crucial. The presence of hierarchical structure and multiple length scales results in nonlinear mechanical characteristics for soft materials. This nonlinearity was addressed by implementing a generic nonlinear viscoelastic model. To simulate the material nonlinearity, a fractal derivative viscoelastic model considering a fractal Maxwell model in parallel with a nonlinear spring was used. The fractal dashpot in combination with the nonlinear spring accounts for the power-law time-dependent rheology of generic soft materials. These two different aspects in the form of nonlinear stiffness and non-Newtonian rheology account for the mechanics of most soft materials. The stress–strain behavior of xerogel, when reinforced with nanocellulose fibers and crystals, is shown in Figure 18.12 [72].

Stochioiuet al. used two rheological models, Zener and Burger to analyze the creep behavior and rheological parameters of flax reinforced in epoxy biocomposites. The fiber orientation for the case was $\pm 45°$. The Zener model consists of spring and dashpot arranged in parallel and a second spring mounted in the series of the two. The Burger model is a dashpot mounted in series with the Zener model. The material parameters identified by Zener and Burger model had an error of 3.6% and 1.8%,

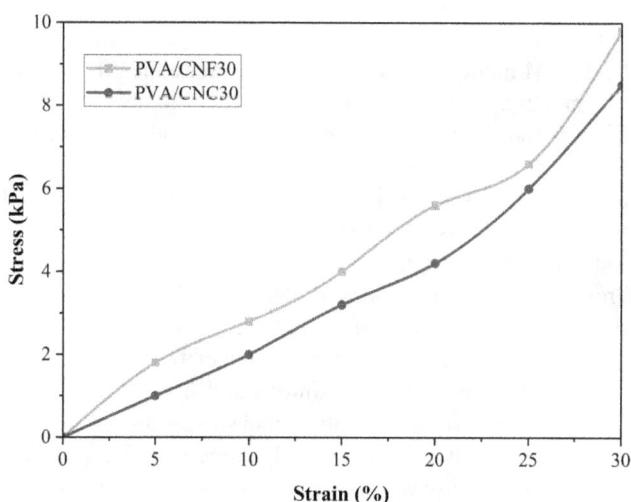

FIGURE 18.12 Stress–strain behavior for PVA xerogels reinforced with nanocellulose fibers (CNF) and crystals (CNC) [72].

respectively. In addition, the creep curve was obtained at different stress for both models. The Burger model exhibited lower error for the case of 5 MPa stress than the Zener model. Both the models were successfully validated in the linear domain [73]. Marklund et al. [74] carried out the tensile test of flax/polypropylene biocomposites to show the nonlinear loading behavior and hysteresis upon unloading. To analyze the creep behavior, the nonlinear model developed by Schapery was used. Further Prony series, power law, and modified power law were used to determine the visco-elastic compliance in Schapery's model. Zapas model was implemented to approximate the viscoelastic response. To compare the models, creep strain at a stress of 24 MPa evaluated indicated that strain for the power-law model was 1.5%, for Prony series also it was 1.5% and for modified power law, the strain was around 1.2% at the period of 1.5 hours [74].

18.3.3 MICROMECHANICAL MODELS

Mourid et al. [75] used analytical homogenization to characterize the viscoelastic behavior of textile composites. Using Mori-Tanaka (MT) and Self-Consistent (SC) models, the framework of elastic homogenization of textile composites was extended to viscoelastic composites with the help of the correspondence principle based on Laplace–Carson transforms. In addition, finite element simulations of 3D models of woven and braided fabrics using 20 node brick elements were performed to obtain the viscoelastic behavior. FEM results were compared with the analytical predictions. It was shown that the homogenization model based on the SC scheme did not predict the properties accurately. In comparison, the MT methods predicted visco-elastic nature more efficiently. By adjusting the aspect ratio of fibers, the MT method's accuracy can be improved even further [75].

Strand-based wood and bamboo composites were analyzed using the microme-chanical approach called morphological approach (MA), which is useful for model-ing the viscoelastic nature of such biocomposites. In addition, numerical analysis using a full-field FEM model was developed to estimate the effectiveness and accu-racy of the MA method. Considering a parallel strand lumber (PSL) beam, an ide-alized unit cell was modeled and micromechanical equations were used to predict elastic and viscoelastic material parameters. It was found that the MA scheme was more suited to predict the viscoelastic nature of biocomposites compared to the clas-sical micromechanical models. MA approach proves to be an efficient and effective multiscale modeling tool that can be used for structural analysis of biocomposites made of different types and orientations of bamboo and wood strands. The effective properties calculated by MA and FEM models are shown in Figure 18.13 [76].

Wu et al. [77] used the shear lag approach under the viscoelastic Kelvin-Voigt to simulate dynamic tensile loading conditions and find out the micromechanical behavior of the tendon. Deformation of fibrils and stress transfer that occurs between these fibrils and interfibrillar matrix under dynamic stretching of the tendon was determined. The microstructure was modeled by considering fibrils embedded in the interfibrillar matrix. The SLM used a microstructure consisting of a unit cell with two adjacent fibrils. The cross-section depicted each fibril neighboring four fibrils. The fibril volume fraction and overlap length were the two main parameters that

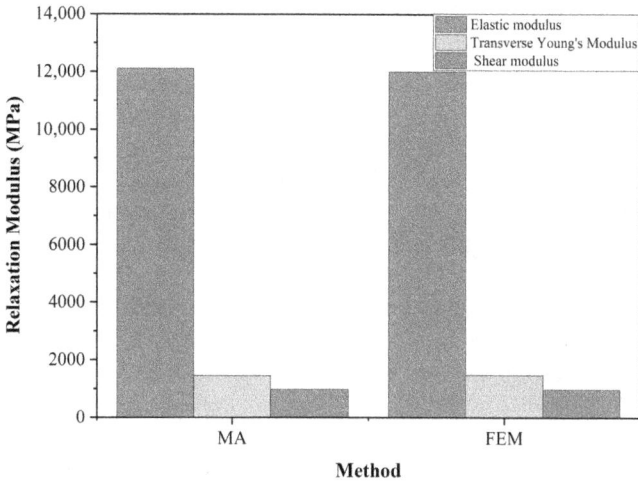

FIGURE 18.13 Various relaxation moduli as predicted by FEM and MA methods [76].

influenced the mechanical properties of tendon and stress transfer characteristics. It was observed that increasing the overlap length by 4 times causes shear stress distribution to reduce from 0.05 to 0.018 MPa. Also, increasing the fibril volume fraction from 0.3 to 0.7 increased Young's modulus from around 40 MPa to 300 MPa for a relative sliding of 180 mm [77].

Hosseini et al. [78] studied flax fiber-reinforced in polyurethane matrix by using the linear viscoelastic model in a micromechanical framework. The mechanical properties of the composite were predicted using the FEM model generated representative volume element (RVE). This RVE was developed if repeating unit cells (RUC) where the RUC was subjected to six loading conditions. This way the characterization of the viscoelastic behavior of the composite was undertaken. The time-history of averaged response was determined in terms of stress and strains. A suitable agreement between experimental and micromechanical models was achieved with the effectiveness of the rate-dependent behavior of flax fibers. For good periodicity in the biocomposites RVE, the proposed model gives accurate and efficient results. Omar et al. [79] conducted FEM-based numerical study along with experimental tests to analyze the micromechanics of oil palm empty fruit bunch fibers reinforced with silica bodies. The RVE was developed in FEM software ABAQUS by considering silica body which was modeled as spherical body and as embedded in the middle of a cube which represented the oil palm fibers. Parallel Rheological Network was the model used for characterizing the viscoelastic material behavior in a finite element framework. This model correctively predicted the stress relaxation and cyclic tensile test of fibers. For fiber with and without silica, the difference in predicted parameters and test parameters was more for the case of shear loading than for tensile or compressive loading. Moreover, the ultrastructure components of oil palm fibers such as cellulose and lignin were found to have more effect on the viscoelastic mechanical behavior than the silica bodies. One can attribute this to the complex interface behavior of these components [79].

Radchenkoet al. [80] studied the creep behavior occurring in the bone tissue by using a mathematically derived model. A microinhomogeneous nonlinear elastic media was considered and instantaneous elastic deformation was studied. Bone tissue being two-phase material, a rheological model similar to the Maxwell model consisting of two structural elements was derived. Both the linear and nonlinear regions of creep were simulated and elastic modulus was determined. A shift in the diagram of elastic strain was observed at creep due to the redistribution of stresses in the structural model. The nonlinear creep behavior of bone tissue is shown in Figure 18.14.

Bedzra et al. [81] generated an RVE of the unidirectional flax fiber/epoxy matrix biocomposites to develop a constitutive model systematically. Subjecting the microstructure to various deformation states, the yield surface and deformation profile for the composite were found out. Developing a viscoelastic–viscoplastic continuum model that encompasses the anisotropic and nonassociative nature of the material, the microstructure was subjected to other loading conditions also. Virtual experiments were carried out for the plain weave flax fiber in epoxy matrix biocomposites using the developed continuum model. The yield surface and deformation behavior of the biocomposites were validated henceforth. The transverse tension relaxation modulus and longitudinal shear relaxation modulus were estimated and found to evolve from 6.1 to 5.84 and 1.85 to 1.81 GPa, respectively, with time [81].

18.3.4 Advanced Computational Tools

Molecular dynamics (MD) tool was used to model the intrinsic viscoelastic mechanical behavior of structural bamboo material, which is a hierarchical biocomposite. This viscoelastic behavior was determined at macro and micro scales and for the individual constituents of the bamboo structure. To simulate this computational

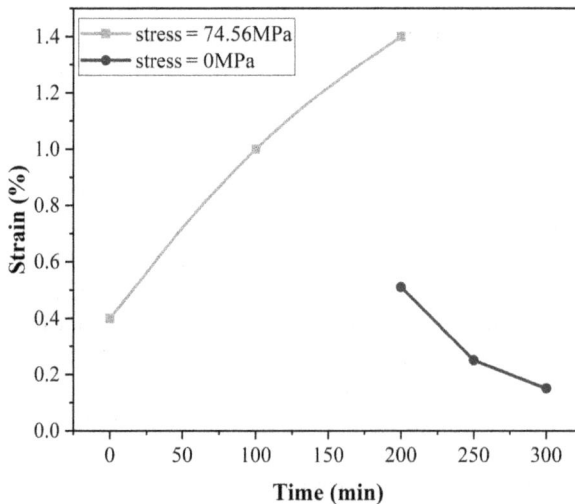

FIGURE 18.14 Strain vs time in nonlinear creep region for bone tissue [80].

creep test was performed using the constant force Steered Molecular Dynamics (SMD) simulations. The macro- and micro-level viscoelastic natures were evaluated using dynamic compression test and nanoindentation tests respectively. The relationship between the microstructure and viscoelastic nature was then developed using the phenomenological model. The key findings were the presence of hemicellulose and lignin matrix greatly affects the microscopic viscoelastic behavior. In addition, increasing the volume fraction of viscous phases causes the viscoelastic damping behavior of bamboo when subjected to dynamic compressive loading [82]. Mlyniec et al. [83] performed atomistic tensile and shear computational tests in the MD framework using the extended AMBER force field. This was done to study the viscoelastic behavior of collagen nanofibrils. The effect of fibril length and loading rate on the failure and viscoelastic behavior of the collagen fibrils was determined. The MD simulations were carried for the case of an aqueous and nonaqueous medium. Intermolecular and intramolecular potentials were used to describe the bonded and nonbonded coarse-grained interactions. It was found out that longer fibril length can cause brittle cracking, whereas ductile behavior was observed for higher loading rates. The viscoelastic behavior was affected by the presence of hydrogen bond networks.

In another study, Mlyniec et al. [84] analyzed the effect of thermal conditions on the viscoelastic behavior of the soft collagenous biomaterials. SMD and Corse-Grained MD simulations using the LAMMPS package were implemented to model the mesoscale collagen microfibril. To simulate the interactions (inter and intramolecular), Lennard-Jones and Coulomb potentials were used. The rheological model for these fibrils was generated by considering the large strain viscoelastic model and stress–strain curves from CG simulation. Ranging from static tensile test to strain rate of $10^9 s^{-1}$, the simulation of the model was carried out and viscoelastic behavior was predicted. The model to simulate viscoelastic behavior consisted of two networks each consisting of one element, where one element comprises of a hyperelastic Arrudae Boyce model, while the other one includes a hyperelastic model with strain rate dependence. The predicted MD results indicated that at the higher temperature the viscoelastic nature is lost at a higher rate (Figure 18.15) [84].

Using a master curve with the Arrhenius model, the relationship between creep behavior and morphology of sisal short fiber-reinforced starch biocomposites was determined. These composites were fabricated using the injection molding technique. The effect of fiber orientation and content on the flexural creep behavior of the biocomposites was studied. Fiber content and temperature effects were also considered, taking into account various methods and equations. Creep power law was used for short time durations and the Arrhenius model was used for longer durations. The results indicated an increase in longitudinal modulus with increasing fiber content. Elastic modulus increased from 2 to 4 GPa on increasing fiber content from 0 to 15 wt.%. The flexural modulus decreased from 7.9 to 4.4 GPa for a similar increase in fiber content [85]. Tirella et al. [86] investigated the viscoelastic behavior of highly hydrated biological materials. Due to hydration, these biological materials become extremely soft and experience pre-stress conditions. A new method epsilon dot method was devised to determine the viscoelastic parameters for such soft materials. This model was then further implemented to determine the viscoelastic behavior of

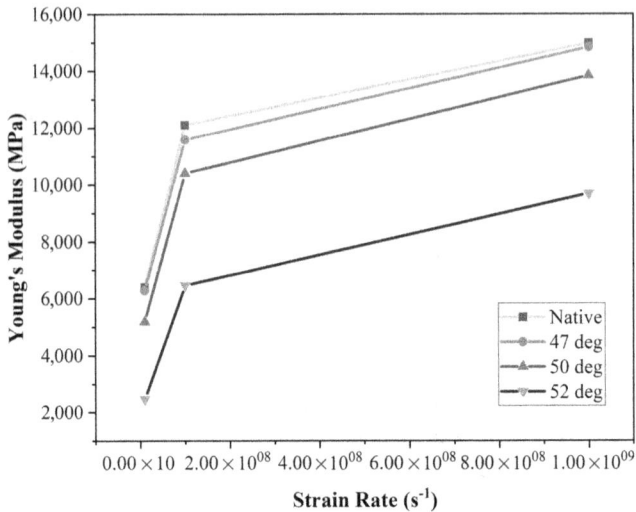

FIGURE 18.15   MD results for Young's modulus degradation with temperature [84].

the porcine liver. Lumped parameter evaluation using Laplace Transform was carried out for the standard linear solid (SLS) and generalized Maxwell models.

Bassir et al. [87] analyzed the nonlinear behavior of laminate composites by undertaking a comparatively newer approach using artificial neural networks (ANNs). This method combined the ANN with a genetic algorithm that uses parallel selection (GAPS) to solve a particular problem at hand. It was basically an identification problem in which the various mechanical behaviors of the composite laminate were considered. The composite laminate shells were studied under elasticity, viscoelasticity, damage, and plasticity conditions, generating a nonlinear model for each case. This hybrid approach of using both GAPS and ANN was used to make use of both the rapid convergence of ANN and the robustness of the GAPS. GAPS was used to provide the required database which was then used by the ANN to carry out the identification process. The cost function, which gives the difference between the predicted and actual value of the material parameter, was related to the mechanical parameters of the composite under consideration using ANN. Using this hybrid approach, the time-consuming phase of GAPS was avoided, and also a good approximation of the mechanical parameters was obtained. The predicted results were in high correlation with the experimental values [87].

18.3.5   BIOMEDICAL APPLICATIONS

Zhang et al. [88] studied the mechanical behavior of blood vessels using viscoelastic modeling. It has been hypothesized that the viscoelasticity of blood vessels causes a reduction in stress and strain induced in the vessel wall. This reduction can cause changes in the fatigue life of the blood vessel. Hence, the impact of these reductions on fatigue life was studied. A quasi-linear viscoelastic model with normalized stress function was adopted The pulsatile behavior of the real blood vessel was simulated by considering wave pulse in Fourier series as suggested by Humphrey [89]. Similar boundary conditions were considered for both the elastic and viscoelastic models and hysteresis behavior of the artery walls was determined. The results indicate the reduction in vibration of artery wall soon considering its viscoelastic nature [88].

Ahmadzadeh et al. [90] carried out a mathematical model-based viscoelastic study of tau proteins which is useful to examine in the case of an axonal stretch injury. The mechanism of breaking of microtubules has been analyzed showing that the viscoelastic nature of tau proteins causes the stretching of microtubules over to their neighbors, leading to their breaking. A microstructural model to computationally study the interaction between the tau proteins and microtubules when subjected to mechanical loading was presented. The axon microstructure consisted of microtubules cross-linked by tau proteins, which were viscoelastically represented using a spring and a dashpot. The viscoelastic SLM was considered and results suggested breaking of microtubules at higher strain rates. At lower strain rates, the microtubules just slide over each other without any inherent damage. A critical strain rate of 22–44 s^{-1} was found out for breaking the microtubules. This was in good agreement with the experimental results. It was also found out that longer microtubules fail faster than shorter ones wherein the failure occurred at the middle of the microtubule span [90]. Shamloo et al. [91] used an SLS model to implement the viscoelasticity of axons in the human brain. The SLS unit was used to connect the discrete mass of axon microtubules to the neighboring masses. The dynamic response under suddenly applied end loads was recorded. The effect of magnitude and rate of the applied force was found out. For a particular rate, increasing the force magnitude caused the rupture of microtubule bundles faster. Failure was found to occur when the non-dimensionalized force was 0.25 (shown in Figure 18.16). This prediction of axonal rupture is essential to study critical brain damage and injury caused by sudden accidents.

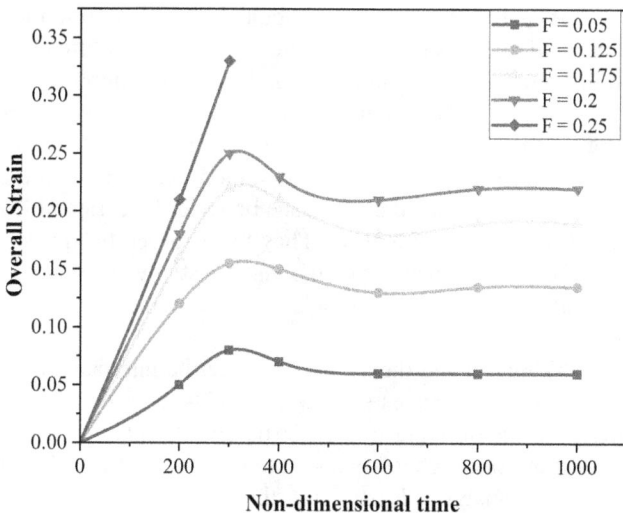

FIGURE 18.16 Viscoelastic response of axon microtubules for different loads [91].

Using the GA optimization process combined with the micromechanical FEM technique, the viscoelastic parameters for axon and extracellular matrix (ECM) in the white matter of the brain were determined. To model the microstructure of the brain tissue, a square periodic RVE was used. To this, periodic boundary conditions (PBC) were applied to ensure the replication of periodicity for neighboring elements. In addition, the physical boundary conditions applied constrained the central node of the RVE for three translational degrees of freedom. A displacement equivalent loading of 1.5% strain was also prescribed. Linear viscoelastic approximation was undertaken and optimization was carried out. The FEA-based approach was successful in optimizing the procedure in 60 iterations. The results predicted that axonal initial elastic modulus was thrice than the elastic modulus of ECM, the predicted numerical value being12.86 kPa. Long-term elastic modulus of ECM was 1.03 kPa and for axon, it was 3.7 kPa. A strong correlation between the FEA and experimental results was observed showing the reliability of the numerical model. A 0.952 concordance correlation coefficient suggested the exactness and efficiency of the developed model. This method can be further applied to analyze the micromechanical feature of the brain tissue and can be used to further study the diffuse axonal injures (DAIs) [92].

Wu et al. [93] used the viscoelastic SLM based on the Kelvin-Voigt rheological approach. They studied the axonal response when subjected to tensile load that varied with time. Mechanical behavior of axons composed of microtubule and tau protein was described by using closed-form analytical equations. The mechanism of stress transfer, failure, and relative sliding between the tau protein and microtubule was investigated using these equations. The major findings were that the failure of axon occurred due to simultaneous failure of microtubule and tau protein in the sliding rupture failure regime. The microtubule-tau protein interface length and stress state were found to significantly affect the response of axon to mechanical loading. In addition, failure mechanisms differed for various loading conditions. Moreover, larger deformations caused the short microtubules to separate from the microtubule bundles, whereas a high-stress rate was the main cause for rupture of long microtubules. The failure predicted was multi-level indicating the importance of a hierarchical study. In another model to characterize the brain tissue, the theory of porous media was used. Comellas et al. [94] implemented finite viscoelasticity along with numerical models of FEM. Their model combined nonlinear poroelasticity with finite viscoelasticity and simulated brain tissue behavior for different loading conditions of shear, compression, and tension. They were successful in representing the interconnectivity between poroelasticity and viscoelasticity and hence signify that a combination of both models is necessary to accurately predict the hysteresis behavior in human brain tissue.

Shelef et al. [95] highlighted the need to analyze the interfacial dynamic modulus of biocomposites. In organisms such as arachnids, insects, and even birds and fishes, the interfacial region lying in their structural morphology can significantly impact the stress transfer, toughness value, and overall mechanical performance of the component. Thus, analytically studying the phenomena to link the loss coefficient and viscoelastic modulus at the interfacial level to the large-scale structure becomes essentially important. It was estimated that a change in the loss coefficient of the interface by 2.5 times could increase the loss coefficient of the composite by around 2 times. Similarly, the change in modulus at the interface by 10 times causes a similar change in the elastic modulus of the biocomposites. More such biomedical applications involve analyzing material nonlinearity such as the study of the human cornea [96] where the presence of an aqueous medium can affect the nonlinear material response.

18.4 FUTURE PROSPECTS

From the wide-scale application of biocomposites and largely nonlinear mechanical behavior exhibited by them, it is evident that viscoelastic modeling offers very crucial information on the performance, behavior, and life of the biocomposites. Till now, mainly analytical models have been used for biocomposites and naturally existing viscoelastic materials. Future studies should try to incorporate such nonlinear material models for damage, impact, and delamination of biocomposites. Attention must also be paid to model the interfacial region and identify its dynamic modulus [95]. This will be beneficial not only from a structural application point of view but the findings can be extended to the biomechanics study of various plant parts, the behavior of mollusks, wasps, and other organisms as well. These interfaces are difficult to model experimentally; hence, numerical modeling will help gather new findings. On a similar line, the need for hierarchical modeling of viscoelastic nature needs to be focused to assess the micro-scale behavior and its impact at the macro scale. Viscoelastic models can be used for capturing the nonlinear and time-dependent characteristics of soft tissues [97]. Their growth and interaction with other biological components such as inorganic molecules can be modeled. For industrial applications, viscoelastic modeling finds scope in studying various shape memory polymers, functionally graded materials, and hydrogels [98]. With the development of nanobiomaterials [99,100], drug delivery systems [101,102], biocomposites damping systems [103], the computational capacity of viscoelastic models needs to be improved to tackle the complexity of such systems. Refining existing models with high-end data analysis tools, use of ML and ANN, image analysis appear to be the future set of goals for an efficient and accurate depiction of viscoelastic phenomena in biocomposites.

18.5 CONCLUSIONS

The study presents an overall holistic view of the recent trends in numerical modeling of the viscoelastic nature of biocomposites. Giving a brief overview of some widely used theories behind the models, the application of computational tools such as FEM, micromechanics, MD has been highlighted. It has also been shown that how the field of biomedical engineering and biomechanics is benefiting from the viscoelastic modeling tools. Biocomposites being multifarious in their origin, composition and behavior present a complex challenge to the numerical modeling framework, which is now being proficiently handled by powerful mathematical techniques. The numerical models are successful in predicting the nonlinear behavior of these materials accurately and are found to concur with the experimental findings. This gives a boost to the application of biocomposites in vibration and damping systems, shock absorbers, and also acoustic equipment, where their viscoelasticity is helpful. Further work can be extended to include damage and fracture behavior in the viscoelastic framework.

REFERENCES

[1] A.K. Mohanty, M. Misra, G. Hinrichsen, Biofibres, biodegradable polymers and biocomposites: an overview, Macromolecular Materials and Engineering, 276–277(1) (2000) 1–24.

[2] O. Faruk, A.K. Bledzki, H. Fink, M. Sain, Biocomposites reinforced with natural fibers: 2000–2010, Progress in Polymer Science, 37(11) (2012) 1552–1596.

[3] M.E. Hoque, Y.L. Chuan, P.M. Meng, 12-Agro-based green biocomposites for packaging applications, In: Naheed Saba, Mohammad Jawaid, Mohamed Thariq (Eds.), Woodhead Publishing Series in Composites Science and Engineering Biopolymers and Biocomposites from Agro-Waste for Packaging Applications, 235–254. Woodhead Publishing, 2021, doi:10.1016/B978-0-12-819953-4.00008-2.

[4] A. Jimenez, R.A. Ruseckaite, Nano-biocomposites for food packaging, Green Energy and Technology (2012), doi:10.1007/978-1-4471-4108-2-15.

[5] H. Park, A study on structural design and analysis of small wind turbine blade with natural fibre(flax) composite, Advanced Composite Materials, 25(2) (2016) 125–142.

[6] S.S. Suhaily, M. Jawaid, H.P.S. Abdul Khalil, A.R. Mohamed, F. Ibrahim, A review of oil palm biocomposites for furniture design and applications: potential and challenges, Bio Resources, 7(3) (2012) 4400–4423.

[7] C. Scarponi, Hemp fiber composites for the design of a Naca cowling for ultra-light aviation, Composites Part B: Engineering, 81 (2015) 53–63.

[8] R. Potluri, N. Chaitanya Krishna, Potential and applications of green composites in industrial space, Materials Today: Proceedings, 22(4) (2020) 2041–2048.

[9] A. Ashori, Wood–plastic composites as promising green-composites for automotive industries!, Bioresource Technology, 99(11) (2008) 4661–4667.

[10] S.C.R. Furtado, A.L. Araújo, A. Silva, C. Alves, A.M.R. Ribeiro, Natural fibre-reinforced composite parts for automotive applications International Journal of Automotive Composites, 1(1) (2014) 18–38.

[11] M. Nagalakshmaiah, S. Afrin, R.P. Malladi, S. Elkoun, M. Robert, M. Ansari, A. Svedberg, Z. Karim, Chapter 9. In: *Biocomposites: Present trends and challenges for the future*, Woodhead Publishing Series in Composites Science and Engineering, Cambridge 2019, Pages 197–215.

[12] K.L. Pickering, M.G.A. Efendy, T.M. Le, A review of recent developments in natural fibre composites and their mechanical performance, Composites Part A: Applied Science and Manufacturing, 83 (2016) 98–112.

[13] T. Gurunathan, S. Mohanty, S.K. Nayak, A review of the recent developments in biocomposites based on natural fibres and their application perspectives, Composites Part A: Applied Science and Manufacturing, 77 (2015) 1–25.

[14] S. Abebayehu, A. Engida, Preparation of biocomposite material with super-hydrophobic surface by reinforcing waste polypropylene with Sisal (Agave sisalana) fibers, *International Journal of Polymer Science*, 2021 (2021) 6642112, doi:10.1155/2021/6642112

[15] A. Singh, S. Halder, A. Kumar, P. Chen, Tannic acid functionalization of bamboo micron fibes: its capability to toughen epoxy based biocomposites, *Materials Chemistry and Physics*, 243 (2020) 122112, doi:10.1016/J.MATCHEMPHYS.2019.122112

[16] Miguel Ángel Hidalgo Salazar, Viscoelastic Performance of Biocomposites, In: Composites from Renewable and Sustainable Materials, Poletto, M (ed.). 2016, Composites from Renewable and Sustainable Materials, IntechOpen, London. 10.5772/62936.

[17] E. Heggset, B. Strand, K. Sundby, S. Simon, G. Chinga-Carrasco, K. Syverud, Viscoelastic properties of nanocellulose based inks for 3D printing and mechanical properties of CNF/alginate biocomposite gels, *Cellulose*, 26(1) (2018) 581–595, doi:10.1007/S10570-018-2142-3

[18] G. George, E.T. Jose, D. Åkesson, M. Skrifvars, E.R. Nagarajan, K. Joseph, Viscoelastic behaviour of novel commingled biocomposites based on polypropylene/jute yarns, *Composites Part A: Applied Science and Manufacturing*, 43(6) (2012) 893–902, doi:10.1016/j.compositesa.2012.01.019

[19] Y.S. Munde, R.B. Ingla, I. Siva, A comprehensive review on the vibration and damping characteristics of the vegetable fibre reinforced composites, *Journal of Reinforced Plastics and Composites*, 38(17) (2019) 822–832, doi:10.1177/0731684419838340

[20] H. Daoud, J.-L. Rebiere, A.E. Mahi, M. Taktak, M. Haddar, Effect of an interleave natural viscoelastic layer on the dynamic behaviour of a bio-based composite, *Advanced Compostite Letters*, 26(6) (2017), doi:10.1177/096369351702600601

[21] S. Kumar, D. Zindani, S. Bhowmik, Investigation of mechanical and viscoelastic properties of flax- and ramie-reinforced green composites for orthopedic implants, *Journal of Materials Engineering and Performance*, 29 (2020) 3161–3171, doi:10.1007/s11665-020-04845-3

[22] A.E. Mahi, H. Daoud, J.-L. Rebiere, I. Gimenez, M. Taktak, M. Haddar, Damage mechanisms characterization of flax fibers–reinforced composites with interleaved natural viscoelastic layer using acoustic emission analysis, *Journal of Composite Materials*, 53(18) (2019) 2623–2637, doi:10.1177/0021998319836236

[23] B. Bayerlein, L. Bertinetti, B. Bar-On, H. Blumtritt, P. Fratzl, I. Zlotnikov, Inherent Role of Water in Damage Tolerance of the Prismatic Mineral–Organic Biocomposite in the Shell of Pinna Nobilis, *Advanced Functional Materails*, 26 (2016) 3663–3669, doi:10.1002/adfm.201600104

[24] Naoki Sasaki, Viscoelastic Properties of Biological Materials, In: Vicente, J (ed.). 2012, Viscoelasticity - From Theory to Biological Applications, IntechOpen, London. 10.5772/3188.

[25] K. Okumura, Strength and toughness of biocomposites consisting of soft and hard elements: a few fundamental models, *MRS Bulletin*, 40(4) (2015) 333–339, doi:10.1557/MRS.2015.66

[26] J.D. Ferry, *Viscoelastic Properties of Polymers*, 4th edition, 1–33. New York: John Wiley & Sons Inc, 1980.

[27] N.Y. Tuna, Finlayson, *Journal of Rheology*, 2879 (1984) 93.

[28] M.J. Crochet, V. Legat, The consistent streamline-upwind/Petrov-Galerkin method for viscoelastic flow revisited, *Journal of Non-Newtonian Fluid Mechanics*, 42 (1992) 283–299. doi:10.1016/0377-0257(92)87014-3

[29] J. Dealy, K. Wissbrun, *Melt Rheology and Its Role in Plastics Processing: Theory and Applications*. Kluwer Academic Publishers, Dordrecht, 1999.

[30] D. Doraiswamy, *The Origins of Rheology*. DuPont iTechnologies, Wilmington, 1988.

[31] R.B. Bird, R.C. Armstrong, O. Hassager, Dynamics of Polymeric Liquids, Fluid Mechanics, vol. 1. Wiley, New York, 1987.

[32] C.L. Tucker III, S.G. Advani, Processing of short fiber system. In: *Flow and Rheology in Polymer Com-posites Manufacturing*, 147–202. Elsevier, Amsterdam, 1994.

[33] T. Grafe, K. Graham, Polymeric Nanofibers and nanofiber webs: a new class of nonwovens, International Nonwovens Journal, 12 (2003) 51–55.

[34] C. Zhou, S. Kumar, Thermal instabilities in spinning of viscoelastic fibers, Journal of Non-Newtonian Fluid Mechanics, 165 (2010) 879–891, doi:10.1016/j.jnnfm.2010.04.009

[35] Hiroshi Murata, Rheology - Theory and Application to Biomaterials, In: Souza, A (ed.). 2012, Polymerization, IntechOpen, London. 10.5772/2750.

[36] R. Cherizol, M. Sain, J. Tjong, Review of non-newtonian mathematical models for rheological characteristics of viscoelastic composites, Green and Sustainable Chemistry, 5(1) (2015) 6–14, doi:10.4236/GSC.2015.51002

[37] Viscoelasticity-Wikipedia, https://en.wikipedia.org/wiki/Viscoelasticity

[38] M.M. Denn, *Polymer Melt Processing: Foundations in Fluid Mechanics and Heat Transfer.* Cambridge Univer-sity Press, New York, 2008. doi:10.1017/CBO9780511 813177

[39] H.M. Krutka, R.L. Shambaugh, D. Papavassiliou, Effects of the polymer fiber on the flow field from a slot melt blowing die, Industrial & Engineering Chemistry Research, 47 (2008) 935–945, doi:10.1021/ie070871i

[40] H. Yaghoobi, A. Fereidoon, Evaluation of viscoelastic, thermal, morphological, and biodegradation properties of polypropylene nano-biocomposites using natural fiber and multi-walled carbon nanotubes, Polymer Composites, 39 (2018) E592–E600, doi:10.1002/PC.24750

[41] K. Haghighi, L. Segerlind, Failure of biomaterials subjected to temperature and moisture gradients using the finite element method: I – thermo-hydro viscoelasticity, Transactions of the ASAE, 31(3) (1988) 930–937, doi:10.13031/2013.30802

[42] R. Ewoldt, A. Hosoi, G. McKinley, Nonlinear viscoelastic biomaterials: meaningful characterization and engineering inspiration, Integrative and Comparative Biology, 49(1) (2009) 40–50, doi:10.1093/ICB/ICP010

[43] S.Vahid, V. Burattini, S. Afshinjavid, A. Dashtkar, Comparison of rheological behaviour of bio-based and synthetic epoxy resins for making ecocomposites, Fluids, 6(1) (2021) 38, doi:10.3390/FLUIDS6010038

[44] W. Zhang, A. Capilnasiu, D. Nordsletten, Comparative analysis of nonlinear viscoelastic models across common biomechanical experiments, Journal of Elasticity, 2021 (2021) 1–36, doi:10.1007/S10659-021-09827-7

[45] M.S. Sukiman, F. Erchiqui, T. Kanit, A. Imad, Design and numerical modeling of the thermoforming process of a WPC based formwork structure, Materials Today Communications, 22 (2020) 100805, doi:10.1016/j.mtcomm.2019.100805

[46] J. Wang, X. Peng, Z. Huang, H. Zhou, A temperature-dependent 3D anisotropic visco-hyperelastic constitutive model for jute woven fabric reinforced poly (butylene succinate) biocomposite in thermoforming, Composites Part B, 208 (2021) 108584, doi:10.1016/j.compositesb.2020.108584

[47] W. Zhang, A. Capilnasiu, G. Sommer, G.A. Holzapfel, D.A. Nordsletten, An efficient and accurate method for modeling nonlinear fractional viscoelastic biomaterials, Computer Methods in Applied Mechanics and Engineering, 362 (2020) 112834, doi:10.1016/j.cma.2020.112834

[48] A.T. Talib, C.C. Jie, M.A.P. Mohammed, A.S. Baharuddin, M.N. Mokhtar, M. Wakisaka, On the nonlinear viscoelastic behaviour of fresh and dried oil palm mesocarp fibres, Biosystems Engineering, 186 (2019) 307–322, doi:10.1016/j.biosystemseng. 2019.08.010

[49] H. Daoud, A.E. Mahi, J.-L. Rebière, M. Taktak, M. Haddar, Characterization of the vibrational behaviour of flax fibre reinforced composites with an interleaved natural viscoelastic layer, Applied Acoustics, 128 (2017) 23–31, doi:10.1016/j.apacoust.2016.12.005

[50] S. Guessasma, M. Sehaki, D. Lourdin, A. Bourmaud, Viscoelasticity properties of biopolymer composite materials determined using finite element calculation and nanoindentation, Computational Materials Science, 44 (2008) 371–377, doi:10.1016/j. commatsci.2008.03.038

[51] F. Fernandes, R. Sousa, M. Ptak, G. Migueis, Helmet design based on the optimization of biocomposite energy-absorbing liners under multi-impact loading, Applied Sciences, 9(4) (2019) 735–761, doi:10.3390/app9040735

[52] F. Trivaudey, V. Placet, V. Guicheret-Retel, M.L. Boubakar, Nonlinear tensile behaviour of elementary hemp fibres. Part II: modelling using an anisotropic viscoelastic constitutive law in a material rotating frame, Composites: Part A, 68 (2015) 346–355, doi:10.1016/j.compositesa.2014.10.020

[53] M.M. Hassani, F.K. Wittel, S. Hering, H.J. Herrmann, Rheological model for wood, Computer Methods in Applied Mechanics and Engineering, 283 (2015) 1032–1060, doi:10.1016/j.cma.2014.10.031

[54] E. Vidal-Sallé, P. Chassagne, Constitutive equations for orthotropic nonlinear viscoelastic behaviour using a generalized Maxwell model application to wood material, Mechanics of Time Dependent Materials, 11 (2007) 127–142, doi:10.1007/s11043-007-9037-2

[55] X. Lv, X. Hao, R. Ou, T. Liu, C. Guo, Q. Wang, X. Yi, L. Sun, Rheological properties of wood–plastic composites by 3D numerical simulations: different components, Forests, 12(4) (2021) 417, doi:10.3390/F12040417

[56] I. Tagiltsev, A. Shutov, Geometrically nonlinear modelling of pre-stressed viscoelastic fibre-reinforced composites with application to arteries, Biomechanics and modelling in Mechanobiology, 20(1) (2020) 323–337, doi:10.1007/S10237-020-01388-3

[57] F. Lachaud, M. Boutin, C. Espinosa, D. Hardy, Failure prediction of a new sandwich panels based on flax fibres reinforced epoxy bio-composites, Composite Structures, 257 (2021) 113361, doi:10.1016/j.compstruct.2020.113361

[58] N. Zobeiry, S. Malek, R. Vaziri, A. Poorsartip, A differential approach to finite element modelling of isotropic and transversely isotropic viscoelastic materials, Mechanics of Materials, 97 (2016) 76–91, doi:10.1016/J.MECHMAT.2016.02.013

[59] C. Xu, S. Lin, Y. Yang, Optimal design of viscoelastic damping structures using layerwise finite element analysis and multi-objective genetic algorithm, Computers & Structures, 157 (2015) 1–8, doi:10.1016/J.COMPSTRUC.2015.05.005

[60] T. Seidlhofer, C. Czibula, C. Teichert, C. Payerl, U. Hirn, M. Ulz, A minimal continuum representation of a transverse isotropic viscoelastic pulp fibre based on micromechanical measurements, Mechanics of Materials, 135 (2019) 149–161, doi:10.1016/J.MECHMAT.2019.04.012

[61] M. Yahyaei-Moayyed, F. Taheri, Experimental and computational investigations into creep response of AFRP reinforced timber beams, Composite Structures, 93(2) (2011) 616–628, doi:10.1016/J.COMPSTRUCT.2010.08.017

[62] A. Masto, F. Trivaudey, V. Guicheret-Retel, V. Placet, L. Boubakar, Nonlinear tensile behaviour of elementary hemp fibres: a numerical investigation of the relationships between 3D geometry and tensile behaviour, Journal of Materials Science, 52, doi:10.1007/s10853-017-0896-x

[63] N.R. Sikame Tagne, D. Ndapeu, D. Nkemaja, G. Tchemou, D. Fokwa, W. Huisken, E. Njeugna, M. Fogue, J.-Y. Drean, O. Harzallah, Study of the viscoelastic behaviour of the Raffia vinifera fibres, Industrial Crops and Products, 124 (2018) 572–581, doi:10.1016/j.indcrop.2018.07.077

[64] J. Wu, H. Yuan, L.-Y Li, Effect of viscoelasticity on interfacial stress transfer mechanism in the biocomposites: a theoretical study of viscoelastic shear lag model, Composites Part B, 164 (2019) 297–308, doi:10.1016/j.compositesb.2018.11.086

[65] B. Ji, H. Gao, Mechanical properties of nanostructure of biological materials, Journal of the Mechanics and Physics of Solids, 52 (2004) 1963–1990, doi:10.1016/j.jmps.2004.03.006

[66] R. Cherizol, M. Sain, J. Tjong, Modeling the rheological characteristics of flexible high-yield pulp-fibre-reinforced bio-based nylon 11 bio-composite, Journal of Encapsulation and Adsorption Sciences, 5 (2015) 1–10, doi:10.4236/jeas.2015.51001

[67] S.H. Hanipah, M.A.P. Mohammed, A.S. Baharuddin, Non-linear mechanical behaviour and bio-composite modelling of oil palm mesocarp fibres, Composite Interfaces, 23(1) (2016) 37–49, doi:10.1080/09276440.2016.1091681

[68] M. Tazi, M.S. Sukiman, F. Erchiqui, A. Imad, T. Kanit, Effect of wood fillers on the viscoelastic and thermophysical properties of HDPE-wood composite, International Journal of Polymer Science, 2016 (2016) 032525, doi:10.1155/2016/9032525

[69] A. Rubio-López, T. Hoang, C. Santiuste, Constitutive model to predict the viscoplastic behaviour of natural fibres based composites, Composite Structures, 155 (2016) 8–18, doi:10.1016/j.compstruct.2016.08.001

[70] E. Kontou, G. Spathis, P. Georgiopoulos, Modeling of nonlinear viscoelasticity-viscoplasticity of bio-based polymer composites, Polymer Degradation and Stability, 110 (2014) 203–207, doi:10.1016/j.polymdegradstab.2014.09.001

[71] K. Wilczyński, K. Buziak, A. Lewandowski, A. Nastaj, Rheological basics for modeling of extrusion process of wood polymer composites, Polymers, 13(4) (2021) 622, doi:10.3390/POLYM13040622

[72] R. Pramanik, F. Soni, K. Shanmuganathan, A. Arockiarajan, Mechanics of soft polymeric materials using a fractal viscoelastic model, Mechanics of Time-Dependent Materials, 2021 (2021) 1–14, doi:10.1007/S11043-021-09486-0

[73] C. Stochioiu, B. Piezel, A. Chettah, S. Fontaine, H. Gheorghiu, Basic modeling of the visco elastic behavior of flax fiber composites, Industria Textila, 70(4) (2019) 331–335, doi:10.35530/IT.070.04.1512

[74] E. Marklund, J. Varna, L. Wallström, Nonlinear viscoelasticity and viscoplasticity of flax/polypropylene composites, Journal of Engineering Materials and Technology, 128(4) (2006) 527–536, doi:10.1115/1.2345444

[75] A.E. Mourid, R. Ganesan, M. Levesque, Comparison between analytical and numerical predictions for the linearly viscoelastic behavior of textile composites, Mechanics of Materials, 53 (2013) 69–83, doi:10.1016/j.mechmat.2012.11.003

[76] S. Malek, C. Nadot-Martin, B. Tressou, C. Dai, R. Vaziri, Micromechanical modeling of effective orthotropic elastic and viscoelastic properties of parallel strand lumber using the morphological approach, Journal of Engineering Mechanics, 145(9) (2019) 04019066, doi:10.1061/(ASCE)EM.1943-7889.0001631

[77] J. Wu, H. Yuan, L. Li, K. Fan, S. Qian, B. Li, Viscoelastic shear lag model to predict the micromechanical behavior of tendon under dynamic tensile loading, Journal of Theoretical Biology, 437 (2018) 202–213, doi:10.1016/j.jtbi.2017.10.018

[78] N. Hosseini, S. Javid, A. Amiri, C.A. Ulven, D.C. Webster, G. Karami, Micromechanical viscoelastic analysis of flax fiber reinforced bio-based polyurethane composites, Journal of Renewable Materials, 3(3) (2015) 205–215, doi:10.7569/JRM.2015.634112

[79] F.N. Omar, S.H. Hanipah, L.Y. Xiang, M.A.P. Mohammed, A.S. Baharuddin, J. Abdullah, Micromechanical modelling of oil palm empty fruit bunch fibres containing silica bodies, Journal of the Mechanical Behaviour of the Biomedical Materials, 62 (2016) 106–118, doi:10.1016/j.jmbbm.2016.04.043

[80] V.P. Radchenko, D.V. Shapievskii, Mathematical model of creep for a microinhomogeneous nonlinearly elastic material, Journal of Applied Mechanics and Technical Physics, 49(3) (2008) 478–483, doi:10.1007/S10808-008-0064-9

[81] R. Bedzra, S. Reese, J. Simon, Hierarchical multi-scale modelling of flax fibre/epoxy composite by means of general anisotropic viscoelastic-viscoplastic constitutive models: Part II – Mesomechanical model, International Journal of Solids and Structures, 202 (2020) 299–318, doi:10.1016/J.IJSOLSTR.2020.05.019

[82] M.K. Habibi, L.-h. Tam, D. Lau, Y. Lu, Viscoelastic damping behavior of structural bamboo material and its microstructural origins, Mechanics of Materials, 97 (2016) 184–198, doi:10.1016/j.mechmat.2016.03.002

[83] A. Mlyniec, L. Mazur, K.A. Tomaszewski, T. Uhl, Viscoelasticity and failure of collagen nanofibrils: 3D coarse-grained simulation studies, Soft Materials, 13(1) (2015) 47–58, doi:10.1080/1539445X.2015.1009549

[84] A. Mlyniec, K.A. Tomaszewski, E.M. Spiesz, T. Uhl, Molecular-based nonlinear viscoelastic chemomechanical model incorporating thermal denaturation kinetics of collagen fibrous biomaterials, Polymer Degradation and Stability, 119 (2015) 87–95, doi:10.1016/j.polymdegradstab.2015.05.005

[85] V. Alvarez, J. Kenny, A. Vazquez, Creep behavior of biocomposites based on sisal fiber reinforced cellulose derivatives/starch blends, Polymer Composites, 25(3) (2004) 280–288, doi:10.1002/PC.20022

[86] A. Tirella, G. Mattei, A. Ahluwalia, Strain rate viscoelastic analysis of soft and highly hydrated biomaterials, Journal of Biomedical Materials Research Part A, 102(10) (2014) 3352–3360, doi:10.1002/JBM.A.34914

[87] D. Bassir, S. Guessasma, L. Boubakar, Hybrid computational strategy based on ANN and GAPS: application for identification of a non-linear model of composite material, Composite Structures, 88(2) (2009) 262–270, doi:10.1016/J.COMPSTRUCT.2008.04.007

[88] W. Zhang, Y. Liu, G.S. Kassab, Viscoelasticity reduces the dynamic stresses and strains in the vessel wall: implications for vessel fatigue, American Journal of Physiology-Heart and Circulatory Physiology, 293(4) (2007) H2355–H2360, doi:10.1152/AJPHEART.00423.2007

[89] J.D. Humphrey, S. Na, Elastodynamics and arterial wall stress, Annals of Biomedical Engineering, 30 (2002) 509–523.

[90] H. Ahmadzadeh, D.H. Smith, V.B. Shenoy, Viscoelasticity of Tau proteins leads to strain rate-dependent breaking of microtubules during axonal stretch injury: predictions from a mathematical model, Biophysics Journal, 106 (2014) 1123–1133, doi:10.1016/j.bpj.2014.01.024

[91] A. Shamloo, F. Manuchehrfar, H. Rafii-Tabar, A viscoelastic model for axonal microtubule rupture, Journal of Biomechanics, 48 (2015) 1241–1247, doi:10.1016/j.jbiomech.2015.03.007

[92] S. Javid, A. Rezaei, G. Karami, A micromechanical procedure for viscoelastic characterization of the axons and ECM of the brainstem, Journal of the Mechanical Behaviour of the Biomedical Materials, 30 (2014) 290–299, doi:10.1016/j.jmbbm.2013.11.010

[93] J. Wu, H. Yuan, L.-y. Li, B. Li, K. Fan, S. Li, K.-N. Lee, Mathematical modeling of microtubule-tau protein transients: insights into the superior mechanical behavior of axon, Applied Mathematical Modelling, 71 (2019) 452–466, doi:10.1016/j.apm.2019.02.030

[94] E. Comellas, S. Budday, J.-P. Pelteret, G.A. Holzapfel, P. Steinmann, Modeling the porous and viscous responses of human brain tissue behavior, Computer Methods in Applied Mechanics and Engineering, 369 (2020) 113128, doi:10.1016/j.cma.2020.113128

[95] Y. Shelef, A. Uzan, O. Braunshtein, B. Bar-On, Assessing the interfacial dynamic modulus of biological composites, Materials, 14(12) (2021) 3428, doi:10.3390/MA14123428

[96] H. Hatami-Marbini, Hydration dependent viscoelastic tensile behavior of cornea, Annals of Biomedical Engineering, 42(8) (2014), doi:10.1007/s10439-014-0996-6

[97] J. Bischoff, E. Arruda, K. Grosh, A rheological network model for the continuum anisotropic and viscoelastic behavior of soft tissue, Biomechanics and Modelling in Mechanobiology, 3(1) (2004) 56–65, doi:10.1007/S10237-004-0049-4

[98] R. Huang, S. Zheng, Z. Liu, T. Ng, Recent advances of the constitutive models of smart materials — hydrogels and shape memory polymers, International Journal of Applied Mechanics, 12 (2020) 2050014, doi:10.1142/S1758825120500143

[99] A. Hari Reddi, J. Becerra, J.A. Andrades, Nanomaterials and hydrogel scaffolds for articular cartilage regeneration, Tissue Engineering Part B: Reviews, 17(5) (2011) 301–305.

[100] D.S. Vara, G. Punshon, K.M. Sales, S. Sarkar, G. Hamilton, A.M. Seifalian, Endothelial cell retention on a viscoelastic nanocomposite vascular conduit is improved by exposure to shear stress preconditioning prior to physiological flow, Artificial Organs, 32 (2008) 977–981, doi:10.1111/j.1525–1594.2008.00659.x

[101] A. Islam, M. Riaz, T. Yasin, Structural and viscoelastic properties of chitosan-based hydrogel and its drug delivery application, International Journal of Biological Macromolecules, 59 (2013) 119–124, doi:10.1016/j.ijbiomac.2013.04.044

[102] H. So, Y.H. So, A.P. Pisano, Refillable and magnetically actuated drug delivery system using pear-shaped viscoelastic membrane, Biomicrofluids, 8 (2014) 044119, doi:10.1063/1.4893912

[103] A. Yassine, R. Benzidane, A. Bendada, S. Zouaoui, Vibration of damaged bio-composite beams reinforced with random short Alfa fibers: experimental and analytical investigations, Advances in Aircraft and Spacecraft Science, 8(2) (2021) 127–149, doi:10.12989/aas.2021.9.2.127

Index

Note: **Bold** page numbers refer to tables; *italic* page numbers refer to figures.

For Product Safety Concerns and Information please contact our EU
representative GPSR@taylorandfrancis.com
Taylor & Francis Verlag GmbH, Kaufingerstraße 24, 80331 München, Germany